MEDICAL RADIOLOGY

Diagnostic Imaging and Radiation Oncology

Springer
Berlin
Heidelberg
New York
Barcelona
Budapest
Hong Kong
London
Milan
Paris
Santa Clara
Singapore
Tokyo

M. Molls · P. Vaupel (Eds.)

Blood Perfusion and Microenvironment of Human Tumors

Implications for Clinical Radiooncology

With Contributions by

J.M. Brown · D.J. Chaplin · B. Endrich · E. Fait · H.J. Feldmann · C. Grau · M. Höckel
M.R. Horsman · R. Jund · M.A. Konerding · J. Overgaard · C. Laubenbacher
R.J. Maxwell · M. Molls · W. Müller-Klieser · M. Parliament · M. Schwaiger
D.W. Siemann · C.W. Song · P. Stadler · F. Steinberg · C. Streffer · M. Stubbs · M.J. Trotter
R. Urtasun · C. Van Ackern · P. Vaupel

Foreword by
L.W. Brady and H.-P. Heilmann

With 99 Figures in 126 Separate Illustrations, Some in Color

 Springer

Professor Dr. med. M. MOLLS
Direktor der Klinik und Poliklinik für Strahlentherapie und Radiologische Onkologie
Klinikum rechts der Isar der Technischen Universität München
Ismaninger Strasse 22
D-81664 Munich
Germany

Professor Dr. med. P. VAUPEL
Leiter des Instituts für Physiologie und Pathophysiologie
Johannes Gutenberg-Universität Mainz
Duesbergweg 6
D-55099Mainz
Germany

MEDICAL RADIOLOGY · Diagnostic Imaging and Radiation Oncology

Continuation of
Handbuch der medizinischen Radiologie
Encyclopedia of Medical Radiology

ISSN 0942-5373
ISBN 3-540-58866-3 Springer-Verlag Berlin Heidelberg New York

Library of Congress Cataloging-in-Publication Data. Blood perfusion and microenvironment of human tumors : implications for clinical radiooncology / M. Molls, P. Vaupel, eds. ; with contributions by J. M. Brown ... [et al.] ; foreword by L. W. Brady and H.-P. Heilmann   p.   cm. -- (Medical radiology)   Includes bibliographical references and index. ISBN 3-540-58866-3 (alk. paper) 1. Tumors--Blood-vessels. 2. Microcirculation. 3. Tumors--Metabolism. 4. Cancer--Radiotherapy. I. Molls, M. (Michael), 1944– . II. Vaupel, Peter. III. Brown, J. M. (J. Martin) IV Series. [DNLM: 1. Neoplasms--blood supply. 2. Microcirculation--physiopathology. 3. Neoplasms--metabolism. 4. Neoplasms--therapy. QZ 200 B6545 1997] RC269.B53 1997 616.99'407--dc21 DNLM/DLC for Library of Congress 97-20657 CIP

Printed in Germany

The use of general descriptive names, registered names, trademarks, etc. in this publication does not imply, even in the absence of a specific statement, that such names are exempt from the relevant protective laws and regulations and therefore free for general use.

Product liability: The publishers cannot guarantee the accuracy of any information about dosage and application contained in this book. In every individual case the user must check such information by consulting the relevant literature.

Cover design: de'blik, Berlin

Typesetting: Best-set Typesetter Ltd., Hong Kong

SPIN: 10524941   21/3155 – 5 4 3 2 1 0 – Printed on acid-free paper

Dedication

This volume is dedicated to the memory of Dr. med. Mildred Scheel (1932–1985). Dr. Scheel, the wife of the former President of the Federal Republic of Germany, was a radiologist and specialized in mammography. To fulfill her noble ambition to contribute to the fight against cancer she founded the German Cancer Aid (Deutsche Krebshilfe) in 1974. From that time on, all her efforts were directed towards encouraging people to contribute money for this crucial purpose. She developed many significant ideas for organizing cancer prevention, early diagnosis and treatment that are applicable on a large scale, and initiated the establishment of the first five cancer centers in our country. She supported the treatment of childhood cancer in many hospitals and initiated the psychosocial after-care of patients and their families. In addition, she aided individuals who experienced financial difficulties because of their cancer. She established a foundation (Dr. Mildred Scheel Stiftung für Krebsforschung) to promote and support cancer research. This foundation provides a fellowship programme for scientists to work and study at institutions abroad.

Over the past 22 years the citizens of Germany have donated a total of 925 million DM for the work of the German Cancer Aid. We are deeply grateful that our investigations and those of other groups concerning the microenvironment of malignant tumors have been so generously supported by the German Cancer Aid.

Munich   MICHAEL. MOLLS
Mainz   PETER VAUPEL

Foreword

Over the past decade, several revolutionary therapy concepts for cancer treatment – including immunological treatments and gene therapy – have been developed. At present, however, many of these concepts are still in their infancy and remain far from bringing widespread improvements in oncological treatment. Therefore, since radiotherapy remains the most important nonsurgical treatment for cancer, the continuation and support of radiooncological research is paramount and should be encouraged.

The emphasis of this book lies in a consideration of the tumor as a pathophysiological entity and the implications that the properties of this entity have on the responsiveness of malignant tumors to radiation therapy and chemotherapy. While there are strong links between the molecular, cellular, histomorphological and physiological levels, a description of tumor molecular biology has been deliberately avoided. Instead, the multiple physiological components of malignant tumors, including blood flow, oxygenation, pH and energy status, are discussed in detail, using current knowledge from both experimental and clinical investigations. In addition, the various possibilities for modification of tumor perfusion and microenvironment with the aim of improving therapeutic efficiency are considered. From this book it becomes clear that tumors are highly complex biological entities which cannot be broadly categorized. With this in mind, the development of individualised therapy concepts targeted at the pathophysiological properties of a given tumor are necessary. Nuclear magnetic resonance and positron emission tomography technologies for use in patients may in future support the development of such therapy concepts by means of assessment of tumor microenvironment prior to and during a course of treatment. Additionally, results of such investigation may allow for a more precise definition of causes of therapy failure as well as success.

This book, by providing an overview of the current research activities in the field of tumor perfusion and microenvironment together with detailed insights into the therapeutic relevance of tumor pathophysiological properties, will be of critical importance to all oncologists involved in the management of patients.

Philadelphia
Hamburg

LUTHER W. BRADY
HANS-PETER HEILMANN

Contents

1 The Impact of the Tumor Microenvironment on Experimental and Clinical Radiation Oncology and Other Therapeutic Modalities

M. Molls[1] and P. Vaupel[2]

Radiotherapy is the most important nonsurgical treatment for cancer (Tobias and Tattersall 1985; Tobias 1992). Today 45%–50% of all cancer patients can be cured. About 70% of those who are cured received radiotherapy, either alone (e.g., Hodgkin's disease, non-Hodgkin's lymphoma, cancer of the cervix uteri, prostate cancer, different types of squamous cell carcinomas including head and neck cancers and anal cancer, skin cancer) or in combination with other treatment modalities (e.g., breast cancer, rectal cancer, soft tissue sarcomas, lymphomas, malignant diseases in children, the entities given above). At least 50% of all cancer patients need radiotherapy as part of either curative or palliative treatments (DeVita et al. 1997; Inter-Society Council for Radiation Oncology 1991; Vermorken and Schermer 1994). Important aspects of the clinical effectiveness of radiotherapy, the assessment of outcome, the resources needed for adequate radiooncological treatment, the costs of radiotherapy, and quality assurance are discussed in a recent comprehensive and very interesting publication (*Critical Issues in Radiotherapy*, SBU rapport no. 130 E, SB Offset, Stockholm 1996, ISBN 91-87890-34-8).

Over the past decade revolutionary concepts of cancer therapy such as immunological treatment and gene therapy have been developed. However, it has become more and more evident that these new modalities are still far from improving oncological treatment results on a widespread scale. It is therefore necessary to continue and even to reinforce the efforts and the support of radiooncological research. Radiotherapy – a local treatment of high precision – has a future as a treatment applied either alone or in combination. In particular, the possibilities for combination with new modalities (e.g., gene therapy, immunological treatment) will be of high interest.

The unique potential of radiooncology is due to the fact that this discipline reaps benefits from physical-technical as well as biological research activities. Concerning physical-technical research activities, a recent significant step forward has been the development of conformal and stereotactic radiotherapy with X-rays, using the linear accelerator. Radiotherapy with neutrons, heavy ions and protons is also based on elaborate accelerator technology.

With regard to radiobiological research, investigations on tumor cell proliferation have led to promising clinical protocols of accelerated radiotherapy in tumors with a short potential doubling time. Different experimental concepts have been developed which aim at a better understanding of the mechanisms of radiation-induced cell death and provide tools which can be used in clinical radiooncology to increase tumor cell kill (combination of radiotherapy with cytostatic drugs, hypoxic radiosensitizers, hyperthermia). One important direction in experimental radiooncology is the investigation of the many factors that determine the radiosensitivity of mammalian cells. The influence of the progression of cells through the G_2 phase of the cell cycle on X-ray sensitivity is but one of these factors (Molls and Streffer 1984).

According to classical radiation biology, the radiosensitivity of tumors is mainly determined by the four Rs: repair, repopulation, reoxygenation and redistribution of cells in the cell cycle (Hall 1994). The molecular events linked to the four Rs are under investigation together with the molecular targets which can be influenced with the aim of modulating tumor cell radiosensitivity.

On the molecular level, ionizing radiations cause the production of free radicals. Free radicals interact with DNA (Hall 1994). The very different molecular lesions of DNA can be repaired by different repair systems (Leadon 1996). In addition, free radicals also interact with macromolecules of the cytoplasm

[1] M. Molls, Prof. Dr. med., Klinik und Poliklinik für Strahlentherapie und Radiologische Onkologie, Klinikum rechts der Isar der Technischen Universität München Ismaninger Strasse 22, D-81675 München, Germany
[2] P. Vaupel, Prof. Dr. med., Institute of Physiology and Pathophysiology, University of Mainz Duesbergweg 6 D-55099 Mainz, Germany

and membranes. These interactions initiate the transduction of signals and the expression of genes. One can speculate that in future the classic radiation modifiers (cytostatic drugs, hypoxic radiosensitizers, hyperthermia) will be supplemented by molecular modifiers of DNA repair and signal transduction (HALLAHAN 1996; HANNA et al. 1996).

What is the significance of investigations concerning the blood supply and the microenvironment of human tumors? Within a defined entity and stage of a solid tumor the response after irradiation, chemotherapy and other nonsurgical treatment modalities may vary from patient to patient. Different factors can be involved in the case of reduced sensitivity. In addition to intrinsic, genetically determined resistance, physiological properties primarily created by inadequate and nonuniform vascular networks constitute the responsiveness of an individual malignant tumor. Factors which are closely linked and which define the so-called metabolic microenvironment encompass circulatory parameters, including blood and nutrient supply, transvascular and interstitial transport, and oxygenation and bioenergetic status, as well as tumor pH (VAUPEL 1994; see Fig. 1.1). The impact of these physiological proper-

ties can be mediated through direct actions. For example, the effectiveness of radiation therapy with photons or some anticancer drugs is directly dependant on the oxygen tension. The influence of indirect mechanisms, which might be even more important, should also be considered. Changes in the metabolic microenvironment modulate the proliferation kinetics of tumor cells and in turn the amount of cell kill following nonsurgical treatment modalities (DURAND 1991; SUTHERLAND 1988; TANNOCK and ROTIN 1989). The effectiveness of irradiation and cytotoxic drugs substantially differs between the different phases of the cell cycle (HALL 1994). As a further indirect mechanism we would like to stress the use of vectors and molecules which in the future will be given to interact with molecular targets of the cell and to sensitize the tumor to nonsurgical treatments. The bioavailability of these compounds is dependent on the micromilieu.

One might argue that with regard to the sensitivity of tumors the investigation of molecular events is more important than studies on factors and processes of the micromilieu. However, we are convinced that oncological research which addresses the biology of solid malignant tumors and their

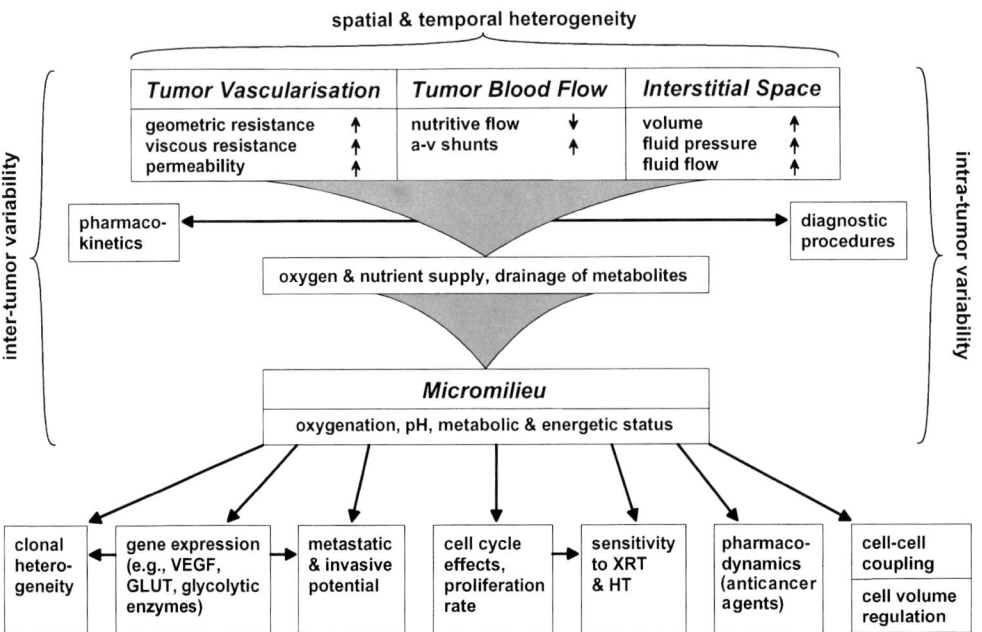

Fig. 1.1. Simplified diagram showing the critical role of the parameters defining the metabolic micromilieu for tumor growth and the responsiveness of tumor cells to nonsurgical treatment modalities, including anticancer drugs, standard radiotherapy (*XRT*) and hyperthermia (*HT*). *GLUT*, glucose transporter; *VEGF*, vascular endothelial growth factor; *a-v*, arteriovenous

individual sensitivity has to deal with the different levels of organization of biological subsystems: the molecular pathways and cascades, the system of organelles and the whole cell, the networks of information between cells (tumor cells to tumor cells, tumor cells to stroma cells), the physiological micromilieu encompassing the parameters described above and the physiological macromilieu with its influences exerted by hormones and mediators.

It is obvious that the whole system of the solid tumor, including its physiological milieu, represent a structure in which interdependences exist between the different levels of biological organization such as the molecular, cellular, histomorphological and physiological levels. The regulation of these interdependences determines not only those events and mechanisms in which classical radiation biology has been interested for decades: cell death, repair of DNA, repopulation, oxygenation and redistribution of cells in the cell cycle. It also exerts an influence on intercellular and cellular signal transduction and on the activities of genes and gene products. Recent experimental findings support our view of the solid tumor as a highly complex biological entity in which systemic conditions may alter molecular functions and vice versa. For example, the abnormal functioning of a normal p53 protein in a disturbed microenvironment has been reported (HAINAUT and MILNER 1993). Activators of protein kinase C mediate cellular cytotoxicity to hypoxic cells but not normoxic cells (KOONG et al. 1994). According to GRAEBER et al. (1996), hypoxia can dramatically alter cellular phenotype and may be important in tumor progression.

Considering the micromilieu as an important influence on the behavior of tumors after non-surgical treatment, we felt it might be of advantage to summarize our current knowledge. Thus the present volume gives an overview of the current research activities in this field and the most recent results.

References

DeVita VT, Hellmann S, Rosenberg SA (1997) Cancer: principles and practice of oncology. Lippincott-Raven, Philadelphia

Durand RE (1991) Keynote Address: The influence of microenvironmental factors on the activity of radiation and drugs. Int J Radiat Oncol Biol Phys 20:253–258

Graeber TG, Osmanian C, Jacks T, et al. (1996) Hypoxia-mediated selection of cells with diminished apoptotic potential in solid tumours. Nature 379:88–91

Hainaut P, Milner J (1993) Redox modulation of p53 conformation and sequence-specific DNA binding in vitro. Cancer Res 53:4469–4473

Hall EJ (1994) Radiobiology for the radiologist. Lippincott, Philadelphia

Hallahan DE (1996) Introduction. Semin Radiat Oncol 6:243–244

Hanna NN, Hallahan DE, Wayne JD, Weichselbaum RR (1996) Modification of the radiation response by the administration of exogenous genes. Semin Radiat Oncol 6:321–328

Inter-Society Council for Radiation Oncology (1991) Radiation oncology in integrated cancer management (The "Blue Book") The American College of Radiology, Philadelphia

Koong AC, Chen EY, Kim CY, Giaccia AJ (1994) Activators of protein kinase C selectively mediate cellular cytotoxicity to hypoxic cells and not aerobic cells. Int J Radiat Oncol Biol Phys 29:259–265

Leadon SA (1996) Repair of DNA damage produced by ionizing radiation: a minireview. Semin Radiat Oncol 6:295–305

Molls M, Streffer C (1984) The influence of G2 progression on X-ray sensitivity of two-cell mouse embryos. Int J Radiat Biol 46:355–365

Sutherland RM (1988) Cell and environment interactions in tumor microregions: the multicell spheroid model. Science 240:177–184

Tannock IF, Rotin D (1989) Acid pH in tumors and its potential for therapeutic exploitation. Cancer Res 49:4373–4384

Tobias J, Tattersall M (1985) Doing the best for the cancer patient. Lancet 1:35–38

Tobias J (1992) Clinical practice of radiotherapy. Lancet 339:159–164

Vaupel P (1994) Blood flow, oxygenation, tissue pH distribution, and bioenergetic status of tumors. In: Ernst Schering Research Foundation Lecture Publication Vol 23. Information and Standards, Medical Scientific Publications, Druckerei Hellmich KG, Berlin ISSN 0940-9300

Vermorken AJM, Schermer FAJM (1994) Towards coordination of cancer research in Europe. IOS Press, Amsterdam

2 Morphological Aspects of Tumor Angiogenesis and Microcirculation

M.A. Konerding[1], C. van Ackern[2], E. Fait[3], F. Steinberg[4], and C. Streffer[5]

CONTENTS

2.1 Introduction

Angiogenesis plays a significant role during normal growth, in physiological conditions (e.g., in the placenta and endometrium), and in pathological conditions such as inflammation, wound healing, and tumor growth. Angiogenesis is thus not a specific phenomenon in tumors or a pathological condition, but instead an integral element of numerous different normal and pathological conditions.

It is widely accepted that no solid tumor can grow beyond a critical size of approximately $1\,mm^3$ with about 10^6 cells without sufficient vascular supply (Folkman 1985; Hirst et al. 1982). However, our knowledge of angiogenesis is not as extensive as might be suggested by the significance of this research front. Thus, for example, the "International Symposium on Angiogenesis" in 1991 in St. Gallen, Switzerland, as well as the meeting "Angiogenesis in Health and Disease" of the NATO Advanced Study Institute in 1993 on Rhodes and in 1995 in Porto Carras, Greece, where the latest experimental results were presented and the current interdisciplinary research on angiogenesis was summarized, showed that basic questions still remain unresolved despite extensive and detailed knowledge in various fields.

This is also true for morphological aspects of tumor angiogenesis: from 1983 to 1994, approximately 1150 papers on angiogenesis were to be found in the Medline data bank under the keyword angiogenesis, but of those, only 8%, or 92 reports, were mainly or at least partially concerned with morphological aspects. Of these, only 26 publications focussed on the morphology of tumor angiogenesis (e.g., Hammersen et al. 1983a; Konerding et al. 1989a,b,c; Paweletz and Knierim 1989). In this respect, it is interesting that numerous studies on angiogenesis were carried out on in vitro and not on suitable in vivo tumor models. Studies on primary tumors, as have been frequently suggested in the past (Jain 1988), are also widely missing.

Most studies have been concerned with the "why" of angiogenesis and paid less attention to the morphological "how." Since the initial extensive studies on capillary sprouting by Thoma in the last century (1893), the prevailing view is that all morphologically tangible events during primary and secondary angiogenesis (including normal growth, inflammation, wound healing, and tumor angiogenesis) are similar and, thus, comparable to each other (for literature see Simpson and Fraser 1983; Paweletz and Knierim 1989). However, Hammersen et al. (1983b) have shown convincingly that angiogenesis in wound healing after trauma and tumor angiogenesis involves different mechanisms.

Numerous studies on tumor vessels are concerned with pattern formations, course, and density, but not with the structure in the morphological sense. This can also be applied to tumor angiogenesis since such important processes as sprout formation, development of the three-dimensional architecture, and vessel regression have only been investigated

[1] M.A. Konerding, Univ.-Prof. Dr. med., [2] C. van Ackern, Dr. med., and [3] E. Fait, Dipl.-Biol. Anatomisches Institut, Johannes Gutenberg-Universität Mainz, D-55099 Mainz, Germany
[4] F. Steinberg, Dr. med. and [5] C. Streffer, Univ.-Prof. Dr. rer. nat. Institut für Medizinische Strahlenbiologie, Universitätsklinikum Essen, Hufelandstrasse 55, D-45122 Essen, Germany

with respect to morphology in exceptional cases despite the biological and clinical relevance.

It would of course be best to study tumor angiogenesis on human primary tumors. However, this is possible only to a very limited extent (WELLER et al. 1977) since the tumor vascular system in surgical specimens is mainly made up of morphologically established vessels and late degenerative forms. Nevertheless, the vascular systems in both primary and experimental tumors show high remodeling rates, so that studies on primary tumors are in principle possible (e.g., MIODONSKI et al. 1980).

For our own studies we have mainly used the model of xenografted human tumors on nude mice, because this model shows many parallels to the human primary tumor and because it allows for sequential studies.

2.2
Methodology of the Presented Studies

For our studies of tumor angiogenesis, we used undifferentiated melanomas (MeWo, B11), sarcomas (MOR), and carcinosarcomas (4197) transplanted onto nude mice, as well as spontaneous murine tumors. The methods of transplantation were described by BUDACH et al. (1986) in detail. For sequential studies, series of tumors were subsequently examined from the 3rd day up to the 78th day after

transplantation. Additionally, seven human primary colonic and rectal carcinomas were studied.

Transmission electron microscopy was used to characterize the structure of newly formed vessels, and serial semithin and ultrathin sections allowed for insights into the arrangement of sprouting endothelial cells and the formation of new vessel buds. Scanning electron microscopy of microvascular corrosion casts was used to demonstrate architectonic features (for details see KONERDING 1991; KONERDING et al. 1989a,b, 1995).

2.3
Morphology of Angiogenesis

2.3.1
Origin of Endothelial Cells Involved in Tumor Angiogenesis

Several authors such as CLARK et al. (1939), ILLIG (1961), or HEIMARK and SCHWARTZ (1988) postulated that secondary angiogenesis in tumors takes place exclusively by sprouting and mitoses of endothelial cells. Others such as CLIFF (1981) and HATA (1981) showed that endothelial cells can also derive from local mesenchymal cells. Transformation of fibrocytes to endothelial cells was shown as early as 1957 by PETRY and HEBERER after implantation of vascular prostheses. According to FEIGL et al.

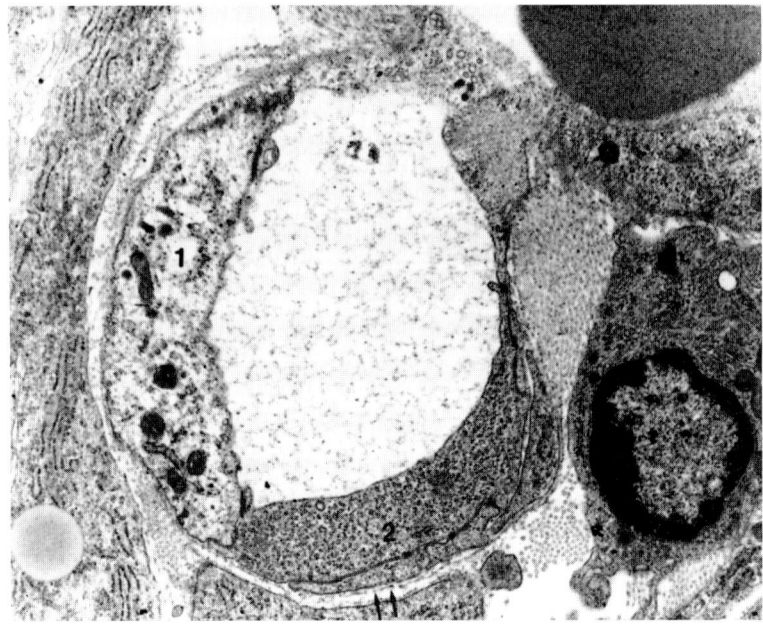

Fig. 2.1. Transmission electron microscopic image of an early tumor vessel in a xenografted carcinoma (day 79). The cell organelle contents of the two endothelial cells (*1, 2*) differs significantly and cell *2* shows more similarities to the adjacent pericytes. The diffuse basal lamina seems to be discontinuous in part (*arrows*). Note the intracellular edema and the varying degrees of degeneration. ×11.435

Fig. 2.2. Advanced intracytoplasmatic (*star*) and extracytoplasmatic lumen formation. The latter lumen is occluded by an erythrocyte (*e*) and shows extensive reflexive overlapping of two ramification (*1, 2*) from the same cell that have developed contacts between themselves (*arrows*). The two lumina are still separated from one another by a wide cytoplasmatic bridge. Fragments of the basal lamina can be seen along the entire circumference, and pericytes are missing. Transmission electron microscopy, melanoma MeWo, 18th day. ×4790

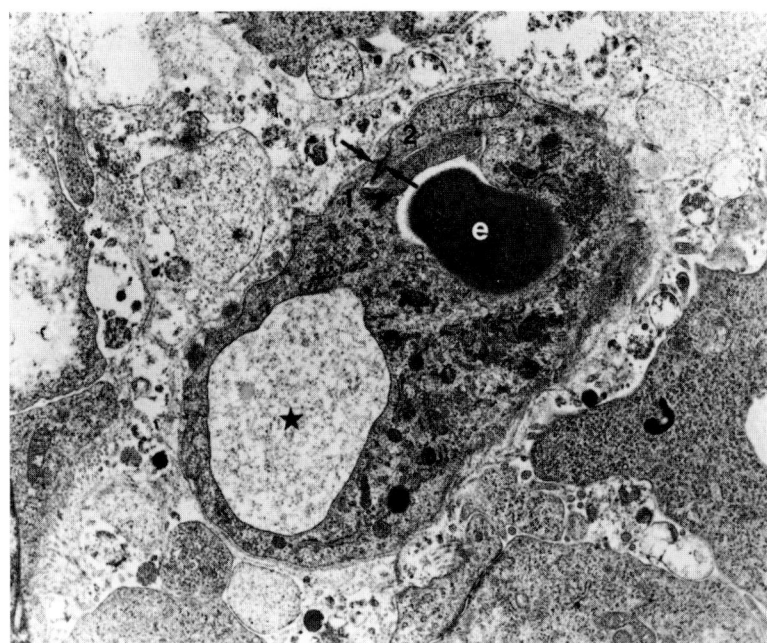

(1985), even mononuclear cells can take part in capillary formation.

In tumor angiogenesis, obviously both the angioblastic and mesenchymal pathways of endothelial cell and new vessel formation are possible: HAMMERSEN et al. (1983) observed in a murine melanotic melanoma mesenchymal cells rich in organelles forming new lumina and wide intracytoplasmatic vacuoles. In our own studies we have frequently seen newly formed vessels made up of cells with extremely different cell organelle contents, indicative of the involvement of different cell races (Fig. 2.1). Sometimes "endothelial cells" could be observed which had engulfed bundles of collagen fibers.

Practically all labeling methods with mono- and polyclonal antibodies against endothelial cells, which can be used reliably in normal tissues, are of limited value in tumor vessels. This may be due not only to a low specificity of the endothelial cells in tumors, as suggested by DENEKAMP (1989), but also to the involvement of different cell races.

2.3.2
Structural Development of the Tumor Vascular System

The first steps in the formation of new vessels involve the stimulation of the endothelial cells of preexisting

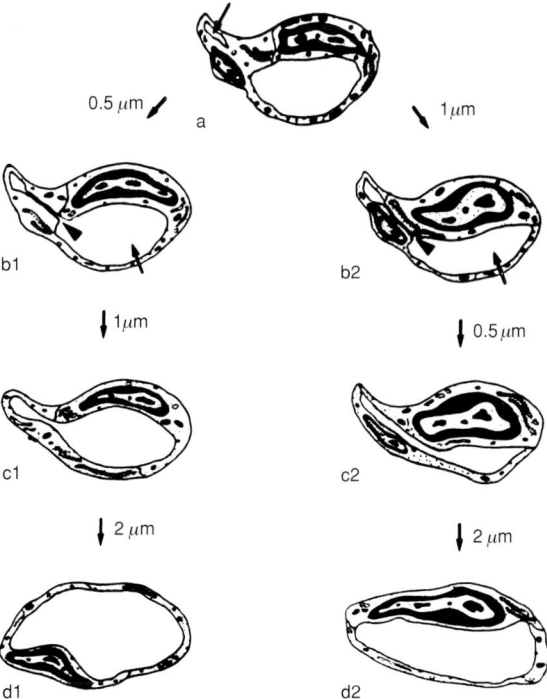

Fig. 2.3. Connection of an intracytoplasmatic, autophagic vacuole to the vessel lumen in order to increase the vascular diameter. Total length = 7 μm. The distance between the single section planes is depicted between the drawings. *a*, Autophagic vacuole (*arrow*); *b1, b2*, connection of the vacuole to the lumen (*arrows*) by a tiny slit (*arrowheads*); *c1, c2*, widening of the connecting slit; *d1, d2*, increasing luminal diameter and flattening of the endothelial lineage. Drawing based on serial sections. MeWo, 8th day. ×4050

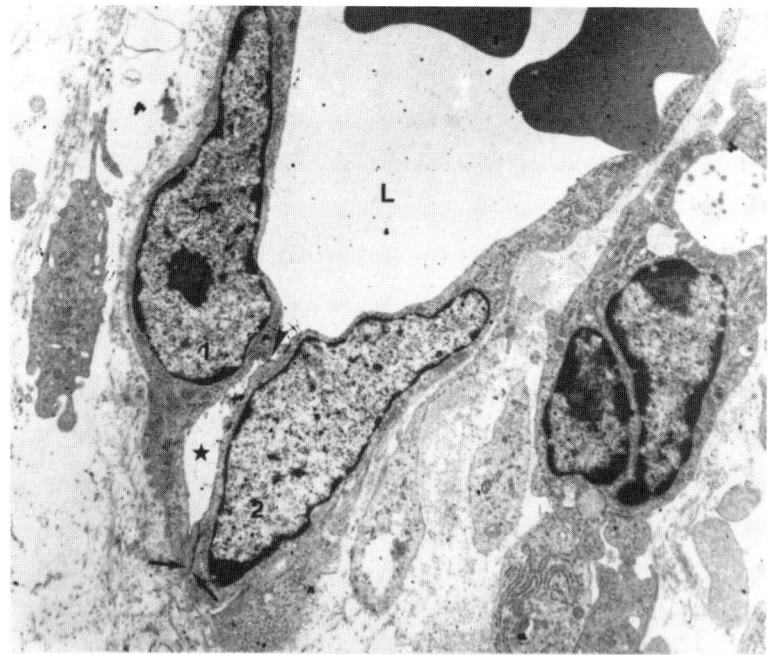

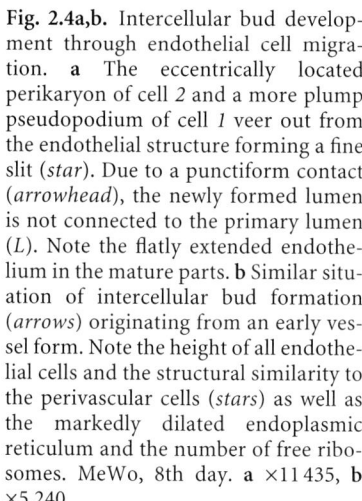

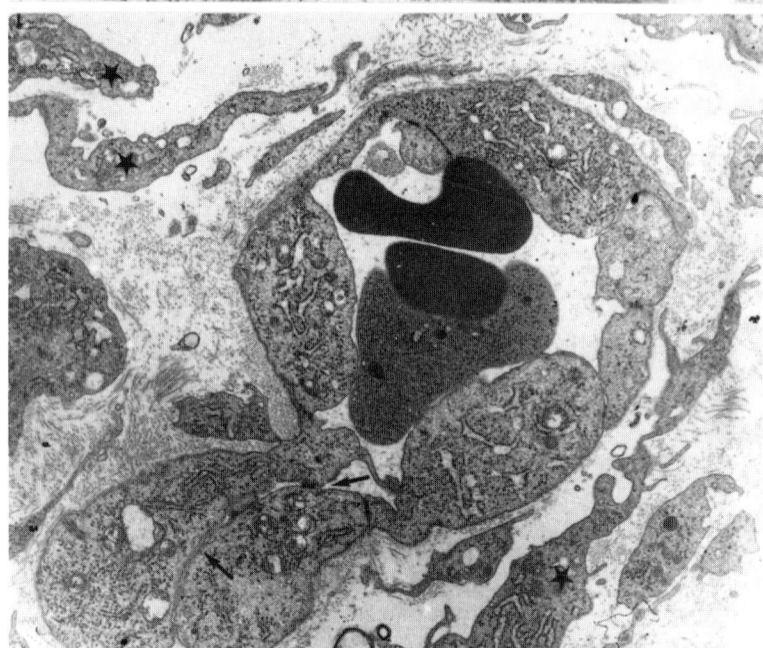

Fig. 2.4a,b. Intercellular bud development through endothelial cell migration. **a** The eccentrically located perikaryon of cell *2* and a more plump pseudopodium of cell *1* veer out from the endothelial structure forming a fine slit (*star*). Due to a punctiform contact (*arrowhead*), the newly formed lumen is not connected to the primary lumen (*L*). Note the flatly extended endothelium in the mature parts. **b** Similar situation of intercellular bud formation (*arrows*) originating from an early vessel form. Note the height of all endothelial cells and the structural similarity to the perivascular cells (*stars*) as well as the markedly dilated endoplasmic reticulum and the number of free ribosomes. MeWo, 8th day. **a** ×11 435, **b** ×5.240

vessels, mainly capillaries and veins, aimed at a degradation of the basal lamina and a migration of endothelial cells (for reviews see AUSPRUNK and FOLKMAN 1977; MAHADEVAN and HART 1990). Small pseudopodes of varying size are indicative of this migration process of endothelial cells (CLIFF 1981). These migrating and sprouting endothelial cells are rich in rough endoplasmic reticulum, mitochondria, and polyribosomes (Fig. 2.2).

The widely held belief that migrating endothelial cells first form a solid cord that subsequently devel-

ops a lumen is based on the early studies of CLIFF (1963), FURUSATO et al. (1985), ILLIG (1961), and even THOMA (1893). However, according to our findings, different mechanisms of sprouting and lumen formation occur during tumor angiogenesis. New lumina can be formed both intracellularly by single endothelial cells and intracellularly involving two or more endothelial cells.

In the first case, a primary vacuole or slit is formed originating from the Golgi apparatus or the endoplasmic reticulum (Fig. 2.2). This form is pre-

Fig. 2.5. Horizontal section through a vascular bud in the periphery of a melanoma. Three endothelial cells form a primitive lumen by means of simple contact structures. Note the sections of the perivascular cells (*stars*) grouped around the bud which is partially surrounded by basal lamina material (*arrows*). The course of perivascular collagen fiber bundles (*k*) is suggestive of a lead function of the connective tissue. MeWo, 25th day. ×9240

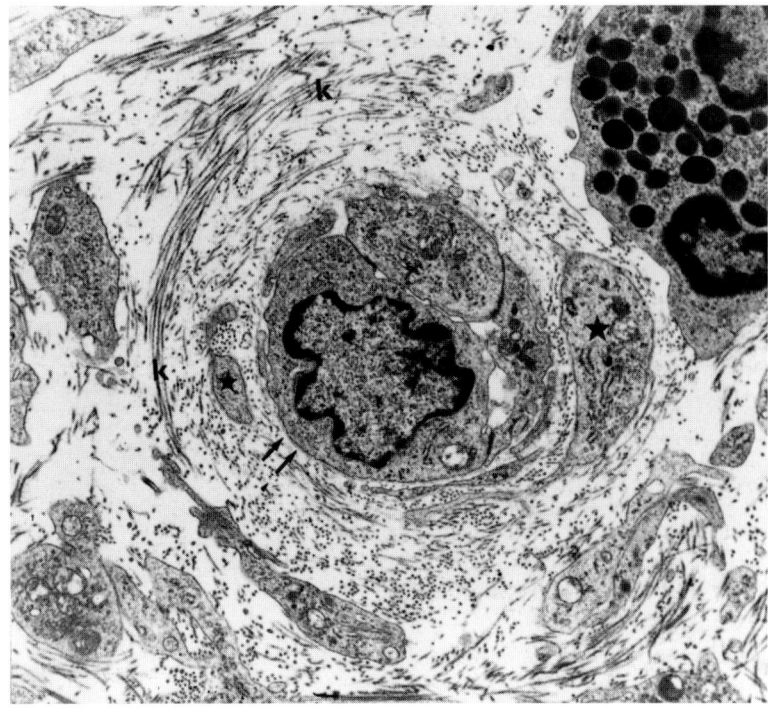

dominantly seen in endothelial cells still integrated into the vessel wall of established vessels, thus facilitating the connection of the newly formed intracytoplasmic lumen to the already established perfused lumen. This feature shows parallels to the development of vacuoles in endothelial cells which fuse to tubes as described by FOLKMAN and HAUDENSCHILD (1980). Reconstructions of serial sections of the amelanotic melanoma MeWo have shown that such intracellular lumina can be observed only over short distances of up to 5 μm. Such intracellular lumina must be discerned from autophagic vacuoles (FUCHS 1977) which are formed in order to increase the vascular luminal diameter (Fig. 2.3). Another form of lumen formation by single endothelial cells lies in the protrusion of one or two cell ramifications with reflexive contact formation between these branches (Fig. 2.2).

Intercellular lumen formation involves active migration of endothelial cells towards an angiogenic stimulus. Here, more plump pseudopodia or the eccentrically situated perikaryon of the endothelial cells involved veer out from the endothelial structure (Fig. 2.4). An intercellular lumen is thus already formed at the start of bud formation and not always, as generally postulated, after formation of a solid endothelial cell cord. Even immature, structurally not yet stabilized vessels characterized by high endothelial cells can form new buds (Fig. 2.4b), thus creating the impression of overhasty new vessel formation.

The endothelial cells of buds are connected to each other by simply structured cell contacts which frequently consist only of cell membrane appositions (Fig. 2.5). Pericytes and a basal lamina can be detected before the onset of blood flow (Fig. 2.5), whereas several authors have postulated no occurrence of pericytes before the fusion and further differentiation of the sprouts (FOLKMAN 1985) because of the assumed inhibitor function of the pericytes (SCHLINGEMANN 1990). Contrary to this, RHODIN and FUJITA (1989) have also described in normal secondary angiogenesis an early formation of pericytes around tiny sprout tips.

Recently, BURRI and TAREK (1990) described intussusception as a novel mechanism of capillary growth in the pulmonary microcirculation involving slender tissue posts extending across the capillary lumina. They described different phases of new vessel formation and postulated this mechanism for other organs, too. However, in extensive studies on serial sections we could not confirm the existence of intussusceptive capillary growth in tumors.

The further development of the sprouts to early vessel forms is characterized by the fusion of the sprout tips. Little is known about this crucially important process (PAWELETZ and KNIERIM 1989). The term "fusion" in itself indicates the lack of a definition of this step. It is more likely that the cell membranes do not fuse, but rather only form cell contacts. Therefore, the term "sprout anastomoses" as inaugurated by RHODIN and FUJITA (1989) should be preferred.

The structure of the early forms (Fig. 2.6), which are defined as vessels after the onset of blood flow, is mainly made up of a high, cell-organelle-rich endothelium. Compared to more mature tumor vessels, the endothelium is two to four times higher. The flattening and thinning of the endothelium sets in with further stabilization and increase in luminal diameter. However, only some of the early forms will persist: many early forms undergo degeneration and destruction (Fig. 2.6b) just like those sprouts that do not fuse with others. No reliable morphometric data are available on the fate of sprouts and early forms. We have the impression that at least two thirds of all formed sprouts will not differentiate into early forms or established vessels.

Frequently, villi-like protrusions of the luminal surfaces indicate the young age of these vessels (Fig. 2.6c). Abluminal protrusions of cell ramifications, again, indicate active migration and new sprouting (Fig. 2.7a). The endothelium is continuous with more or less vertically orientated cell contacts. Only in exceptional cases have we observed differentiations such as fenestrations shut by diaphragms (Fig. 2.7b).

Apart from these exceptions, no real further differentiation takes place. The wall of the vast majority of tumor vessels is made up only of an endothelial lining and a basal lamina. The endothelium becomes flattened and the vascular diameters increase, and there is no formation of a medial layer as in arteries or veins.

2.3.3
Architectural Development and Remodeling of the Tumor Vascular System

Microvascular corrosion casts of tumor vessels allow three-dimensional imaging of the above-mentioned sprouting and budding activity. However, new vessel formation can be observed in subcutaneously growing human xenografts, albeit not before the 4th day after grafting. Budding initially takes place on venules, smaller veins, and capillaries; arterial sprouting was never observed by us. This finding was reported by others, too. On principle, however, arteries or arterioles must also be involved in tumor angiogenesis, because otherwise no connection to

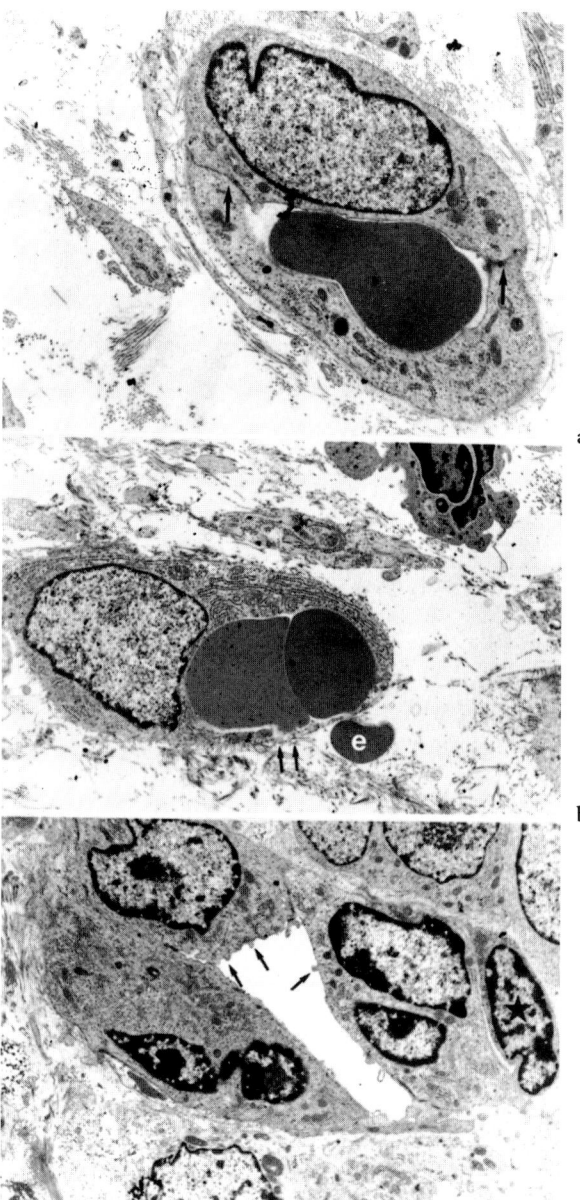

Fig. 2.6a–c. Sections of early vessel forms in the tumor periphery. **a** Already perfused early form with typical, still high endothelium. Note the vertically extended contact structures (*arrows*). **b** Early vessel with severe degeneration. In part, the vessel wall is destroyed (*arrows*). Extravascular erythrocytes (*e*) hint at more severe damage. **c** Villi-like luminal cell protrusions (*arrows*) indicative of early vessel forms. Note the marked perivascular cell (*star*) which shows the same morphological properties as the lumen-confining cells. MeWo, 8th day. **a** ×4970, **b** ×6620, **c** ×4970

the arterial blood flow would be possible. In other forms of secondary angiogenesis, sprouting of arterial endothelial cells was demonstrated, e.g., in the case of new myocardial capillary formation after internal thoracic artery grafting (BAIRD 1969).

Fig. 2.7a,b. Lowly differentiated, newly formed vesels. **a** Abluminal protrusions and pseudopodes (*arrows*) indicative of migration. **b** Partial compression of an early form by an extravascular cell (*star*) which can best be assigned to a tumor cell and which results in a thinning of the endothelium. *Inset*: fenestrated areas are superimposed by continuous endothelium. MeWo, 8th day. **a** ×8680, **b** ×5240, *inset* ×18 180

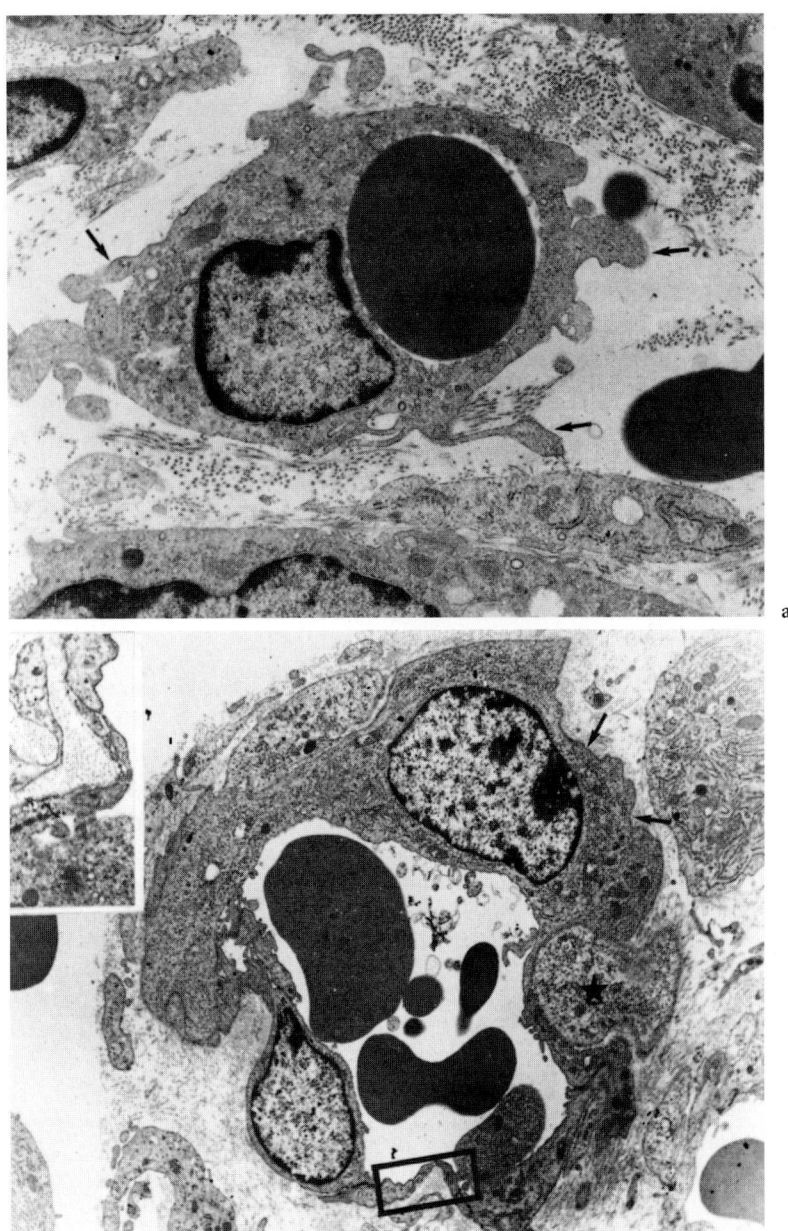

The newly formed vascular stretches do not show any hierarchy (Fig. 2.8). Abrupt changes of the diameters and extravasates due to leaky sprout tips are common features (Fig. 2.8a, c). Numerous fungiform and plump protrusions, bulges and flaps represent new sprouts originating from these new vascular stretches (Fig. 2.8a–c). These features must, however, be differentiated from artifacts. GRUNT et al. (1986) have presented a detailed classification of the morphological appearance of sprouts and buds based on corrosion casts of Lewis lung carcinomas showing many parallels to our own findings.

Tortuous courses and elongations as well as vascular compressions will be retained and will become aggravated in the course of time (Fig. 2.9). Features such as absence of hierarchy are more obvious in the more centrally located sinusoidal vessels (Fig. 2.9b) than at the borderline to the normal tissue. Especially in the xenografts, a vascular envelope (GRUNT et al. 1986) of sprouting and preexistent host vessels

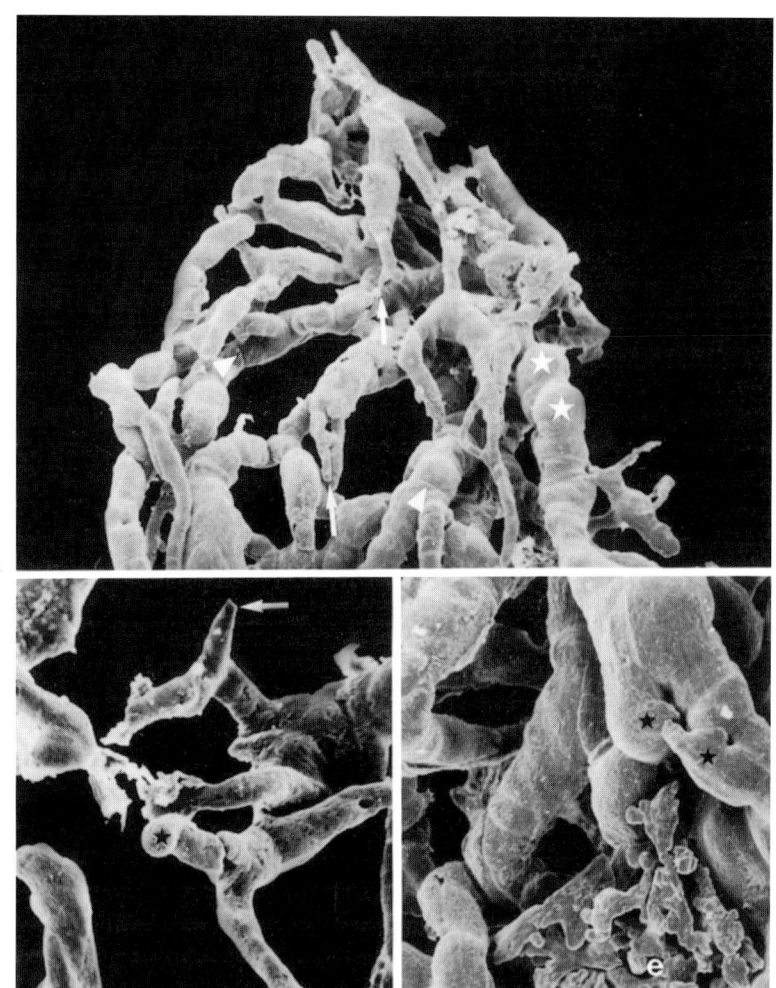

Fig. 2.8a–c. Scanning electron micrographs of a corrosion cast specimen of a melanoma 8 days after transplantation. a Numerous bud formations with globular shape (*stars*) as well as tiny tips (*arrows*). Note the circumferential furrows (*arrowheads*) and constrictions. b Detail of Fig. 8a. The *arrow* marks sprout tip, the *stars* fungiform protrusions. c Extravasates (*e*) direct vicinity of newly formed loop (*stars*). a ×310, b ×540, c ×540

frequently retains a certain hierarchy (Fig. 2.9c). This is not true for human primary tumors (MIODONSKI et al. 1980), since the invasive growth prevents formation of such a vascular and/or fibrous tissue capsule.

However, even though a tumor vascular envelope is formed in many experimental tumors, the intratumoral vascularization is obviously not better than in human primary tumors. Only few vessels run centripetally from this peritumoral plexus into the tumor center (Fig. 2.10a). Comparative morphometry shows an extreme heterogeneity of vascular densities both in human primary and in experimental tumors. Basic features of the tumor microcirculatory unit such as elongated sinusoids originating from and drained by large-caliber veins (Fig. 2.10b) as well as glomeruloidal arrangements of peripheral vessels (Fig. 2.10c) can be equally observed in all tumors irrespective of origin (for details see KONERDING et al. 1995). However, there is strong evidence that individual tumors express an individual microvascular architecture. Parameters defining the microvascular unit, such as intervascular distance, interbranching distance, and branching angles show tumor-type-

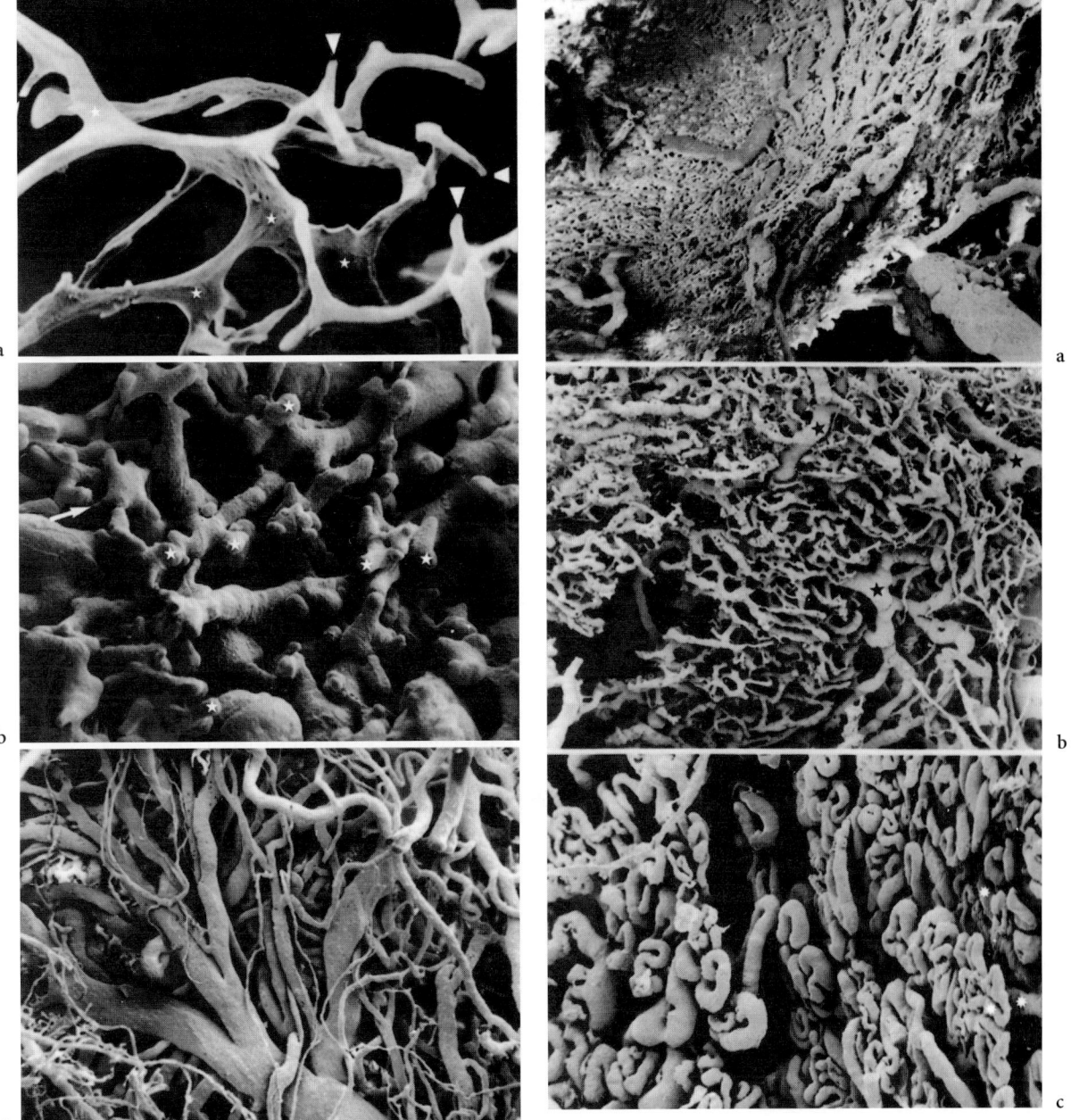

Fig. 2.9a–c. Different forms of tumor vessels within the same melanoma (25th day). **a** Peripheral, mainly compressed vessels (*stars*). *Arrowheads* indicate cul-de-sac vessels. **b** Sinusoidal plexus in the center with changes in diameter (*arrow*) and numerous globular outpouchings (*stars*). **c** Nearly normal hierarchy of venous vessels in the peripheral tumor vascular envelope. **a** ×380, **b** ×425, **c** ×110

Fig. 2.10a–c. Tumor sinusoids at different locations and ages. **a** Day 28: tumoral side of the tumor vascular envelope. Note the flattening of all major vessels (*star*). **b** Day 35: more centrally located sinusoids void of any hierarchy which are supplied and drained by large-caliber veins (*stars*). **c** Day 79: glomeruloidal arrangement of tortuous sinusoids in the periphery (*stars*). **a** ×110, **b** ×110, **c** ×140

dependent significant differences as demonstrated by MALKUSCH et al. (1995).

As already seen by transmission electron microscopy, vascular corrosion casts show that there is no further differentiation of newly formed tumor vessels into structurally complete arteries or veins. Sinusoidal capillary plexuses prevail. A real remodeling of the vasculature, as seen in wound healing or inflammatory angiogenesis, is widely missing, too (Fig. 2.11).

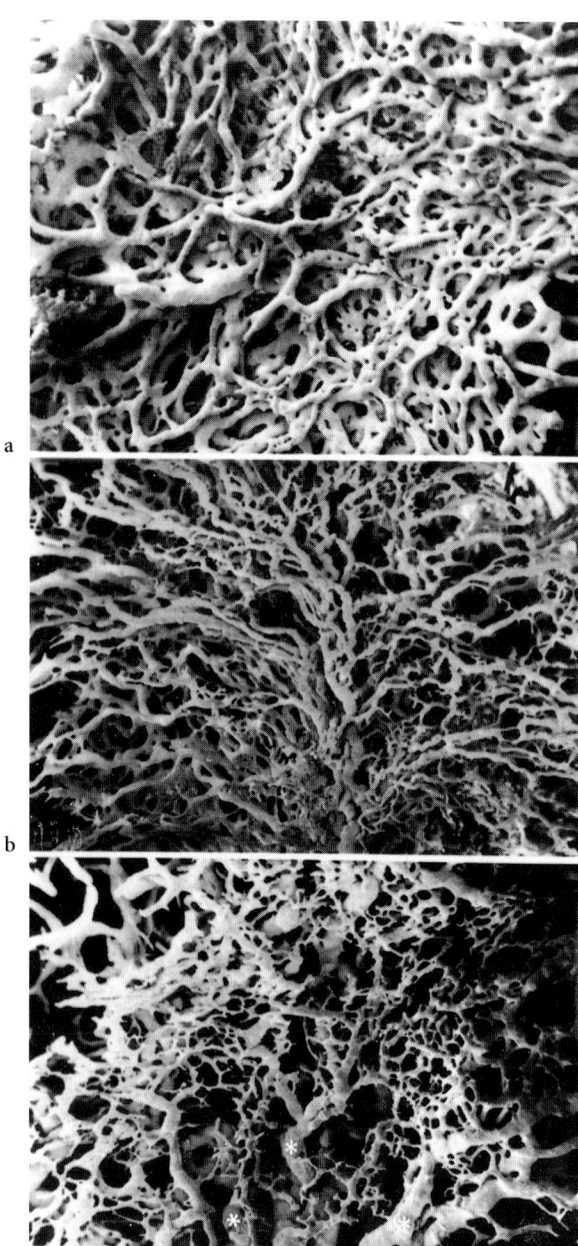

Fig. 2.11. Plexus formation in the periphery (**a**) and the center (**b, c**) of a melanoma on day 79. Note the absence of an hierarchy, the vessel elongation and the changes in diameter. *Asterisks* depict draining veins. **a** ×185, **b** ×42, **c** ×115

2.3.4
The Vascular Architecture of Human Primary Tumors

As pointed out earlier, most work on the tumor vascular system has involved experimental tumors, i.e., either chemically induced, inoculated, or xenografted ones of various origin (for a review see KONERDING et al. 1995). Only few spontaneously occurring human tumors were studied. Comparisons of the vascular architecture of primary tumors with that of experimental tumors, however, are necessary in order to determine to what extent results of experimental studies on tumor vascularity can be extrapolated to the human primary tumor. In this context it should also be remembered that the tumor growth rates, volume doubling times, and probably also the induction of the vascular system in human primary and experimental tumors show significant differences.

In contrast to the subcutaneous xenotransplantation of small tumor tissue slices, primary tumors, e.g., carcinomas, grow in tissue with a well-defined microvascular architecture which offers from the very beginning of tumor growth a sufficient nutritive blood flow. The microvasculature of the great intestine, for example, is characterized by a flatly extended mucosal capillary plexus with a hexagonal honeycomb pattern around the tissue of the mucosal glands (Fig. 2.12a,b). This subepithelial plexus is supplied by arteries that divide within the submucosa and is drained by venules originating immediately under the mucosal surface (Fig. 2.12b).

This honeycomb pattern is not retained in the tumor but replaced and/or modulated by tumor-induced vessels (Fig. 2.13). In the tumor periphery, dilated and elongated vessels take a tortuous course, as seen in the xenografts. However, there is no evidence for a vascular envelope surrounding the tumor (Fig. 2.13a). At the macroscopic border of tumor and normal mucosa, vascular density is not increased. However, the microvascular pattern is obviously altered.

The intratumoral vessels show all of the characteristics already seen in the xenografts and other experimental tumors (KONERDING et al. 1995). Sinusoidal plexuses with numerous flattenings, changes in diameter, bulges, and blind ends are seen throughout the tumor (Fig. 2.13b) The vascular density in central tumor areas is nearly as high as in the surrounding normal tissue, whereby highly and poorly vascularized areas can be located directly next to each other.

In summary, qualitative comparison of primary and experimental tumors shows a high degree of similarity in the vasculature, at least in the tumor center, and quantitative analysis of microvascular casts reveals similar ranges of parameters such as diameters and intervascular distances.

Fig. 2.12a,b. Microvascular corrosion cast of the ascending colon of a 75-year-old male: the mucosal plexus 10 cm distal of a colonic carcinoma. The mucosal capillary plexus (*c*) is arranged in a honeycomb pattern and shows numerous intercapillary connections (*arrows*). The supplying arterioles (*a*) and draining veins (*v*) take a straight course from the underlying submucosal vessels (*stars*). **a** ×145, **b** ×200

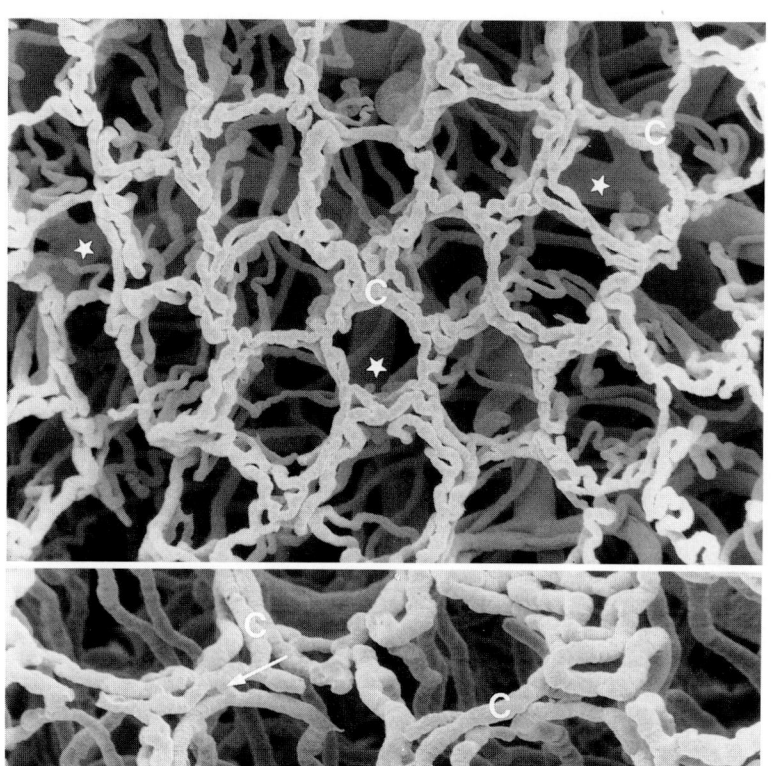

a

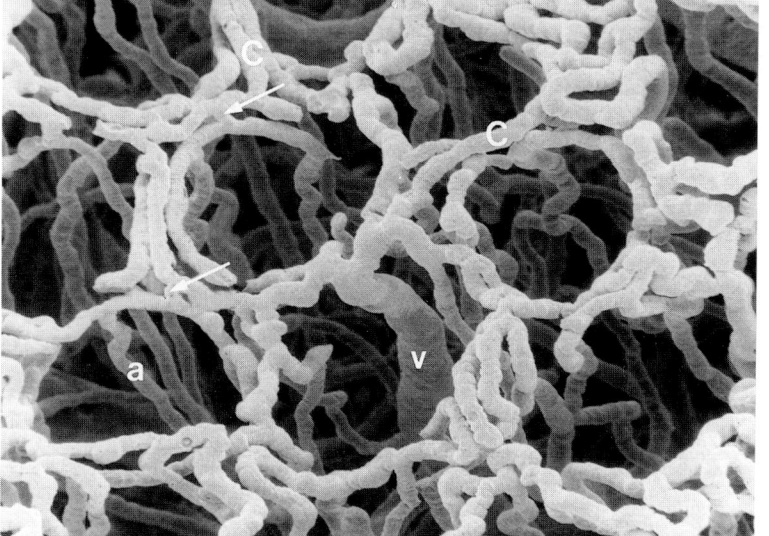

b

References

Ausprunk DH, Folkman J (1977) Migration and proliferation of endothelial cells in preformed and newly formed blood vessels during tumour angiogenesis. Microvasc Res 14:53–65

Baird RJ, Manktelow RT, Shah PA (1969) Pressure in a vascular implant in the myocardium during systole. Circulation 39:175–181

Budach V, Bamberg M, Donhuisen K, Schnidt U, van Beuningen D, Stuschke M (1986) Serial xenotransplantation of a human embryonal carcinoma in experimental urology. J Urol 136:1143–1147

Burri PH, Tarek MR (1990) A novel mechanism of capillary growth in the rat pulmonary microcirculation. Anat Rec 228:35–45

Clark ER, Clark EL, Williams RG (1939) Microscopic observations on the growth of blood capillaries in the living mammal. Am J Anat 64:251–301

Cliff WJ (1963) Observations on healing tissues: a combined light and electron microscopic investigation. Philos Trans R Soc Lond [Biol] 246:305–325

Cliff WJ (1981) Endothelial structure and ultrastructure during growth, development and aging. In: Schwartz CJ, Werthessen NT, Wolf S (eds) Structure and function of the circulation, vol 2. Plenum, New York, pp 695–718

Denekamp J (1989) Review: angiogenic attack as a therapeutic strategy. In: Michael B, Hance M (eds) Gray laboratory annual report 1989. Cancer Research Campaign, London, pp 6–13

Feigl W, Leu HJ, Lintner F, Pedio G, Susani M (1985) Neue Befunde zur Angiogenese im Rahmen von Organisationsprozessen. Vasa 14:371–378

Folkman J (1985) Tumor angiogenesis. Adv Cancer Res 43:175–203

Folkman J, Haudenschild C (1980) Angiogenesis in vitro. Nature 288:551–556

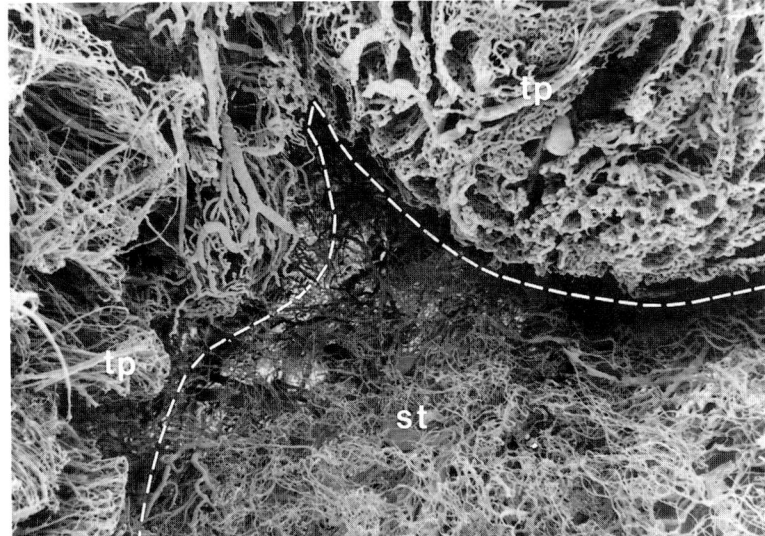

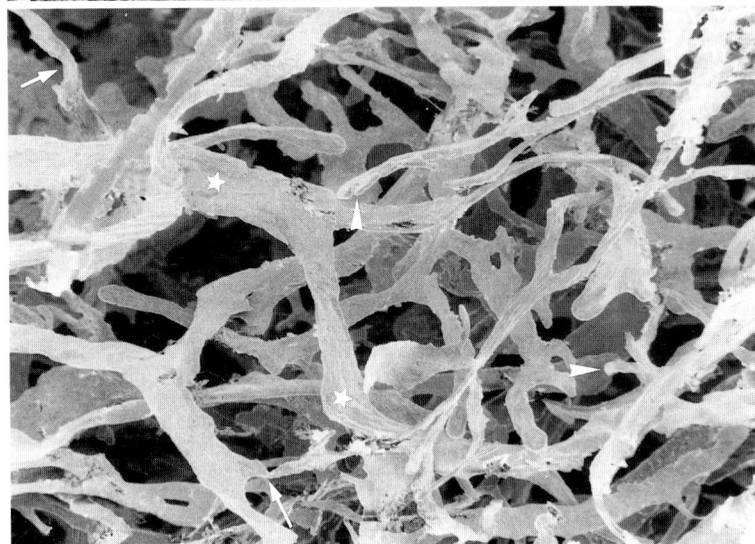

Fig. 2.13a,b. Vasculature of a colonic carcinoma (pT3, pN0, pMx; G2). Same patient as in Fig. 2.12. **a** In the tumor periphery (*tp*) numerous vessels are elongated and take a tortuous course. The normal hexagonal mucosal architecture is no longer visible. Even in the surrounding tissue (*st*) the original microvascular architecture is not retained. The *dashed line* marks the macroscopic tumor borderline. **b** Deeper vessels form a sinusoidal plexus without hierarchy with numerous blind ends (*arrow-heads*), flattenings (*stars*), and variations in diameter (*arrows*). **a** ×20, **b** ×130

Fuchs U (1977) Morphologische Reaktionsmusterd derterminalen Strombahn. In: Meesen H (ed) Mikrozirkulation. Springer, Berlin Heidelberg New York, pp 477–632 (Handbuch der allgemeinen Pathologie, 3rd edn, vol 7)

Furusato M, Shimoda T, Yokota K, et al. (1985) Angiogenesis of juvenile hemangioma. In: Tsuchiya M, Asano M, Mishima Y (eds) Microcirculation annual. Elsevier, Amsterdam, pp 101–107

Glaser BM, D'Amore PA, Seppa H, Seppa S, Schiffmann E (1980) Adult tissues contain chemoattractants for vascular endothelial cells. Nature 288:483–484

Glaser BM, Kalebic T, Garbisa S, Connor TB Jr, Liotta LA (1983) Degradation of basement membrane components by vascular endothelial cells: role in neovascularization. In: Development of the vascular system. Pitman, London, pp 150–162 (Ciba foundation symposium 100)

Grunt TW, Lametschwandtner A, Karrer K (1986) The characteristic structural features of the blood vessels of the Lewis lung carcinoma. Scanning Electron Microsc 2: 575–589

Hammersen F, Hammersen E, Osterkamp-Baust U (1983a) Bau und Funktion endothelialer Zellen. Eine Einführung. In: Messmer K, Hammersen F (eds) Structure and function of endothelial cells. Karger, Basel (Mikrozirkulation in Forschung und Klinik/Progress in applied microcirculation, vol 7)

Hammersen F, Osterkamp-Baust U, Endrich B (1983b) Ein Beitrag zum Feinbau terminaler Strombahnen und ihrer Entstehung in bösartigen Tumoren. In: Vaupel P, Hammersen F (eds) Mikrozirkulation in malignen Tumoren. Karger, Basel, pp 15–51 (Mikrozirkulation in Forschung und Klinik/Progress in applied microcirculation, vol 2)

Hammersen F, Endrich B, Messmer K (1985) The fine structure of tumor blood vessels. I. Participation of non-endothelial cells in tumor angiogenesis. Int J Microcirc Clin Exp 4:31–43

Hata J (1981) Morphological studies on tumor angiogenesis. Microvasc Res 23:131–147

Heimark RL, Schwartz SM (1988) Endothelial morphogenesis. In: Simionescu N, Simionescu M (eds) Endothelial cell biology in health and disease. Plenum, New York, pp 123–143

Hirst DG, Denekamp J, Hobson B (1982) Proliferation kinetics of endothelial and tumour cells in three mouse mammary carcinomas. Cell Tissue Kinet 15:251–261

Illig L (1961) Die terminale Strombahn. Capillarbett und Mikrozirkulation. In: Hegglin R, Leuthardt F, Schoen R, Schwiegk H, Zollinger HU (eds) Pathologie und Klinik in Einzeldarstellungen, vol 10. Springer, Berlin Göttingen Heidelberg

Jain RK (1988) Determinants of tumor blood flow: a review. Cancer Res 48:2641–2658

Konerding MA (1991) Scanning electron microscopy of corrosion casting in medicine. Scanning Microsc 5:851–865

Konerding MA, Steinberg F, Budach V (1989a) The vascular system of xenotransplanted tumors – scanning and light microscopic studies. Scanning Microsc 3:327–336

Konerding MA, Steinberg F, Streffer C (1989b) The vasculature of xenotransplanted human melanomas and sarcomas. 1. Vascular corrosion casting studies. Acta Anat (Basel) 136:21–26

Konerding MA, Steinberg F, Streffer C (1989c) The vasculature of xenotransplanted human melanomas and sarcomas. 2. Scanning and transmission electron microscopic studies. Acta Anat (Basel) 136:27–32

Konerding MA, Miodonski AJ, Lametschwandtner A (1995) Microvascular corrosion casting in the study of tumor vascularity: a review. Scanning Microsc 9:1233–1244

Köpf-Meier P, Kestenbach U (1989) Host-supplied connective tissue as a guide for proliferating cells in human tumor xenografts. Acta Anat (Basel) 135:289–295

Mahadevan V, Hart IR (1990) Metastasis and angiogenesis. Acta Oncol 29:97–103

Malkusch W, Konerding MA, Klapthor B, Bruch J (1995) A simple and accurate method for 3-D measurements in microcorrosion casts illustrated with tumour vascularization. Ann Cell Pathol 9:69–81

Miodonski A, Kus J, Olszewski E, Tyrankiewicz R (1980) Scanning electron microscopic studies on blood vessels in cancer of the larynx. Arch Otolaryngol Head Neck Surg 106:321–332

Paweletz N, Knierim M (1989) Tumor-related angiogenesis. Crit Rev Oncol Hematol 9:197–242

Petry G, Heberer G (1957) Die Neubildung der Gefäßwand auf der Grundlage synthetischer Arterienprothesen. Langenbecks Arch Chir 286:249–257

Rhodin JAG, Fujita H (1989) Capillary growth in the mesentery of normal young rats. Intravital video and electron microscope analyses. J Submicrosc Cytol Pathol 21:1–34

Schlingemann RO (1990) Vascular markers in tumor biology. Königliche Bibliothek, Den Haag, pp 11–37

Simpson JG, Fraser RA (1983) Angiogenese in malignen Tumoren. In: Vaupel P, Hammersen F (eds) Mikrozirkulation in malignen Tumoren. Karger, Basel, pp 1–32 (Mikrozirkulation in Forschung und Klinik/Progress in applied microcirculation, vol 2)

Thoma R (1893) Untersuchungen über die Histogenese und Histomechanik des Gefäß-Systems. Enke, Stuttgart

Weller RO, Foy M, Cox S (1977) The development and ultrastructur of the microvasculature in malignantgliomas. Neuropathol Appl Neurobiol 3:307–322

3 The Role of the Microcirculation in the Treatment of Malignant Tumors: Facts and Fiction

B. Endrich[1] and P. Vaupel[2]

3.1 Introduction

Cancer is characterized by progressive growth of cells that have lost their proliferative control. These cells ultimately destruct tissue and metastasize to organs distant from the primary site. In general, only one of three cancer victims can be cured by a single treatment modality, usually surgery, less frequently radio- or chemotherapy. If, however, the chosen modality fails, the cell mass will again reach a volume of only approximately $1 \, mm^3$ (ca. 10^6 cells) without "nutritive deficiency" (Ausprunk and Folkman 1977; Folkman 1986; Folkman et al. 1971; Gullino 1991), despite its loss of proliferative control. Beyond this "preclinical growth stage," diffusional supply and waste removal are inadequate. Further tumor growth depends on nutrient supply via a network of microvessels (Denekamp 1993; Denekamp et al. 1983; Endrich 1988; Endrich and Götz 1991; Jain 1988; Vaupel et al. 1989a; Vaupel 1992).

Therefore, the vascularization of solid tumors is a prerequisite if a clinically relevant size is to be reached. This review (a) summarizes today's understanding of the microanatomy and pathophysiology of microscopic blood vessels in tumors and (b) looks at the possible relevance of these findings for clinical diagnosis and treatment as well as for the patient's outcome.

In contrast to some recent clinical work on tumor microcirculation (Obermair et al. 1994a, b; Weidner et al. 1992; Weidner 1995), a distinction should be made between new, sprouting blood vessels and the "mature" tumor microcirculation. On the basis of the data presented, we will provide good reasons for doing so while discussing the possible impact of various factors in clinical oncology.

3.2 Vascular Anatomy of Tumors During Early Growth

3.2.1 Vascular Supply/Arterial Vessels

A neoplasm incorporates arterial vessels of all categories during growth because the walls of arteries exhibit a striking "resistance" to cancer cell invasion and keep their normal vessel wall structure for quite some time (Endrich 1988; Intaglietta et al. 1977; Reinhold et al. 1977; Reinhold 1979; Vaupel and

[1] B. Endrich, Dr. med., Krankenhaus St. Elisabeth, Chirurgische Abteilung, Ziegelstrasse 38, D-89407 Dillingen/Donau, Germany
[2] P. Vaupel, Prof. Dr. med., Institute of Physiology and Pathophysiology, University of Mainz, Duesbergweg 6, D-55128 Mainz, Germany

Dedicated to Prof. Dr. med., Dr. med., h.c. K. Messmer on the occasion of his 60th birthday

MENKE 1989; VAUPEL et al. 1988). When the blood supply to small tumors is traced from its inception, a great number of thin-walled channels are seen that have developed from capillary sprouts, finally forming a chaotic microvascular network (for reviews see DENEKAMP et al. 1983; ENDRICH 1988; GULLINO 1991; JAIN 1988; VAUPEL 1977; VAUPEL et al. 1989a; SKINNER et al. 1990).

3.2.2
Neovascularization

As the neoplastic mass enlarges, the host tissue must provide building blocks for vascular stroma as already suggested in the past century when PAGET (1989) presented his "seed and soil" hypothesis of metastatic growth. The nonrandom pattern of visceral metastases suggested to Paget a special affinity of tumor cells for the "milieu" of specific organs. Verifying this notion, GULLINO and GRANTHAM (1961) found already 35 years ago that a hepatoma cell culture was unable to form collagen. However, when transplanted into a host, tumors developed with a collagen content equal to that obtained by implanting small fragments of whole hepatoma tissue.

In the early 1970s the presence of an angiogenic stimulus was suggested, since the generation of new tumor blood vessels was found to be critical for tumor expansion. Subsequently, a number of angiogenesis factors were identified (FOLKMAN 1971; FOLKMAN et al. 1971). As a specific feature of malignant growth, however, many tumors impose modifications on a microvascular bed that are different from angiogenesis induced by nonmalignant cells (for a review see BREM et al. 1972). For instance, a histologic cross section of a capillary in the normal brain reveals one or two endothelial cells per lumen; in brain tumors such as the glioblastoma, however, five to ten endothelial cells may occupy a single lumen. Most tumor microvessels are dilated and they sometimes contain tumor cells within the endothelial lining (BLOOD and ZETTER 1990; FOLKMAN 1990; HAMMERSEN et al. 1985; HÖCKEL et al. 1993b; KLAGSBRUN and D'AMORE 1991; KONERDING et al. 1989a,b,c; STEINBERG et al. 1990; YAMAURA and SATO 1974; YAMAURA et al. 1976). Although microvessels in tumors are usually pericyte-poor and do not regain their normal density of pericytes until capillary growth has ceased, capillaries in some tumors contain excessive numbers of pericytes. Why such enormous differences exist and whether they

are of clinical significance (WURSCHMIDT et al. 1990; VERHOEVEN and BUYSSENS 1988), is far from being understood at present. In recent years, however, great efforts have been made to elucidate biochemical events during angiogenesis. The evolution of new minute blood vessels has received increasing attention among clinicians, particularly because of the capacity of endothelial cells to respond in a specific way to a specific stimulus.

3.3
Clinical Implications of Angiogenesis in Malignant Tumors

3.3.1
Angiogenic Factors

Stimulation of new vessel growth allows tumor cells to tap into a source from which not only blood-borne nutrients but also "growth regulating" molecules can be released. The angiogenic capacity of cellular or transcellular fluid, as well as of excretions without such a capacity in benign tissue, could be of diagnostic value. The deliberate search for such diagnostic tools revealed media that became angiogenic in vivo upon contact with a growing tumor. Some examples are:

a) Brain tumors of children in which angiogenesis and the immunoreactivity of basic fibroblast growth factor (bFGF) in cerebrospinal fluid are correlated with the likelihood of relapse (LI et al. 1994); in fact, many cell lines derived from pediatric tumors (e.g., retinoblastoma, neuroblastoma, rhabdomyosarcoma, and Ewing's sarcoma) contain this angiogenic factor (SCHWEIGERER 1993)

b) A fair number of carcinomas in which urinary bFGF levels decrease with tumor eradication but increase with tumor relapse (NGUYEN et al. 1994), suggesting that this parameter is suitable for assessing a tumor's response

c) Tumors in children such as Wilms' tumor, osteosarcoma, or Ewing's sarcoma, in which NMR imaging illustrates dense vascularization during early tumor growth; vascularization, to which angiogenesis factors including bFGF may contribute, permits visualization, spatial assignment, and demarcation of the tumor (SCHWEIGERER et al. 1987a)

d) The neovascular membrane observed on the iris that is quite frequently associated with a locally infiltrating retinoblastoma; again, it seems likely that angiogenesis factors such as bFGF are released by the tumor, thus contributing to retinoblastoma-associated neovascularization (SCHWEIGERER et al. 1987b,c)

e) Suborbital hematomas sometimes suggesting in pediatric oncology the presence of an orbital metastatic neuroblastoma; this tumor is furnished with leaky blood vessels, possibly as a result of enhanced vascular endothelial expression (LEVY et al. 1989)

f) Cerebrospinal fluid with respect to glioblastomas and meningiomas, aqueous humor with respect to ocular tumors, and urine with respect to bladder carcinomas suggesting that the release of an angiogenic stimulus into the extracellular compartment might be useful in early diagnosis (CHODAK et al. 1981; GLASER et al. 1980; GULLINO 1991; LÓPEZ-POUSA et al. 1981; TAPPER et al. 1981)

At present, however, such angiogenesis tests cannot be applied on a broad basis in clinical practice (CHODAK et al. 1981; GLASER et al. 1980; GULLINO 1991; LÓPEZ-POUSA et al. 1981; SCHWEIGERER 1995; TAPPER et al. 1981).

Other approaches related to microvascular research have also experienced a revival in recent years. BREM et al. (1992) found fibroblast growth factor predominantly present in surgical specimens from 52 human brain tumors. This suggests a cellular depot for this potent growth factor that mediates angiogenesis and tumorigenesis. Based on these findings, BREM et al. (1992) concluded not only that fibroblast growth factor might play a role in the transition from the benign to the malignant phenotype, but also identified microvascular proliferation as a possible target for anticancer therapy.

Similar observations were made by LI et al. (1994) in brain tumors of children. Such studies also created a great deal of enthusiasm in specifically targeting the microcirculation in tumor therapy.

3.3.2
Antiangiogenesis

Only a few years ago, it was proposed that solid malignant tumors should be primary targets for *antiangiogenic treatment*; in fact, three options have been considered so far:

1. Inhibition of the turnover from an avascular primary tumor into a fully vascularized tumor (quite frequently, this will not be feasible because avascular primary tumors have a diameter of only a

Table 3.1. Preclinical studies with antiangiogenic substances (for further details see HAYES 1994; LICHTENBELD et al. 1996; SCHWEIGERER 1995)

Compound	Trial phase	Tumor type	Institution
1. AGM 1470 (TNP-470) (angioinhibin)	I	AIDS associated Kaposi's sarcoma	Natl. Cancer Inst., USA
2. See 1	I	Metastatic cancer of prostate or cervix	M.D. Andersen Hospital, USA
3. See 1	I	Solid tumors	Dana Farber Cancer Institute, Boston, USA
4. Carboxy-aminoimidazole	I	Solid tumors	Natl. Cancer Inst., USA
5. Metalloproteinase inhibitor BB-94 (balimastat)	I	Female patients with ascites due to malignant disease	Western General Hospital, Edinburgh, UK; St. John's Hosp., Howden, UK
6. Peptidoglycan DS 4152	I	Relapsed malignancy of breast, lung, head and neck, soft tissue	See 5
7. Platelet factor 4	I/II	Kidney and colon cancer, melanoma, Kaposi's sarcoma	Various cancer centers in USA
8. Polysaccharide CM 101	I	Kaposi's sarcoma, various malignomas with and without metastases	University Hospital, Vanderbilt, Tennessee, USA

few millimeters and will usually escape clinical detection)

2. Slowdown of tumor progression by preventing a tumor from becoming highly vascularized
3. Prevention of neovascularization of distant metastases (SCHWEIGERER 1995)

In the very near future, pharmacologic antiangiogenesis might be introduced into clinical practice; antiangiogenic agents are currently being tested in at least eight clinical trials (Table 3.1).

Further developments of inhibitors of angiogenesis are listed in Table 3.2; most of them attenuate signals that stimulate angiogenesis (AUERBACH and AUERBACH 1994; GASPARINI and HARRIS 1994).

Some of these compounds, however, were found to prevent neovascularization in chronic inflammatory diseases such as arthritis as well as psoriasis.

A summary of recent findings in this field has been provided by FOLKMAN and BREM (1992) with a few promising approaches related to clinical oncology:

a) Steroid-heparin combinations were found to dissolve the basement membrane of growing capil-

laries (INGBER et al. 1986). One application presently questioned among clinicians is the treatment of corneal neovascularization which is resistant to conventional therapy. It also seems possible to suppress neovascularization in mast cell-rich inflammatory lesions or hemangiomas by corticosteroids alone because of the high levels of endogenous heparin.

b) A number of studies suggest that components of the basal lamina such as fibronectin, collagen type IV, and laminin undergo a more rapid turnover in growing capillaries. Consequently, agents that interfere with the synthesis of these components or stimulate their degradation enhance destruction of the basement membrane, thus inhibiting angiogenesis. In particular, interferon alpha has received marked attention during recent years. This agent has been used for successful treatment of at least 20 children suffering from pulmonary capillary hemangiomatosis as well as systemic hemangiomatosis in which steroid therapy had already failed (WHITE et al. 1989). However, beyond this rare disease with its excessive growth of capillaries, interferon alpha

Table 3.2. Selection of widely known angiogenesis inhibitors classified according to their presumed action (for a more detailed summary see SCHWEIGERER 1995; SCOTT and HARRIS 1994)

Substance	Category	Presumed mechanism	Author
Interferon alpha	Inhibition of compound delivery	Interference with production or export of angiogenic stimuli	SCHWEIGERER 1995
Specific antibodies vs bFGF or VEGF	Intransit inhibitors	Inactivation of angiogenic factors after their release from tumor cells or extracellular matrix	HORI et al. 1991; KIM et al. 1993; KONDO et al. 1993
Protamine Platelet factor 4 Suramine	Inhibition of receptor action	Blocking of binding to the specific receptor	NEUFELD and GOSPODAROWICZ 1987; BROWN and PARISH 1994; COFFEY et al. 1987
Angiostatin D-Penicillamine Gold compounds Tamoxifen Thrombospondin Tumor necrosis factor alpha	Inhibition of endothelial cell proliferation	Multiple	O'REILLY et al. 1994; MATSUBARA and ZIFF 1987; MASTSUBARA et al. 1989; GAGLIARDI and COLLINS 1993; BAGAVANDOSS and WILKS 1990
Cartilage derived inhibitor Dexamethasone Medroxyprogesterone Retinoid acid	Invasion inhibitors	Prevention of degradation of the endothelial cell basement membrane	MOSES et al. 1990; WOLFF et al. 1993; BLEI et al. 1993; PEPPER et al. 1994
Tumor necrosis factor alpha	Endothelial cell killers	Stimulation of apoptosis of endothelial cells	ROBAYE et al. 1991

has not been established as a potent agent in medical oncology (FOLKMAN 1989).

This strongly suggests that the effectiveness of nearly all of the antiangiogenic agents listed above may still be restricted to specific capillary beds, since some tumors are more angiogenic than others (ENDRICH and VAUPEL 1995a,b; TAKAMIYA et al. 1993).

Two *endogenous angiogenesis inhibitors* deserve further attention, namely, angiostatin and thrombospondin. It was assumed that a tumor mass containing 10^9 cells could produce these inhibitors in larger quantities (FOLKMAN 1995) before they appear in the circulation. At least one of these inhibitors of endothelial cell proliferation (angiostatin) remains detectable for approximately 5 days after removal of the primary tumor. Since angiogenesis factors are eliminated from the circulation in less than 1h, this time lapse might explain why metastases, which were unable to induce angiogenesis and remained dormant in the presence of the primary tumor, appeared suddenly vascularized and grew rapidly after the primary tumor was surgically removed (FOLKMAN 1995).

According to McCARTY (1996), it also seems appropriate to consider nutritional factors for a role in modulating angiogenesis. Fish oil (long chain omega-3 polyunsaturated) was proposed in particular to impede angiogenesis by down-regulating protein kinase C and modulating eicosanoid production. Based on a fair number of studies in rats and mice, it was felt that the application of supplementary fish oil in human cancer therapy should be an urgent priority.

There is still considerable concern about establishing antiangiogenic therapies in the clinical setting. It is widely believed that reduction of vessel growth or elimination of tumor microcirculation might compromise the therapeutic impact of other modalities. With a reduced number of tumor blood channels, it is conceivable that cytostatic drugs could be delivered in smaller quantities to the tumor.

The opposite, however, might be true: Cytostatic agents achieve high concentrations at the blood/endothelial interface where the rapidly proliferating endothelium of the tumor microcirculation is particularly sensitive to agents such as methotrexate (HIRATA et al. 1989), vincristine (BAGULEY et al. 1991), bleomycin (SCHWEIGERER 1995), and nitrosourea (LAZO 1986). As a result, endothelial growth will be inhibited and tumor blood flow reduced due to the decreased number of perfused capillaries. At the same time, the high intratumoral interstitial pressure (see also Sect. 3.4.2) may decrease and some "decompression" of microvessels within the tumor is seen. In fact, it has been proposed for quite some time that lowering interstitial tumor pressure might offer a means of improving chemotherapeutic efficacy (JAIN 1989, 1990). This has already been achieved by irradiation (LAZO 1986), hyperthermia (ENDRICH 1988; OTTE 1988), and, more recently, antiangiogenesis (SCHWEIGERER 1995). Recent studies have supported the notion that angiogenesis inhibitors increase blood flow by "unpacking" tumor cells and reduce interstitial tumor pressure (FOLKMAN 1995). Such data are corroborated by the finding that antiangiogenic treatment per se can potentiate chemotherapy (KATO et al. 1994; MONTI and SINHA 1994; TEICHER et al. 1992, 1993, 1994).

Another possible drawback is that tumor hypoxia can be aggravated by antiangiogenesis, thus compromising the therapeutic efficacy of standard radiotherapy. Again, there seems to be sound evidence that this novel treatment potentiates radiotherapy (MILAS et al. 1991; TEICHER et al. 1992) by affecting the reoxygenation of tumor cells after lowering interstitial tumor pressure. Consequently, due to the decompression of blood channels, oxygenation might be improved, rendering the tumor more susceptible to radiotherapy with sparsely ionizing radiation.

Even so, there are still a number of open questions regarding potential *side effects of antiangiogenic treatment*. For instance, the most significant side effect will be on the menstrual cycle, where this treatment could eventually cause secondary amenorrhea. Another undesirable reaction seems to be the impairment of wound healing or realignment of vessels in fractured bones after antiangiogenic treatment. Such possible unwanted effects on "physiological angiogenesis," as outlined above, must be considered prior to a general use of antiangiogenesis in clinical practice.

Increased attention must also be paid to the work of GAGLIARDI and COLLINS (1993). Their novel findings showed that antiestrogens are effective inhibitors of angiogenesis in breast cancer. Moreover, the fact that angiostatic activity is not altered in the presence of excess estrogens suggests that this activity is exerted via mechanisms other than the inhibition of estrogen action. They also suggested that this angiostatic activity may contribute to the therapeutic effect of antiestrogens in estrogen receptor-negative tumors. Although microscopic counting is very time-consuming, medical oncology as well as other subspecialities could benefit from these new microvascular techniques as well (FOLBERG et al. 1992).

3.3.3
Microvascular Density as a Prognostic Indicator in Clinical Oncology

In recent years, angiogenesis in malignant tumors has frequently been proposed as a prognostic tool. Immunohistochemistry for staining endothelial cells and microvascular imaging as well as modern computer techniques were introduced in an attempt to identify high risk groups and correlate disease stage with the patient's prognosis. Such human studies have been reported for breast, cervix, colon, and prostatic carcinomas as well as for tumors of the oral cavity.

As shown in Tables 3.3 and 3.4, some authors provide clinical evidence of a relation between tumor cell proliferation and vascularity; this hypothesis has been particularly enforced by the use of the parameter "microvascular density" as a prognostic indicator in human mammary carcinoma in which a correlation is believed to exist between increased intratumoral vessel density, the risk of metastasis, and patient's survival (WEIDNER et al. 1992; SNEIGE et al. 1992; SAHIN et al. 1993; PAGE and DUPONT 1992). In addition, in a greatly increasing number of publications, such a relation was demonstrated for prostatic carcinomas (BIGLER et al. 1993; BRAWER et al. 1994; FEGENE et al. 1993; VESALAINEN et al. 1994; WEIDNER et al. 1993; WAKUI et al. 1992). Further evidence exists for head and neck carcinomas (GASPARINI et al. 1993; MIKAMI et al. 1991; ALBO et al. 1994), rectal carcinomas (SACLARIDES et al. 1994),

Table 3.3. Prognostic value of intratumoral microvessel density – review of the literature with particular reference to mammary carcinomas

Authors	Year	Tumor/patients	Method of staining Mean follow-up (years)	Prognostic value
BEVILACQUA et al.	1995	Node-negative breast cancer in 211 patients	Anti-CD 31 antibody, 6 y follow-up	+
GOULDING et al.	1995	165 breast cancer patients, primarily node-negative	QBEnd/10 and JC 709 antibody, 15 y follow-up	–
HALL et al.	1992	87 breast cancer patients, mainly node-negative ($n = 50$)	F8RA/vWF antibody, 1.5 y follow-up	–
HORAK et al.	1992	103 breast cancer patients	JC 70 antibody, 2.5 y follow-up	+
LAURIA et al.	1995	1408 breast cancer patients (T_1–T_{3a}, N_0–N_2, M_0)	Study of blood vessel invasion, no specific staining mentioned, 5 y follow-up	–
MILIARAS et al.	1995	42 breast cancer patients (T_1–T_{3a}, N_0–N_1, M_0)	F8RA, 4 y follow-up	–
OBERMAIR et al.	1994b	106 patients with various stages of breast cancer	F8RA, 7 y follow-up	+
WEIDNER et al.	1992	165 breast cancer patients in a prospective blind study	F8RA, 5 y follow-up	+
TOI et al.	1993	125 breast cancer patients (T_1–T_3, N_0–N_2, M_0)	F8RA, 5 y follow-up	+
VAN HOEF et al.	1993	93 node-negative breast cancer patients	F8RA/vWF antibody	–
VISSHER et al.	1993	58 patients with various stages of breast cancer	Anti-type IV collagen, 4.3 y follow-up	+

F8RA, factor VIII-related antigen; vWF, von Willebrand factor.

Table 3.4. Role of microvessel density at other primary tumor sites – review of the literature (for further details see WEIDNER 1995)

Authors	Year	Tumor/patients	Method of staining	Prognostic value
SARBIA et al.	1996	150 patients with esophagus carcinomas at various stages	QBEnd/10 antibody	−
SHPITZER et al.	1996	25 patients with oral tongue T_1 cancer	F8RA	+
TAHAM and STEIN	1995	41 patients with lip carcinomas	F8RA	−
MAEDA et al.	1995	108 patients with gastric carcinomas at various stages	F8RA	+
TANIGAWA et al.	1996	110 patients with gastric carcinomas at various stages	Anti-CD 34 antibody	+
MACCHIARINI et al.	1992	87 patients with non-small cell lung carcinomas (T_1, N_0, M_0)	F8RA, PC 10 antibody	+
MATTERN et al.	1995	204 patients with non-small cell lung carcinomas at various stages	F8RA	−
BOSSI et al.	1995	178 patients with colorectal cancer at various stages	Anti-CD31 antibody	−
FRANK et al.	1995	105 patients with colorectal cancer at various stages	F8RA	+

testicular tumors (OLIVAREZ et al. 1994), ovarian cancer (HOLLINGSWORTH et al. 1994), and urothelial carcinomas (JAEGER et al. 1994). However, as demonstrated in Tables 3.3 and 3.4, conclusive and unanimous evidence will not be available as long as staining methods, the individual bias of selecting microvessel areas, and the counting procedure itself are still questioned (DAVIDSON et al. 1994; ENDRICH and VAUPEL 1995b). As a result, a lack of correlation between microvessel density and patient's prognosis was reported among others by BOSSI et al. (1995) for colorectal tumors and by SARBIA et al. (1996) for carcinoma of the esophagus.

From the viewpoint of a clinician, it seems questionable that an angiogenesis index alone will ever reasonably identify those patients who will finally suffer from distant metastases. There are two reasons for this statement:

1. At the time of diagnosis, human tumors are heterogeneous and consist of subpopulations of cells with different biological properties that include invasion and metastasis (FIDLER 1990).
2. The process of tumor growth is sequential and selective and consists of a whole series of independent steps that include angiogenesis, motility, invasion, survival in the circulation, adhesion, extravasation, and proliferation. To generate metastases, tumor cells must go through all of these steps (ELLIS and FIDLER 1995; FIDLER 1990). The complexity of this cascade suggests that prediction of metastatic potential and prog-

nosis of a patient requires a multiparametric approach.

Moreover, experimental data in xenografted systems show that only 20%–85% of tumor microvessels are perfused at a given time. Therefore, the morphological aspect of the surgical specimen at the moment of fixation might not reflect the "true or functional vascular density" of the tumor under study and prognostic evaluation.

Other experimental studies have revealed that the mitotic index was high in the vicinity of microvessels or at high oxygen tensions in the tissue and decreased with increasing distance from blood vessels. According to the radial O_2 tension gradient from the vessel into the tissue, these results merely confirm the role of oxygen as a limiting substrate for the proliferation of tumor cells. It should also be noted that there is an intravascular O_2 tension gradient from the arterial to the venous end. This implies substantial differences of proliferative indices not only as a function of distance from the microvessels but also as a function of the position with respect to the arterial or venous end of the microcirculation.

GABBERT et al. (1982) reported experiments dealing with cell proliferation in chemically induced colon carcinomas of rats. These tumors were characterized by different degrees of differentiation but equally developed vascularization. Since significant differences in proliferation pattern were found, the authors concluded that the proliferation behavior of a tumor may depend primarily on inherent properties of tumor cell proliferation, while vascularization merely provides the conditioning for tumor enlargement.

In human tumors, similar studies are scarce: Using six different human melanoma xenograft lines, LYNG et al. (1992) concluded that blood supply (number of vessels) per viable tumor cell was likely to be the key to the proliferative activity of tumor cells. Studying cell proliferation and vascularization in human breast carcinomas, MONSCHKE et al. (1991) found that highly vascularized tumors exhibited greater proliferative activity than tumor specimens with a lower vascular density.

To summarize undoubtedly promising activities, much more work needs to be done particularly to solve the discrepancy presently existing between clinical and experimental data.

3.4
Pathophysiology of the Mature ("Established") Tumor Microcirculation

As a tumor grows, "tumorous" blood channels form a chaotic network of microvessels ("established" tumor microcirculation; ENDRICH et al. 1979). Such vessels are characterized by a lack of smooth muscle cells and nervous components as well as tortuosity, arteriovenous anastomoses, irregular vessel branching patterns, and vessels with little or no endothelial lining (for a summary see HAMMERSEN et al. 1985). In addition, a functioning lymphatic system is absent in most solid tumors. Thus, interstitial fluid drainage is limited even though the total number of venous vessels is much larger in tumor tissue than in its benign counterpart. Therefore, it is not surprising that interstitial pressures as high as 45 mmHg have been reported for human tumors (BOUCHER et al. 1990; LESS et al. 1992). Such hydrostatic pressures were identical to arteriolar pressures reported for tumors in the rat and in the hamster (ENDRICH 1988; ENDRICH and GÖTZ 1991; BOUCHER and JAIN 1992; PETERS et al. 1980). Quite recently, simultaneous measurements of interstitial and venular pressures on the surface of a mammary adenocarcinoma in the rat also showed more or less identical values (BOUCHER and JAIN 1992). However, local changes in pressure and diameter might also be elicited by swelling and destruction of endothelial cells and adhesion of leukocytes, platelets, or even cancer cells. It is evident that all of the factors mentioned above may contribute to make the flow of blood through any malignant tumor more chaotic, sluggish, hetero-

geneous, and ultimately inadequate (ENDRICH 1988; ENDRICH and GÖTZ 1991; JAIN 1988; VAUPEL 1977; VAUPEL et al. 1989a,b,c).

For many years, the disorganization of the microvascular network of tumors associated with a poor delivery of blood, oxygen (see Chap. 6), nutrients, or anticancer agents has been considered a major drawback in substantially improving radiation and chemotherapy regiments (for reviews see JAIN 1991; VAUPEL 1990; VAUPEL et al. 1991). To illustrate the enormous extent of dysfunction in any tumor microcirculation, "transport" and "exchange" functions must be considered separately in malignant tumors.

3.4.1
Transport Function in Tumor Microvessels

Blood flow in the normal microcirculation as well as through the microvascular network of a tumor is assumed to be proportional to the pressure difference between arterial and venous vessels, and inversely proportional to viscous and geometric resistance (JAIN 1991). Arteriolar pressures in tumors are almost identical for both tumor and normal microcirculation (ENDRICH 1988; PETERS et al. 1980). Venular pressures, however, are significantly lower in a tumor, a fact that can be explained by a much higher number of outflow vessels (ENDRICH 1988; PETERS et al. 1980).

The viscous resistance (apparent viscosity) of blood depends on the number of blood cells, their flow behavior, and, to a much lesser extent, on plasma viscosity. In animal experiments, red blood cell rigidity was increased by hypoxia, reducing pH and cytotoxic hyperthermia or raising plasma glucose concentration. It should be noted that due to their larger capillary diameter, tumors may reveal a less pronounced Fahraeus and Fahraeus-Lindqvist effect, which could in turn lead to an increased number of microvessels with low or no perfusion as long as red blood cells are less flexible than normal. Therefore, information on geometric resistance, though still limited, should include the number of vessels (capillary density) and their length and diameter, as well as the branching pattern (for a review see JAIN 1991).

For unknown reasons, malignant cells rarely invade vessels of the arterial tree. As a result of the well-maintained contractile and nervous apparatus, one would assume that these vessels might respond to physical, pharmacological, and chemical stimuli, and might reveal the phenomenon of vasomotion.

Even though the earliest studies of the "normal" microcirculation referred to the presence of rhythmic diameter changes of arterioles associated with a modulation of capillary flow, vasomotion in tumors and its impact on tumor microvessels have widely been neglected. One possible explanation might be that vasomotion has only rarely been seen in tumors (ENDRICH 1988; INTAGLIETTA et al. 1977; REINHOLD et al. 1977; VAUPEL and MENKE 1989).

The significance of this phenomenon becomes apparent in clinical angiology. Periodic activity shows pronounced alterations when arterial stenosis of the lower extremity is treated by transluminal angioplasty. When using laser Doppler flowmetry in such patients, normal microcirculatory function was found to be highly nonlinear. Tissue perfusion, exchange of molecules, fluid balance, and peripheral vascular resistance are fundamentally different if microvessel diameter behaves in an oscillatory fashion instead of being steady, even though the average value might be the same (INTAGLIETTA 1989a). Quite recently, it was suggested that the patterns of diameter and flow variation in the microcirculation represent features of chaotic dynamics that become periodic only if pathological changes such as hemorrhage or hypotension require some reaction from the microcirculation (INTAGLIETTA 1989a,b; SLAAF et al. 1989). One hypothesis describes this reaction as a result of competing control mechanism, with the specific condition leading to the prevalence of one of the control mechanisms.

An entirely different situation is found in tumors for which the term "chaos" has been used for a long time to describe the irregular arrangement of microvessels as well as the flow pattern in tumor capillaries (GULLINO 1975; REINHOLD et al. 1977; TANNOCK 1968; VAUPEL 1977). Moreover, vasomotion is nearly eliminated even though tumor arteries and arterioles, with their original morphological features, are incorporated by the tumor tissue. Tumor arterioles in particular might experience a condition in which a presumed constrictor stimulus from a local pacemaker is inadequate in producing oscillatory or even chaotic behavior (INTAGLIETTA 1989b; MEYER et al. 1987, 1988; SLAAF et al. 1989). The significant impact on homeostasis can further be substantiated by the calculations of TSAI and INTAGLIETTA (1993). They demonstrated in a mathematical model that under flowmotion (cyclic time-varying fluctuations of capillary blood flow), tissue which under steady-state conditions would be anoxic becomes oxygenated. Therefore, this slow wave flowmotion observed in the normal microcirculation appears to be a physiological "reserve" mechanism to improve tissue oxygenation in situations in which oxygen supply is reduced; this capability to compensate oxygen deficiency is also lacking in tumors, to an extent as yet unknown (for further details see VAUPEL and MENKE 1989).

3.4.2
Exchange of Fluid and Solutes in the Tumor Microvasculature

Considering the lack of vasomotion as one of the significant differences between normal and tumor microcirculation, it is not surprising that most exchange parameters show differences when compared to the normal tissue "microenvironment." These alterations were summarized in recent years in a number of excellent reviews (BAXTER and JAIN 1989; CURTI et al. 1993; GULLINO 1975; JAIN 1987, 1989, 1990, 1991; REINHOLD et al. 1977; VAUPEL and MÜLLER-KLIESER 1983).

3.4.2.1
Transport Across the Microvascular Wall

Anticancer agents are carried over long distances within the body by convection. Upon reaching exchange vessels, substances extravasate and meet their target through the interstitium by two primary modes, diffusion and convection. *Diffusion* is defined as the transport of dissolved particles on the basis of their thermokinetic energy. A prerequisite for the motion of particles is the existence of a concentration gradient. Diffusion will depend on the diffusion medium, the molecular weight of the diffusing particles, and temperature. In general, transport of low molecular weight substances is diffusion-dominated, while convective transport in the microcirculation becomes important at higher molecular weights. For larger molecules, and particularly for those with polar groups, the microvascular wall can be a considerable obstacle to diffusion.

In malignant tumors, however, where most of the exchange vessels arise from the expansion of the vascular network by angiogenesis, the vessels have wide (water-filled) interendothelial gaps with discontinuous or absent basement membranes (HAMMERSEN et al. 1985; HÖCKEL et al. 1993b; KONERDING et al. 1989a,b,c; STEINBERG et al. 1990; YAMAURA and SATO 1974). This feature alone suggests that there should be a much higher diffusive exchange rate than in vessels of most normal tissues. In contrast, the average microvascular surface area decreases with tumor growth: Hence, one would expect reduced diffusive exchange in bulky tumors compared to smaller ones.

Convective fluid flow through the microvascular wall is defined as filtration. Morphologic characteristics of tumor microvessels suggest that the filtration coefficient should be higher than in most normal tissues. Extravasation of macromolecules, however, is poor in tumors. This is due to high interstitial fluid pressures that limit fluid extravasation (JAIN 1987). Recent studies of intratumor pressure gradients show that although interstitial pressure is elevated throughout the tumor, it drops precipitously to normal levels in the tumor's periphery (JAIN 1991). As the tumor grows, interstitial fluid pressure often rises, presumably due to proliferation of tumor cells in a confined space, high vascular permeability, and the absence of lymphatic vessels (JAIN 1991).

In normal tissue, the oncotic pressure difference between vascular and interstitial space is approximately 5–20 mmHg. Oncotic pressure in the tumor interstitium is most likely higher than in normal tissues due to high vascular permeability, and thus higher concentrations of plasma proteins in the tumor interstitium than in the interstitial compartment of normal tissue. Furthermore, small amounts of macromolecules are transported across endothelial cells by coated vesicles (transcytosis or cytopempsis).

3.4.2.2
Transport Within the Interstitial Space

Once an agent has passed the endothelium, its interstitial transport occurs by diffusion and convection. The interstitial space in most tumors is large compared with that in host normal tissue; hyaluronate and proteoglycan are present in lower concentrations, and the relative volume of free fluid is high (for reviews see GULLINO 1975; VAUPEL and MÜLLER-KLIESER 1983). These features suggest a relatively high diffusion coefficient in malignant tumors. Diffusion coefficients for macromolecules, an order of magnitude higher than those of several normal tissues (VAUPEL and MÜLLER-KLIESER 1983), should favor diffusion of large-sized molecules. The opposite, however, is true. Since diffusion distances are significantly increased in many malignant tumors, agents need more time to traverse the distances to reach their targets. For instance, the time required

for the diffusion of IgG in tumors is of the order of 0.5 h for a distance of 100 µm; its Fab fragment would take about 0.2 h to pass the same distance in the tumor's interstitium (JAIN 1989).

Based on the characteristics of the interstitial space in tumors (high relative volume of free water, interstitial hypertension, low levels of glycosaminoglycan, increase in fractional volume of interstitial compartment), convective transport is significant in tumors, especially for substances of larger molecular size. Bulk flow of free fluid in the interstitial compartment is typical for malignant tumors. Whereas in normal tissues convective currents in the interstitial space are estimated to be 0.5%–1.0% of plasma flow, bulk flow of free fluid in tumors can reach 14% of the plasma flow rate, provoking a significant hemoconcentration during the passage of blood through the tumor (VAUPEL and KALLINOWSKI 1987). Since interstitial fluid pressure is high in the center of the tumor and decreases towards the invasion front at the tumor's periphery, interstitial fluid motion is from the center towards the periphery, from where it will ooze out into the surrounding tissue, contributing to the formation of peritumoral edema. This fluid leakage leads to an interstitial fluid velocity of 0.1–25 µm/s directed radially outward (YOUNG et al. 1950; BUTLER et al. 1979). The outward convection has to be overcome by macromolecules diffusing from peripheral to central areas.

To sum up the pros and cons of tumor characteristics for the delivery of anticancer agents (especially macromolecules), the physiological "benefits" include high values of vascular permeability, diffusion coefficient, and hydraulic conductivity, whereas problems arise from a heterogeneous microcirculation, elevated interstitial pressure, the enlarged volume of the interstitial space, large intercapillary distances, and a radially directed bulk flow of free water towards the tumor periphery. This aspect, which is of great clinical relevance in chemotherapy, is often neglected. It might still greatly limit this approach (see also Chap. 4).

3.5
Measures Related to the (Mature) Microcirculation in Tumors

3.5.1
Diagnostic Procedures

New possibilities of monitoring a tumor's response to therapy performed using imaging modalities have received increased attention during recent years. Tumor size is still the main criterion, although the reduction of tumor size after therapy will only become evident after several weeks. NMR spectroscopy could express earlier effects in quantitative terms, so the time gain could be used to optimize tumor therapy (for reviews see MOLLS and FELDMANN 1991; SEMMLER et al. 1991; OKUNIEFF et al. 1989, 1991; VAUPEL et al. 1989b). Recently, blood-flow-related data were obtained using dynamic CT and dynamic magnetic resonance imaging (ADLER et al. 1993; EVELHOCH et al. 1991). In pelvic tumors, FELDMANN et al. (1993) were able to differentiate between areas of low and high perfusion rates. Encouraged by the possibility of having an additional tool for the prediction of tumor response to thermoradiotherapy and radiation, they started to evaluate changes in tumor perfusion by positron emission tomography (PET). LAMMERTSMA et al. (1991) also reported some of their experience in women suffering from breast carcinomas while utilizing PET for in vivo measurements of tumor perfusion. PET may have great potential in oncology for the staging and follow-up of proliferation of tumors, as well as for the evaluation of the distribution of labelled chemical agents in tumor metastases. It should, however, be noted that due to perfusion heterogeneities, a quantitative estimate of oxygen utilization is still not possible with this technique. In contrast, "metabolic imaging" using bioluminescence in cryosections of tissue renders feasible data on the relation between the tumor's histology and the spatial distribution of compounds such as ATP, glucose, and lactate (VAUPEL 1990; KROEGER et al. 1991). This novel technique and its clinical potential have been evaluated directly in specimens of human tumors such as the cervix carcinoma (SCHWICKERT et al. 1995).

Color-coded duplex sonography seems limited in its suitability for staging breast tumors. Based on a clinical study of 149 patients, measurement of blood flow provided some additional diagnostic information only in T_1 mammary carcinoma. According to FIEDLER et al. (1996), however, this technique is reliable in detecting recurrences in an older scar after primary surgery.

3.5.2
Alterations of Tumor Blood Flow After Drug Application

Despite the problems concerning tumor microcirculation discussed earlier, a great number of studies on blood flow in human and experimental tumors have been reported. In addition, a reasonable number of drugs has been tested for their effect on blood flow in experimental tumors (for a review see CHAPLIN and TROTTER 1991). The compound most extensively evaluated is *angiotensin II* (JIRTLE 1988; BURTON et al. 1985, 1988; SUZUKI et al. 1981; KUROIWA et al. 1987; TAKEMATSU et al. 1985; EKELUND and LUNDERQUIST 1974), a potent vasoconstrictor that induces hypertension upon administration. This agent was found to restore flow in regions within the tumor where interstitial pressure was higher than intravascular pressure. From the clinical point of view, a short-term infusion of angiotensin II could be beneficial during radiation or chemotherapy (EKELUND and LUNDERQUIST 1974). Even though this drug has been shown to improve the efficiency of chemotherapy, it is not widely used in medical oncology. The same is true for diagnostic imaging, although blood vessel visualization is greatly improved by angiotensin II. One reason for its very limited use might be the extensive variation of the dose-response relationship among individuals as well as different organs.

The use of another group of drugs might be of more clinical significance. To improve the cytotoxicity of antineoplastic drugs, *calcium antagonists* have been tested under in vitro and in vivo conditions. Moreover, these agents may be capable of inhibiting the respiration rate of malignant cells in vitro (VAUPEL and MENKE 1989). There is also some experimental evidence that verapamil and nifedipine inhibit spontaneous and experimental tumor metastasis (HONN et al. 1983; TSURUO et al. 1985). At the microcirculatory level, calcium antagonists can lead to an increase in peripheral blood flow at constant perfusion pressure via dilation of vessels feeding the tumor. VAUPEL and MENKE (1989) have summarized their findings in the rat DS carcinosarcoma and have found that among the drugs tested, only flunarizine caused a 20% increase in tumor flow that lasted more than 30 min. It should again be mentioned that they observed rhythmic oscillations of red blood cell flux in approximately 15% of the subepidermal tumors. The average time for these oscillations was nearly 3 min with approximately 21 cycles per hour (VAUPEL et al. 1988). Upon application of all of the calcium antagonists tested, these rhythmic variations slowed down significantly to a mean of more than 5 min with 11 cycles per hour. Comparing these data with results from the literature (for review see VAUPEL and MENKE 1989), the flow increase seems to be a rather consistent finding. Presently, however, one can only speculate on the clinical significance of these agents as an adjuvant measure in diagnosis and anticancer treatment. They may modulate pharmacokinetics and pharmacodynamics of cytostatic agents, increase radiosensitivity, or improve the delivery of monoclonal antibodies and biological response modifiers.

3.5.3
Effects of Chemotherapy on the Normal and the Tumor Microcirculation

Significant tumor regression is seen in many instances following chemotherapy. However, only a few studies are available on vascular effects of cytostatic agents. MURRAY et al. (1987) demonstrated a pronounced reduction of blood flow after employing melphalan and misonidazole. A combination of both agents had an even better and more prolonged effect. Vinca alkaloids were also found to affect blood flow (STEPHENS and PEACOCK 1978) in a B16 melanoma in mice. A similar observation was reported by BAGULEY et al. (1991) for colon carcinomas in the same animal species.

Two agents, namely, flavone acetic acid and phoquidone, were recently reported to cause vascular-mediated injury in almost all solid tumors, but not in leukemia or lymphomas (CORBETT et al. 1986; DENEKAMP 1993). Upon application, a rapidly developing hemorrhagic necrosis was visible within a few hours. This was associated with a reduction of tumor perfusion (FINLAY et al. 1988; HILL et al. 1989).

Apparently, chemotherapy can cause both systemic and peripheral vascular toxicity, the latter affecting the microcirculation directly. This might be beneficial, but there are also some reports on acute and late vascular complications (GERL et al. 1993; TEUTSCH et al. 1977). Using vinblastine and/or bleomycin, Raynaud's phenomenon was reported as early as 1977 (TEUTSCH et al. 1977). Subsequently, a number of studies confirmed this observation in a varying proportion of patients ranging from 2.6% to 41%. The combination of cisplatin and bleomycin seems to bear an inherent risk of the development of low peripheral perfusion (AASS et al. 1990; BISSETT et al. 1990; HANSEN and OLSEN 1989; HEIER et al. 1991; SCHEULEN and SCHMIDT 1982; VOGELZANG et al. 1981). Moreover, it should be noted that only gradual resolution of symptoms was seen in 50% of the patients (HEIER et al. 1991).

3.5.4
Tumor Blood Flow upon Localized Hyperthermia

Over the past decade, hyperthermia has been introduced into clinical oncology as an important supplement to the conventional treatment of cancer. The interest in this modality increased in particular with increasing understanding of the pathophysiology of the tumor microcirculation. Although hyperthermia is well understood in this respect, great problems are still present concerning the uniform heating of deep-seated tumors. Today, it seems certain that malignant tumors can be destroyed by a thermal dose innocuous to normal tissues. Moreover, small changes of flow at the microcirculatory level could have major effects on the microenvironment of tumors, which in turn could modify the response to therapeutic modalities combined with hyperthermia. It should, however, be noted that the question of whether normal and malignant cells have a different thermosensitivity is still being discussed controversially.

There are a number of reviews and books available that deal with hyperthermia (ACKER et al. 1990; ANGHILERI and ROBERT 1986; ENDRICH 1988; ENDRICH and HAMMERSEN 1986; HINKELBEIN et al. 1988; ISSELS and WILMANNS 1988; KALMUS et al. 1990; MAYER et al. 1992; REINHOLD and ENDRICH 1986; STREFFER 1987; VAUPEL 1990; VAUPEL et al. 1989c; WATERMAN et al. 1991). In one review (VAUPEL 1990), a diagram (Fig. 3.1) showing the normal vascular response to heat and the effect of hyperthermia on tumor tissue vasculature was designed on the basis of a literature search of clinical and experimental studies of blood flow upon heating. Due to its low perfusion, a tumor creates its own heat reservoir with the convective heat dissipation being lower in fast-growing tumors. As a result, the tumor temperature will be higher than in normal tissue at comparable energy input. In contrast, the microcirculation of malignant tumors appears to have little tolerance to elevated tissue temperatures. This vascular shutdown permits, at least experimentally, a "selective" treatment of tumors. The mechanisms involved are rather complex; they are discussed in several publications (ENDRICH 1988; ENDRICH and VAUPEL 1995a; SEEGENSCHMIEDT et al. 1993; LYNG et al. 1993; VAUPEL 1990). The hyperthermia-induced dysfunction of the microcirculation will not only be advantageous for a selective heating of the tumor as a whole, but can sensitize

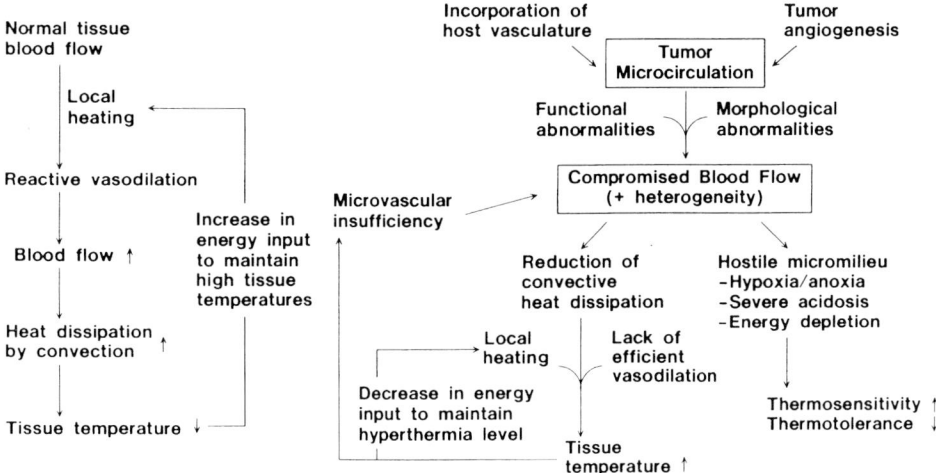

Fig. 3.1. The normal vascular reaction to heat (*left*) and the effect of hyperthermia on the vasculature in tumors (*right*). Upon temperature elevation in normal tissues (e.g., subcutis), vasodilation occurs, leading to increased heat dissipation by convection. In many tumors, efficient heat transfer does not take place. Vascular damage leads to a further temperature rise, rather than a drop in temperature as is expected in normal tissues. (For a review see VAUPEL, 1990)

tumor cells to (subsequent) hyperthermic treatment by changing the "milieu interne" of the tumor. However, since these phenomena have only been observed experimentally in fast growing animal tumors, clinical relevance needs to be verified during clinical application of hyperthermia. Particularly for isolated extremity perfusion of melanomas and soft tissue sarcomas, a fair number of clinical studies are available (for reviews see ENDRICH 1988; GILLY et al. 1993; HENNEKING et al. 1993; SCHLAG and KETTELHACK 1993; ISSELS et al. 1993). Some clinicians have also begun to realize the tremendous impact of changes in tumor microcirculation upon regional hyperthermia, which affects the response of the cytostatic agents added and thus subsequent tumor growth (SCHLAG and KETTELHACK 1993; ISSELS et al. 1993).

It should also be mentioned that comparable microvascular damage has been reported upon photodynamic and shock-wave therapy (for reviews see ENDRICH and GÖTZ 1991; LOWDELL et al. 1993; GAMARRA et al. 1993). Based on encouraging experimental data, photodynamic therapy has also been used clinically (LOWDELL et al. 1993).

3.5.5
Microvascular Changes After Radiation: Implications for Therapy

As early as 1927, changes in the tumor microvasculature upon radiation were discussed. MERWIN and

ALGIRE (1950) were the first to report a pronounced narrowing of microvessels upon application of doses of 20–50 Gy in single fractions. They also observed, as an effect of radiation, that tumor blood flow was variable even for the same tumor and an identical dose of radiation. From a number of clinical and experimental studies, it appeared that a single large dose of radiotherapy would destroy the microvasculature and lead to parenchymal cell death, while fractionated radiation in a clinically relevant dose range could temporarily improve tumor oxygenation, thus facilitating subsequent radiation treatment (BAKER and KROCHAK 1989; DEWHIRST 1991; ZYWIETZ et al. 1994).

Besides directly evaluating the response of the microvasculature to radiation, many factors have been considered as potentially useful predictors of tumor response to radiotherapy (for a review see WEST 1994). Clonogenic and growth assays, the evaluation of repair capacity to predict intrinsic radiosensitivity, proliferative indices, ploidy, tissue oxygenation, and vascularity have all received attention as valid predictors. Vascular density and intercapillary distances have been correlated with the outcome of radiotherapy in carcinomas of the cervix (AWWAD et al. 1986; KOLSTADT 1968; RÉVÉSZ et al. 1989; SIRACKA et al. 1982, 1988; SIRACKY et al. 1988), the larynx, and the nasopharynx (DELIDES et al. 1988). In these studies, small intercapillary distances and/or high vascular density were associated with better local control of the disease or even longer overall survival. In contrast, in a small series of 26

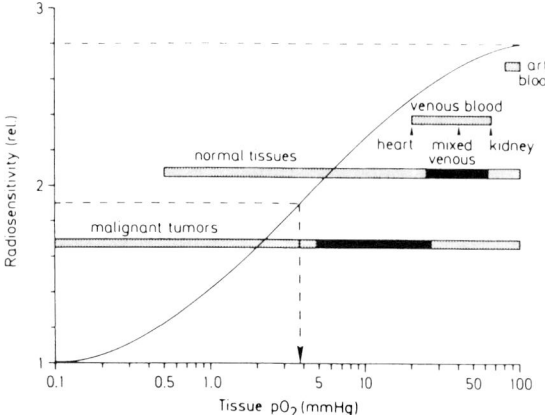

Fig. 3.2. Relative radiosensitivity as a function of O_2 tensions (pO_2) in the cellular environment, *Shaded bars*: range of pO_2 values usually found in blood, normal tissues, and malignant tumors. *Black bars*: range of median O_2 partial pressures in normal tissues and malignancies. The *arrow* indicates the pO_2 value at which the sensitizing effect is half-maximal (3–4 mmHg). (Adapted from VAUPEL et al. 1989)

squamous cell carcinomas of the oral cavity, better local control was observed in poorly vascularized tumors (LAUK et al. 1989). An explanation for this striking difference cannot be provided at present.

Basically, these studies highlight the role of tumor oxygenation in determining a response to radiotherapy (see Fig. 3.2), particularly in cervix carcinomas. Since measurements of vascularity are laborious and do not always describe the functional status of microvessels (not all microvessels detected and counted by morphometric techniques are perfused at any given time), the results obtained indicate again that alternative, more rapid and direct methods for evaluation of tumor oxygenation could have a significant impact on individual radiotherapy (LAUK et al. 1989). In a more recent study on cervix cancers treated with radiotherapy, this has been explicitly stated because tumor oxygenation predicted both local control and overall survival (HÖCKEL et al. 1993a).

3.6
Future Perspectives

Tumor therapy depends on being able to remove the neoplasm as a whole or destroy the clonogenic activity of tumor cells still in place while limiting the damage to normal tissue. A variety of factors are needed to explain why chemical or physical treatments quite frequently fail. As discussed in this

chapter, it becomes increasingly clear that physiological parameters are of utmost importance for a sufficient outcome in cases in which surgery needs to be supplemented (Table 3.5). As a result of the nonuniform vascular network and its inadequate function, global blood flow and microcirculation, transcapillary and interstitial transport, and tissue oxygen and nutrient supply as well as the bioenergetic status will markedly change any therapeutic response beyond a genetically determined intrinsic resistance of tumor cells.

In fact, as in vitro data become readily and increasingly available, methods such as intravital microscopy employed in intact, nonanesthetized animals are of utmost importance to assess a tumor's response in a preclinical model (MENGER and LEHR 1993). Knowledge of microcirculatory and microenvironmental parameters will be relevant to clinical oncology in three ways:

1. As a diagnostic aid when new methods such as MRI, PET, and quantitative bioluminescence are utilized to evaluate tumor pathophysiology in clinical oncology. With a broader use of these novel techniques, a new generation of clinical trials might be initiated using the increasing amount of preclinical data as a guideline for designing clinical protocols.
2. Clinical attempts reportedly stimulate angiogenesis successfully. In a patient suffering from limb ischemia, an increase in collateral vessels was seen upon application of vascular endothelial growth factor (ISNER et al. 1996). A temporary stimulation of angiogenesis could, for instance,

Table 3.5. Microcirculation of malignant tumors and its apparent clinical usefulness at the end of 1996

Facts	Fiction
Chaotic morphology of tumor vessels as compared to normal microcirculation	Homogeneous perfusion of tumor microvessels
Angiogenesis is biologically stimulated	Vasoactive drugs improve tumor therapy
Angiogenic tests are available, but cannot be used on a broad clinical basis	The extent of angiogenesis has some prognostic significance in patients
MRI provides an additional tool to illustrate dense vascularization of a malignant tumor	Antiangiogenic treatment has been introduced as routine measure in clinical oncology
Hyperthermia and radiation affect flow and morphology of the tumor microcirculation	Predictive indicators on a tumor's response after hyperthermia and/or radiation are readily available

lead to earlier detection of a tumor's remnant. However, such attempts have not yet been undertaken.

3. As a prognostic indicator derived from a surgical specimen. Some data on vascular density, obtained primarily in mammary carcinomas at the time of surgery, seem to open a new horizon in prognosis and patient selection (VAN HOEF et al. 1993). Some encouraging results, however, need to be confirmed not only in a much larger number of patients but should also confirm the huge amount of experimental data presently existing on tumor pathophysiology; the development of computerized image analysis might help to provide instant data at low cost (BARBARESCHI et al. 1995; MESSMER et al. 1984; ZEINTL et al. 1986) and avoid the more or less intentional (subjective) evaluation only of areas with high vascular density.

References

Aass N, Kaasa S, Lund E, Kaalhus O, Heier MS, Fossa SD (1990) Long-term somatic side effects and morbidity in testicular cancer patients. Br J Cancer 61:151–155

Acker JC, Dewhirst MW, Honoré GM, Samulski TV, Tucker JA, Oleson JR (1990) Blood perfusion measurements in human tumours: evaluation of laser Doppler methods. Int J Hyperthermia 6:287–304

Adler LP, Crowe JP, Alkaisi NK, Sunshine JL (1993) Evaluation of breast masses and axillary lymph nodes with (F-18)-2-Deoxy-2-fluoro-D-glucose PET. Radiology 187:743

Albo D, Granick MS, Jhala N, Atkinson B, Solomon MP (1994) The relationship of angiogenesis to biological activity in human squamous cell carcinomas of head and neck. Ann Plast Surg 32:588–594

Anghileri LJ, Robert J (1986) Hyperthermia in cancer treatment, vols I–III. CRC Press, Boca Raton

Auerbach W, Auerbach R (1994) Angiogenesis inhibition. A review. Pharmacol Ther 63:265–311

Ausprunk DH, Folkman J (1977) Migration and proliferation of endothelial cells in preformed and newly formed blood vessels during tumor angiogenesis. Microvasc Res 14:53–65

Awwad HK, Naggar M, Mocktar N, Barsoum M (1986) Intercapillary distance measurement as an indicator of hypoxia in carcinoma of the cervix uteri. Int J Radiat Oncol Biol Phys 12:1329–1333

Bagavandoss P, Wilks JW (1990) Specific inhibition of endothelial cell proliferation by thrombospondin. Biochem Biophys Res Commun 170:867

Baguley BC, Holdaway KM, Thomsen LL, Zhuang L, Zwi LJ (1991) Inhibition of growth of colon 38 adenocarcinoma by vinblastine and colchicine: evidence for a vascular mechansim. Eur J Cancer 27:482–487

Baker DG, Krochak RJ (1989) The response of the microvascular system to radiation: a review. Cancer Invest 7:287–294

Barbareschi M, Gasparini G, Morelli L, Forti S, Dalla Palma P (1995) Novel methods for the determination of the angiogenic activity of human tumors. Breast Cancer Res Treat 36:181–192

Bevilacqua P, Barbareschi M, Verderio P, et al. (1995) Prognostic value of intratumoral microvessel density, a measure of tumor angiogenesis in node-negative breast carcinoma – results of a multiparametric study. Breast Cancer Res Treat 36:205–217

Bigler SA, Deering RE, Brawer MK (1993) Comparison of microscopic vascularity in benign and malignant prostate tissue. Hum Pathol 24:220–226

Bissett D, Kunkeler L, Zwanenburg L, et al. (1990) Long-term sequelae of treatment for testicular germ cell tumors. Br J Cancer 62:655–659

Blei F, Wilson EL, Mignatti P, Rifkin DB (1993) Mechanism of action of angiostatic steroids: suppression of plasminogen activator activity via stimulation of plasminogen activator inhibitor synthesis. J Cell Physiol 144:568

Blood CH, Zetter BR (1990) Tumor interactions with the vasculature: angiogenesis and tumor metastasis. Biochim Biophys Acta 1032:89

Bossi P, Viale G, Lee AKC, Alfano RM, Coggi G, Bosari S (1995) Angiogenesis in colorectal tumors: microvessel quantitation in adenomas and carcinomas with clinicopathological correlations. Cancer Res 55:5049–5053

Boucher Y, Jain RK (1992) Microvascular pressure is the principal driving force for interstitial hypertension in solid tumors: implications for vascular collapse. Cancer Res 52:5110–5114

Boucher Y, Baxter LT, Jain RK (1990) Interstitial pressure gradients in tissue-isolated and subcutaneous tumors: implications for therapy. Cancer Res 50:4478–4484

Brammer I, Zywietz F, Jung H (1979) Changes of histological and proliferative indices in the Walker carcinoma with tumour size and distance from blood vessel. Eur J Cancer 15:1329

Brawer MK, Deering RE, Brown M, Preston SD, Bigler SA (1994) Predictors of pathologic stage in prostate carcinoma. Cancer 73:678–697

Brem S, Cotran R, Folkman J (1972) Tumor angiogenesis: a quantitative method for histological grading. J Natl Cancer Inst 4:347–356

Brem S, Tsanaclis AMC, Gately S, Gross JL, Herblin WF (1992) Immunolocalization of basic fibroblast growth factor to the microvasculature of human brain tumors. Cancer 70:2673–2680

Brown KJ, Parish CR (1994) Histidine-rich glycoprotein and platelet factor 4 mask heparan sulfate proteoglycans recognized by acidic and basic fibroblast growth factor. Biochemistry 33:13918

Burton MA, Gray BN, Self GW, Heggie JC, Townsend PS (1985) Manipulation of experimental rat and rabbit liver tumor blood flow with angiotensin II. Cancer Res 45:539

Burton MA, Gray BN, Coletti A (1988) Effect of angiotensin II on blood flow in the transplanted sheep squamous cell carcinoma. Eur J Cancer 45:54

Butler TP, Grantham FH, Gullino PM (1979) Bulk transfer of fluid in the interstitial compartment of mammary tumors. Cancer Res 35:3084

Chaplin DJ, Trotter MJ (1991) Chemical modifiers of tumor blood flow. In: Vaupel P, Jain RK (eds) Tumor blood supply and metabolic microenvironment. Fischer, Stuttgart, p 65–85

Chodak GW, Scheiner CJ, Zeter BR (1981) Urine from patients with transitional cell carcinoma stimulates migration of capillary endothelial cells. N Engl J Med 305:869–874

Coffey RJ Jr, Leof EB, Shipley GD, Moses HL (1987) Suramin inhibition of growth factor receptor binding and mitogenicity in AKR-2B cells. J Cell Physiol 132:143

Corbett TH, Bissery MC, Wozniak A (1986) Activity of flavone acetic acide (NSC.347512) against solid tumors in mice. Invest New Drugs 4:207

Curti BD, Urba WJ, Alvord WG, Janik JE, Smith JW II, Madara K, Longo DL (1993) Interstitial pressure of subcutaneous nodules in melanoma and lymphoma patients: changes during treatment. Cancer Res 53:2204

Davidson SE, Ngan R, Wilks DP, Moore JV, West CML (1994) A comparison of four methods for assessing tumour vascularity in carcinoma of the cervix. Int J Oncol 5:639–645

Delides GS, Venizelos J, Révész L (1988) Vascularization and curability of stage III and IV nasopharyngeal tumours. J Cancer Res Clin Oncol 114:321

Denekamp J (1993) Angiogenesis, neovascular proliferation and vascular pathophysiology as targets for cancer therapy. Br J Radiol 66:181–196

Denekamp J, Hill SA, Hobson B (1983) Vascular occlusion and tumor cell death. Eur J Cancer 19:271–275

Dewhirst MW (1991) Microvascular changes induced by radiation exposure: implications for therapy. In: Vaupel P, Jain RK (eds) Tumor blood supply and metabolic environment. Fischer, Stuttgart, p 109–122

Ekelund L, Lundequist A (1974) Pharmacoangiography with angiotensin. Radiology 110:533–540

Ellis LM, Fidler IJ (1995) Angiogenesis and breast cancer metastasis. Lancet 346:388–389

Endrich B (1988) Hyperthermie und Tumormikrozirkulation – eine kritische Wertung experimenteller und klinischer Befunde. Contributions to oncology, vol 31. Karger, Basel

Endrich B, Götz A (1991) Tumor microcirculation: scope and clinical applicability. In: Vaupel P, Jain RK (eds) Tumor blood supply and metabolic microenvironment Fischer, Stuttgart, p 37–52

Endrich B, Hammersen F (1986) Morphologic and hemodynamic alterations in capillaries during hyperthermia. In: Anghileri LJ, Robert J (eds) Hyperthermia in cancer treatment, vol II. CRC Press, Boca Raton, p 17–47

Endrich B, Vaupel P (1995a) Malignant tumors and the microcirculation. In: Barker JH, Andersen GL, Menger MD (eds) Clinically applied microcirculation research. CRC Press, Boca Raton, p 107–126

Endrich B, Vaupel P (1995b) Tumorale Gefäßdichte bei Mammatumoren und ihr Einfluß auf das rezidivfreie Überleben. Chirurg 66:76–77

Endrich B, Reinhold HS, Gross JF, Intaglietta M (1979) Tissue perfusion inhomogeneity during early tumor growth in rats. J Natl Cancer Inst 62:387–395

Evelhoch JL, Larcombe-McDouall JB, McCoy CL, Mattiello J, Simpson NE, Sayedsadr M (1991) Deuterium NMR imaging of regional tumor blood flow: effect of size in RIF-1 murine tumors. In: Vaupel P, Jain RK (eds) Tumor blood supply and metabolic microenvironment. Fischer, Stuttgart, p 293–304

Feldmann HJ, Sievers K, Fuller J, Molls M, Löhr E (1993) Evaluation of tumor blood perfusion by dynamic MRI and CT in patients undergoing thermoradiotherapy. Eur J Radiol 16:224–229

Fidler IJ (1990) Critical factors in the biology of human cancer metastasis: twenty-eighth GHA Clowes memorial award lecture. Cancer Res 50:6130

Fiedler V, Neubauer K-D, Schneiders A, Herzig P (1996) Stellenwert der farbkodierten Duplexsonographie (FKDS) bei der Dignitätsbestimmung von Mammatumoren. Fortschr Röntgenstr 165:159–165

Finlay GJ, Smith GP, Fray LM, Baguley BC (1988) Effect of flavone acetic acid on Lewis lung carcinoma: evidence for an indirect effect. J Natl Cancer Inst 80:241–245

Folberg R, Peer J, Gruman LM, et al. (1992) The morphologic characteristics of tumor blood vessels as a marker of tumor progression in primary human uveal melanoma – a matched case-control study. Hum Pathol 23:1298–1305

Folkman J (1971) Tumor angiogenesis: therapeutic implications. N Engl J Med 285:1182–1186

Folkman J (1986) How is blood vessel growth regulated in normal and neoplastic tissue? Cancer Res 46:467

Folkman J (1989) Successful treatment of an angiogenic disease. N Engl J Med 320:1211

Folkman J (1990) What is the evidence that tumors are angiogenesis dependent? J Natl Cancer Inst 82:4–6

Folkman J (1995) Angiogenesis inhibitors generated by tumors. Mol Med 1:120–122

Folkman J, Brem H (1992) Angiogenesis and inflammation. In: Gallin JI, Goldstein IM, Snyderman R (eds) Inflammation – basic principles and clinical correlates. Raven, New York, p 82

Folkman J, Merler E, Abernathy C, Williams G (1971) Isolation of a tumor factor responsible for angiogenesis. J Exp Med 133:275–288

Frank RE, Saclarides TJ, Leurgans S, Speziale NJ, Drab EA, Rubin DB (1995) Tumor angiogenesis as a predictor of recurrence and survival in patients with node-negative colon cancer. Ann Surg 222:695–699

Fregene TA, Khanuja PS, Noto AC, Gehani SK, van Egmont EM, Luz DA, Pienta KJ (1993) Tumor-associated angiogenesis in prostate cancer. Anticancer Res 13:2377–2381

Gabbert H, Wagner R, Höhn P (1982) The relation between tumor cell proliferation and vascularization in differentiated and undifferentiated colon carcinomas in the rat. Virchows Arch [B] 41:119

Gagliardi A, Collins DC (1993) Inibition of angiogenesis by antiestrogens. Cancer Res 53:533–535

Gamarra F, Spelsberg F, Kuhnle GEH, Goetz AE (1993) High-energy shock waves induce blood flow reduction in tumors. Cancer Res 53:1590

Gasparini G, Harris AL (1994) Does improved control of tumour growth require an anti-cancer therapy targeting both neoplastic and intratumoral endothelial cells? Eur J Cancer 30:201–206

Gasparini G, Weidner N, Bevilacqua P, et al. (1993) Intratumoral microvessel density and p 53 protein: correlation with metastasis in head-and-neck squamous-cell carcinoma. Int J Cancer 55:739–744

Gerl A, Clemm C, Wilmanns W (1993) Acute and late vascular complications following chemotherapy for germ cell tumors. Onkologie 16:88–92

Gilly FN, Carry PY, Sayag AC, et al. (1993) Gastric cancer with peritoneal carcinomatosis – does hyperthermia constitute a new therapeutic approach? Ann Chir 47:414

Glaser BM, D'Amore PA, Lutty GA, Fenselau AH, Michels RG, Patz A (1980) Chemical mediators of intraocular neovascularization. Trans Ophthalmol Soc 100:369

Goulding H, Abdul Rashid NFN, Robertson JF, Bell JA, Elston CW, Blamey RW, Ellis IO (1995) Assessment of angiogenesis in breast carcinoma: an important factor in prognosis? Hum Pathol 26:1196–1200

Gullino PM (1975) Extracellular compartments of solid tumors. In: Becker FF (ed) Cancer, vol 3. Plenum, New York, p 327–354

Gullino PM (1991) Tumor angiogenesis 1990: comments on the state of the art. In: Vaupel P, Jain RK (eds) Tumor blood supply and metabolic microenvironment. Fischer, Stuttgart, p 11–26

Gullino PM, Grantham FH (1961) Studies on the exchange of fluids between host and tumor. II. The blood flow of hepatomas and other tumors in rats and mice. J Natl Cancer Inst 27:1465–1491

Hall NR, Fish DE, Hunt N, Goldin RD, Guillou PJ, Monson JR (1992) Is the relationship between angiogenesis and metastasis in breast cancer real? Surg Oncol 1:223–229

Hammersen F, Endrich B, Osterkamp-Baust U (1985) The fine structure of tumor blood vessels: I. Participation of non-endothelial cells in tumor angiogenesis. Int J Microcir Clin Exp 4:31–43

Hansen SW, Olsen N (1989) Raynaud's phenomenon in patients treated with cisplatin, vinblastine, and bleomycin for germ cell cancer: measurement of vasoconstrictor response to cold. J Clin Oncol 7:940–942

Hayes DF (1994) Angiogenesis and breast cancer. Breast Cancer 8:51–71

Heier MS, Nilsen T, Graver V, Aass N, Fossa SD (1991) Raynaud's phenomenon after combination chemotherapy of testicular cancer, measured by laser Doppler flowmetry. A pilot study. Br J Cancer 63:550

Henneking K, Binder J, Weyers W, Schwemmle K (1993) Chirurgische Behandlung und regionale Chemotherapie beim Extremitätenmelanom. Chirurg 64:134–138

Hill SA, Williams KB, Denekamp J (1989) Vascular collapse after flavone acetic acid: a possible mechanism of its antitumour action. Eur J Clin Oncol 25:1419–1424

Hinkelbein W, Bruggmoser G, Engelhardt R, Wannenmacher M (1988) Preclinical hyperthermia. Recent results in cancer research, vol 109. Springer, Berlin Heidelberg New York

Hirata S, Matsubara T, Saura R, Tateishi H, Hirohata K (1989) Inhibition of in vitro vascular endothelial cell proliferation and in vivo neovascularization by low dose methotrexate. Arthritis Rheum 32:1065

Hirst DG, Denekamp J (1979) Tumour cell proliferation in relation to the vasculature. Cell Tissue Kinet 12:31–42

Höckel M, Knoop C, Schlenger K, et al. (1993a) Intratumoral pO_2 predicts survival in advanced cancer of the uterine cervix. Radiother Oncol 26:45

Höckel M, Schlenger K, Doctrow S, Kissel T, Vaupel P (1993b) Therapeutic angiogenesis. Arch Surg 128:423–429

Hollingsworth HC, Steinberg SM, Kohn E, Bryant B, Merino MJ (1994) Tumor angiogenesis in advanced stage ovarian cancer. Mod Pathol 7:89a

Honn KV, Onoda JM, Diglio CA, Sloane BF (1983) Calcium channel blockers: potential antimetastatic agents. Proc Soc Exp Biol Med 174:16

Horak ER, Leek R, Klenk N, et al. (1992) Angiogenesis, assessed by platelet endothelial cell adhesion molecule antibodies, as indicator of node metastases and survival in breast cancer. Lancet 340:1120–1124

Hori A, Sasada R, Matsutani E, Naito K, Sakura Y, Fujita T, Kozai Y (1991) Suppression of solid tumor growth by immunoneutralizing monoclonal antibody against human basic fibroblast growth factor. Cancer Res 51:6180

Ingber DE, Madri JA, Folkman J (1986) A possible mechanism for the inhibition of angiogenesis. Endocrinology 119:1768–1775

Intaglietta M (1989a) Reactivation of the microcirculation in ischemia. Prog Appl Microcir 13:27–37

Intaglietta M (1989b) Vasomotion as normal microvascular activity and a reaction to impaired homeostasis. Prog Appl Microcir 15:1–15

Intaglietta M, Myers RR, Gross JF, Reinhold HS (1977) Dynamics of microvascular flow in implanted mouse mammary tumours. Bibl Anat 15:273–276

Isner JM, Piecek A, Schainfeld R, et al. (1996) Clinical evidence of angiogenesis after arterial gene transfer of $phVEGF_{165}$ in patient with ischaemic limb. Lancet 348:370

Issels RD, Wilmanns W (1988) Application of hyperthermia in the treatment of cancer. Recent results in cancer research, vol 107. Springer, Berlin Heidelberg New York

Issels RD, Bosse D, Starck M, Abdel-Rahman S, Jauch KW, Schildberg FW, Wilmanns W (1993) Weichteiltumoren: Indikation und Ergebnisse der Hyperthermie. Chirurg 64:461–467

Jaeger TM, Weidner N, Chew K, Moore DH, Kerschmann RL, Waldman FM, Carroll PR (1994) Tumor angiogenesis and lymph node metastases in invasive bladder carcinoma. J Urol 151:348a

Jain RK (1987) Transport of molecules across tumor vasculature. Cancer Metastasis Rev 6:559–594

Jain RK (1988) Determinants of tumor blood flow: a review. Cancer Res 48:2641–2658

Jain RK (1989) Delivery of novel therapeutic agents in tumors: physiological barriers and strategies. J Natl Cancer Inst 81:570–576

Jain RK (1990) Vascular and interstitial barriers to delivery of therapeutic agents in tumors. Cancer Metastasis Rev 9:253

Jain RK (1991) Haemodynamic and transport barriers to the treatment of solid tumors. Int J Radiat Biol 60:85

Jirtle RL (1988) Chemical modification of tumour blood flow. Int J Hyperthermia 4:355

Kalmus J, Okunieff P, Vaupel P (1990) Dose-dependent effects of hydralazine on microcirculatory function and hyperthermic response of murine FSall tumors. Cancer Res 50:14

Kato T, Sato K, Kakinuma H, Matsuda Y (1994) Enhanced suppression of tumor growth by combination of angiogenesis inhibitor O-(chloroacetyl-carbamoyl) fumagillol (TNP-470) and cytotoxic agents in mice. Cancer Res 54:5143

Kim KJ, Li B, Winer J, Armanini M, Gillet N, Phillips HS, Ferrara N (1993) Inhibition of vascular endothelial growth factor-induced angiogenesis suppresses tumour growth in vivo. Nature 367:576

Klagsbrun M, D'Amore PA (1991) Regulators of angiogenesis. Annu Rev Physiol 53:217

Kolstadt P (1968) Intercapillary distance, oxygen tension and local recurrence in cervix cancer. Scand J Clin Lab Invest 106:145

Kondo S, Asano M, Suzuki H (1993) Significance of vascular endothelial growth factor/vascular permeability factor for solid tumor growth, and its inhibition by the antibody. Biochem Biophys Res Commun 194:1234

Konerding MA, Steinberg F, Budach V (1989a) The vascular system of xenotransplanted tumors – scanning electron and light microscopic studies. Scanning Microsc 3:327–336

Konerding MA, Steinberg F, Streffer C (1989b) The vasculature of xenotransplanted human melanomas and sarcomas on nude mice. I. Vascular corrosion casting studies. Acta Anat (Basel) 136:21–26

Konerding MA, Steinberg F, Streffer C (1989c) The vasculature of xenotransplanted human melanomas and sarcomas on nude mice. II. Scanning and transmission electron microscopic studies. Acta Anat (Basel) 136:27

Kroeger M, Walenta S, Rofstad EK, Müller-Klieser W (1991) Imaging of structure and function in human tumor xenografts. In: Vaupel P, Jain RK (eds) Tumor blood supply and metabolic microenvironment. Fischer, Stuttgart, p 305–318

Kuroiwa T, Aoki K, Taniguchi S, Hasuda K, Baba T (1987) Efficacy of two route chemotherapy using cis-diammechloroplatinum (II) and its antidote sodium thiosulphate in combination with angiotensin II in a rat limb tumor. Cancer Res 47:3618

Lammertsma AA, Wilson CB, Jones T (1991) In vivo physiological studies in human tumors using positron emission tomography. In: Vaupel P, Jain RK (eds) Tumor blood supply and metabolic microenvironment. Fischer, Stuttgart, p 319–325

Lauk S, Skates S, Goodman M, Suit HD (1989) A morphometric study of the vascularity of oral squamous cell carcinomas and its relation to outcome of radiation therapy. Eur J Cancer Clin Oncol 25:1431

Lauria R, Perrone F, Carlomagno C, et al. (1995) The prognostic value of lymphatic and blood vessel invasion in operable breast cancer. Cancer 76:1772–1778

Lazo JS (1986) Endothelial injury caused by antineoplastic agents. Biochem Pharmacol 35:1919–1923

Less JR, Posner MC, Boucher Y, Borochovitz D, Wolmark N, Jain RK (1992) Interstitial hypertension in human tumors. 4. Interstitial hypertension in human breast and colorectal tumors. Cancer Res 52:6371

Levy AP, Tamargo R, Brem H, Nathans D (1989) An endothelial cell growth factor from the mouse neuroblastoma cell line NB 41. Growth Factors 2:9

Li VW, Folkerth RD, Watanabe H, et al. (1994) Microvessel count and cerebrospinal fluid basic fibroblast growth factor in children with brain tumors. Lancet 344:82–86

Lichtenbeld HHC, van Dam-Mieras MCE, Hillen HFP (1996) Tumor angiogenesis: pathophysiology and clinical significance. Neth J Med 49:42–51

López-Pousa S, Ferrier I, Vich JM, Domenech-Mateu J (1981) Angiogenic activity of CSF in human malignancies. Experientia 37:413

Lowdell CP, Ash DV, Driver I, Brown SB (1993) Interstitial photodynamic therapy. Clinical experience with diffusing fibres in the treatment of cutaneous and subcutaneous tumours. Br J Cancer 67:1398–1403

Lyng H, Skretting A, Rofstad EK (1992) Blood flow in six human melanoma xenograft lines with different growth characteristics. Cancer Res 52:584–592

Lyng H, Monge OR, Sager EM, Rofstad EK (1993) Prediction of treatment temperatures in clinical hyperthermia of locally advanced breast carcinoma – the use of contrast enhanced computed tomography. Int J Radiat Oncol Biol Phys 26:451–457

Macchiarini P, Fontanini G, Hardin MJ, Hardin MJ, Squartini F, Angeletti CA (1992) Relation of neovasculature to metastasis of non-small-cell lung cancer. Lancet 340:145

Maeda K, Chung Y-S, Takatsuka S, et al. (1995) Tumour angiogenesis and tumor cell proliferation as prognostic indicators in gastric carcinoma. Br J Cancer 72:319–323

Matsubara T, Ziff M (1987) Inhibition of human endothelial cell proliferation by gold compounds. J Clin Invest 79:1440

Matsubara T, Saura R, Hirohatak, Ziff M (1989) Inhibition of human endothelial cell proliferation in vitro and neovascularization in vivo by D-penicillamine. J Clin Invest 83:158

Mattern J, Koomägi R, Volm M (1995) Vascular endothelial growth factor expression and angiogenesis in non-small cell lung carcinomas. Int J Oncol 6:1059–1062

Mayer WK, Stohrer M, Krüger W, Vaupel P (1992) Laser Doppler flux and tissue oxygenation of experimaental tumours upon local hyperthermia and/or hyperglycaemia. J Cancer Res Clin Oncol 118:523

McCarty MF (1996) Fish oil may impede tumour angiogenesis and invasiveness by down-regulating protein kinase C and modulating eicosanoid production. Med Hypotheses 46:107–115

Menger MD, Lehr H-A (1993) Scope and perspectives of intravital microscopy – bridge over from in vitro to in vivo. Immunol Today 14:519–522

Merwin R, Algire GH (1950) Transparent-chamber observations of the response of a transplantable mouse mammary tumor to local roentgen irradiation. J Natl Cancer Inst 10:593

Messmer K, Funk W, Endrich B, Zeintl H (1984) The perspectives of new methods in microcirculation research. Prog Appl Microcirc 6:77–90

Meyer JU, Lindbom L, Intaglietta M (1987) Coordinated diameter oscillations at arteriolar bifurcations in skeletal muscle. Am J Physiol 253:H568–H573

Meyer JU, Borgström P, Lindbom L, Intaglietta M (1988) Vasomotion patterus in skeletal muscle arterioles during changes in arterial pressure. Microvasc Res 35:193–203

Mikami Y, Tsukuda M, Mochimatsu I, Kokatsu T, Yage T, Sawaki S (1991) Angiogenesis in head and neck tumors. Nippon Jibiinkoka Gakkai Kaiho 96:645–650

Milas L, Hunter N, Furuta Y, Nishiguchi I, Runkel S (1991) Antitumour effects of indomethacin alone and in combination with radiotherapy: role of inhibition of tumour angiogenesis. Int J Radiat Biol 60:65

Miliaras D, Kamas A, Kalekou H (1995) Angiogenesis in invasive breast carcinoma: is it associated with parameters of prognostic significance? Histopathology 26:165–169

Molls M, Feldmann HJ (1991) Clinical investigations of blood flow in malignant tumors of the pelvis and the abdomen in patients undergoing thermoradiotherapy. In: Vaupel P, Jain RK (eds) Tumor blood supply and metabolic microenvironment. Fischer, Stuttgart, p 143–153

Monschke F, Müller WU, Winkler U, Streffer C (1991) Cell proliferation and vascularization in human breast carcinomas. Int J Cancer 49:812

Monti E, Sinha BK (1994) Antiproliferative effect of genistein and adriamycin against estrogen-dependent and -independent human breast carcinoma cell lines. Anticancer Res 14:1221

Moses MA, Sudhalter J, Langer R (1990) Identification of an inhibitor of neovascularization from cartilage. Science 248:1408

Murray JC, Randhawa VS, Denekamp J (1987) The effects of melphalan and misonidazole on the vasculature of a murine sarcoma. Br J Cancer 55:233–238

Neufeld G, Gospodarowicz D (1987) Protamine sulfate inhibits mitogenic activities of the extracellular matrix and fibroblast growth factor, but potentiates that of epidermal growth factor. J Cell Physiol 132:287

Nguyen M, Watanabe H, Budson AE, Richie JP, Hayes DF, Folkman J (1994) Elevated levels of an angiogenic peptide, basic fibroblast growth factor, in the urine of patients with a wide spectrum of cancers. J Natl Cancer Inst 86:356

Obermair A, Czerwenka K, Kurz P, Buxbaum P, Schemper M, Sevelda P (1994a) Influence of tumoral microvessel density on the recurrence-free survival in human breast cancer: primary results. Onkologie 17:44–49

Obermair A, Czerwenka K, Kurz P, Kaider A, Sevelda P (1994b) Tumorale Gefäßdichte bei Mammatumoren und ihr Einfluß auf das rezidivfreie Überleben. Chirurg 65:611–615

Okunieff P, Vaupel P, Sedlacek R, Neuringer LJ (1989) Evaluation of tumor energy metabolism and microvascular blood flow after glucose or mannitol administration using ^{31}P nuclear magnetic resonance spectroscopy and laser Doppler flowmetry. Int J Radiat Oncol Biol Phys 16:1493

Okunieff P, Singer S, Vaupel P (1991) Magnetization transfer ^{31}P-NMR to measure metabolic states, dynamic changes, and enzyme kinetics. In: Vaupel P, Jain RK (eds) Tumor blood supply and metabolic microenvironment. Fischer, Stuttgart, p 267–292

Olivarez D, Ulbright T, De Riese W, Foster R, Reister T, Einhorn L, Sledge G (1994) Neovascularization in clinical stage A testicular germ cell tumor: prediction of metastatic disease. Cancer Res 54:2800–2802

O'Reilly MS, Holmgren L, Shing Y, et al. (1994) Angiostatin: a novel angiogenesis inhibitor that mediates the suppression of metastases by a Lewis lung carcinoma. Cell 79:315

Otte J (1988) Hyperthermia in cancer therapy. Eur J Pediatr 147:560

Page DL, Dupont WD (1992) Breast cancer angiohistogenesis – through a narrow window. J Natl Cancer Inst 84:1850–1851

Paget S (1889) The distribution of secondary growths in cancer of the breast. Lancet 1:571

Pepper MS, Vassalli J-D, Wilks JW, Schweigerer L, Orci L, Montesano R (1994) Modulation of bovine microvascular endothelial cell proteolytic properties by inhibitors of angiogenesis in vitro. J Cell Biochem 55:419

Peters W, Teixeira M, Intaglietta M, Gross JF (1980) Microcirculatory studies in rat mammary carcinoma. I. Transparent chamber method, development of microvasculature, and pressures in tumor vessels. J Natl Cancer Inst 65:631–642

Reinhold HS (1971) Improved microcirculation in irradiated tumours. Eur J Cancer 7:273–280

Reinhold HS (1979) In vivo observations of tumor blood flow. In: Peterson HI (ed) Tumor blood circulation. CRC Press, Boca Raton, p 115

Reinhold HS, Endrich B (1986) Tumour microcirculation as a target for hyperthermia. Int J Hyperthermia 2:111–137

Reinhold HS, Blachewiecz B, Blok A (1977) Oxygenation and reoxygenation in "sandwich" tumours. Bibl Anat 15:270–272

Révész L, Siracka E, Siracky J, Delides G, Pavlaki K (1989) Variation of vascular density within and between tumors of the uterine cervix and its predictive value for radiotherapy. Int J Radiat Oncol Biol Phys 16:1161

Robaye B, Mosselmans R, Fiers W, Dumant JE, Galand P (1991) Tumor necrosis factor induces apoptosis (programmed cell death) in normal endothelial cells in vitro. Am J Pathol 138:447

Saclarides TJ, Speziale NJ, Drab E, Szeluga DJ, Rubin DB (1994) Tumor angiogenesis and rectal carcinoma. Dis Colon Rectum 37:921–926

Sahin AA, Sneige N, Ordonez GN (1993) Tumor angiogenesis detected by Ulex europaeus agglutinin 1 lectin (UEA1) and factor VIII immunostaining in node-negative breast carcinoma (NNBC) treated by mastectomy: prediction of tumor recurrence. Mod Pathol 6:19a

Sarbia M, Bittinger F, Porschen R, Dutkowski P, Willers R, Gabbert HE (1996) Tumor vascularization and prognosis in squamous cell carcinomas of the esophagus. Anticancer Res 16:2117–2122

Scheulen ME, Schmidt CG (1982) Raynaud-Syndrom nach kombinierter zytostatischer Behandlung von Patienten mit malignen Hodentumoren. Dtsch Med Wochenschr 107:1640–1644

Schlag PM, Kettelhack C (1993) Weichteilsarkome: die isolierte Extremitätenperfusion. Chirurg 64:455–460

Schweigerer L (1993) Basic fibroblast growth factor, angiogenesis and therapeutic antiangiogenesis. In: Dengler HJ, Meuer SC (eds) Zellbiologie und klinische Pharmakologie. Fischer, Stuttgart, p 91

Schweigerer L (1995) Antiangiogenesis as a novel therapeutic concept in pediatric oncology. J Mol Med 73:497–508

Schweigerer L, Fotsis T (1992) Angiogenesis and angiogenesis inhibitors in pediatric diseases. Eur J Pediatr 151:472–476

Schweigerer L, Neufeld G, Friedman J, Abraham JA, Fiddes JC, Gospodarowicz D (1987a) Capillary endothelial cells express basic fibroblast growth factor, a mitogen that promotes their own growth. Nature 325:257–259

Schweigerer L, Neufeld G, Gospodarowicz D (1987b) Basic fibroblast growth factor is present in cultured human retinoblastoma cells. Invest Ophthalmol Vis Sci 28:1838–1843

Schweigerer L, Neufeld G, Gospodarowicz D (1987c) Basic fibroblast growth factor as a growth inhibitor for cultured human tumor cells. J Clin Invest 80:1516–1520

Schwickert G, Walenta S, Sundfor K, Rofstad EK, Mueller-Klieser W (1995) Correlation of high lactate levels in human cervical cancer with incidence of metastasis. Cancer Res 55:4757–4759

Scott PAE, Harris AL (1994) Current approaches to targeting cancer using antiangiogenesis therapies. Cancer Treat Rev 20:393–412

Seegenschmiedt MH, Sauer R, Miyamoto C, Chalal JA, Brady LW (1993) Clinical experience with interstitial thermoradiography for localized implantable pelvic tumors. Am J Clin Oncol 16:210–222

Semmler W, Bachert P, van Kaick G (1991) Human tumor response to therapy monitored by magnetic resonance spectroscopy. In: Vaupel P, Jain RK (eds) Tumor blood supply and metabolic microenvironment. Fischer, Stuttgart, p 257–266

Shpitzer Th, Chaimoff M, Gal R, Stern Y, Feinmesser R, Segal K (1996) Tumor angiogenesis as a prognostic factor in early oral tongue cancer. Arch Otolaryngol Head Neck Surg 122:865–868

Siracka E, Siracky J, Pappova N, Révész L (1982) Vascularization and radiocurability in cancer of the uterine cervix. A retrospective study. Neoplasma 29:183

Siracka E, Révész L, Kovac R, Siracky J (1988) Vascular density in carcinoma of the uterine cervix and its predictive value for radiotherapy. Int J Cancer 41:819

Siracky J, Siracka E, Kovac R, Révész L (1988) Prognostic significance of vascular density and a malignancy grading in radiation treated uterine cervix carcinoma. Neoplasma 35:289

Skinner SA, Tutton PJM, O'Brien PE (1990) Microvascular architecture of experimental colon tumors in the rat. Cancer Res 50:2411–2417

Slaaf DW, Oude Vrielink HHE, Tangelder GJ, Reneman RS (1989) Vasomotion under altered perfusion conditions. Prog Appl Microcirc 15:75–86

Sneige N, Singletary E, Sahin A (1992) Multiparameter analysis of potential prognostic factors in node negative breast cancer patients. Modern Pathol 5:18a

Steinberg F, Konerding MA, Streffer C (1990) The vascular architecture of human xenotransplanted tumors: histological, morphometrical, and ultrastructural studies. J Cancer Res Clin Oncol 116:517

Stephens TC, Peacock JH (1978) Cell yield and cell survival following chemotherapy of the mice B 16 melanoma. Br J Cancer 38:591–598

Streffer C (1987) Hyperthermia and the therapy of malignant tumors. Recent results in cancer research, vol 104. Springer, Berlin Heidelberg New York

Suzuki M, Hori K, Abe I, Saito S, Sato H (1981) A new approach to cancer chemotherapy: selective enhancement of tumor blood flow with angiotensin II. J Natl Cancer Inst 67:663

Taham SR, Stein AL (1995) Angiogenesis in invasive squamous cell carcinoma of the lip: tumor vascularity is not an indicator of metastatic risk. J Cutan Pathol 22:236–240

Takamiya Y, Friedlander RM, Brem H, Malick A, Martuza RL (1993) Inhibition of angiogenesis and growth of human nerve-sheath tumors by AGM-1470. J Neurosurg 78:470–476

Takematsu H, Tomita Y, Kato T (1985) Angiotensin-induced hypertension and chemotherapy for multiple lesions of malignant melanoma. Br J Dermatol 113:463

Tanigawa N, Amaya H, Matsumura M, Shimomatsuya T, Horiuchi T, Muraoka R, Iki M (1996) Extent of tumor vascularization correlates with prognosis and hematogenous metastasis in gastric carcinomas. Cancer Res 56:2671–2676

Tannock IF (1968) The relation between cell proliferation and the vascular system in a transplanted mouse mammary tumour. Br J Cancer 22:258

Tapper D, Scheiner C, Frissora H, Zetter B (1981) The stimulation of capillary endothelial cell migration by aqueous humor. J Surg Res 30:262

Teicher BA, Alvarez Sotomayor E, Huang ZD (1992) Antiangiogenic agents potentiate cytotoxic cancer therapies against primary and metastatic disease. Cancer Res 52:6702

Teicher BA, Holden SA, Ara G, Northey D (1993) Response of the FSall fibrosarcoma to antiangiogenic modulators plus cytotoxic agents. Anticancer Res 13:2101

Teicher BA, Holden SA, Ara G, Alvarez Sotomayor E, Huang ZD, Chen Y-N, Brem H (1994) Potentiation of cytotoxic cancer therapies by TNP-470 alone and with other antiangiogenic agents. Int J Cancer 57:920

Teutsch C, Lipton A, Harvey HA (1977) Raynaud's phenomenon as a side effect of chemotherapy with vinblastine and bleomycin for testicular carcinoma. Cancer Treat Rep 61:925–926

Thomlinson RH, Gray LH (1955) The histological structure of some human lung cancers and the possible implications for radiotherapy. Br J Cancer 9:539–547

Toi M, Kashitani J, Tominaga T (1993) Tumor angiogenesis is an independent prognostic indicator in primary breast carcinoma. Int J Cancer 55:371–374

Tsai AG, Intaglietta M (1993) Evidence of flow motion induced changes in local tissue oxygenation. Int J Microcirc Clin Exp 12:75

Tsuruo T, Iida H, Makishima F, Yamori T, Kawabata H, Tsukagoshi S (1985) Inhibition of spontaneous and experimental tumor metastasis by the calcium antagonist verapamil. Cancer Chemother Pharmacol 14:30

Van Hoef M, Knox WF, Dhesis SS, Howell A, Schor AM (1993) Assessment of tumor vascularity as a prognostic factor in lymph node negative invasive breast cancer. Eur J Cancer 29:1141–1145

Vaupel P (1977) Hypoxia in neoplastic tissue. Microvasc Res 13:399–408

Vaupel P (1990) Pathophysiological mechanisms of hyperthermia in cancer therapy. In: Gautherie M (ed) Biological basis of oncologic thermotherapy. Springer, Berlin Heidelberg New York, p 73

Vaupel P (1992) Physiological properties of malignant tumors. NMR Biomed 5:220–225

Vaupel P, Kallinowski F (1987) Hemoconcentration of blood flowing through human tumor xenografts. Int J Microcirc Clin Exp 6:72

Vaupel P, Menke H (1989) Effect of various calcium antagonists on blood flow and red blood cell flux in malignant tumors. Prog Appl Microcirc 14:88–103

Vaupel P, Müller-Klieser W (1983) Interstitieller Raum und Mikromilieu in malignen Tumoren. Mikrozirk Forsch Klin 2:78–90

Vaupel P, Kluge M, Ambroz MC (1988) Laser Doppler flowmetry in subepidermal tumours and in normal skin of rats during localized ultrasound hyperthermia. Int J Hyperthermia 4:307

Vaupel P, Kallinowski F, Okunieff P (1989a) Blood flow, oxygen and nutrient supply, and metabolic microenvironment of human tumors: a review. Cancer Res 49:6449–6465

Vaupel P, Okunieff P, Kallinowski F, Neuringer LJ (1989b) Correlations between ^{31}P-NMR spectroscopy and tissue O_2 tension measurements in a murine fibrosarcoma. Radiat Res 120:477–493

Vaupel P, Okunieff P, Kluge M (1989c) Response of tumour red blood cell flux to hyperthermia and/or hyperglycaemia. Int J Hyperthermia 5:199

Vaupel P, Schlenger K, Höckel M (1991) Blood flow and oxygenation of human tumors. In: Vaupel P, Jain RK (eds) Tumor blood supply and metabolic environment. Fischer, Stuttgart, p 165–185

Verhoeven D, Buyssens N (1988) Desmin-positive stellate cells associated with angiogenesis in a tumour and non-tumour system. Arch Biol Cell Pathol 54:263

Vesalainen S, Lipponen P, Talja M, Alhava E, Syrjanen K (1994) Tumor vascularity and basement membrane structure as prognostic factors in T_{1-2} M_0 prostatic adenocarcinomas. Anticancer Res 14:709–714

Vissher DW, Smilanetz S, Drozdowicz S, Wykes SM (1993) Prognostic significance of image morphometric microvessel enumeration in breast carcinoma. Anal Quant Cytol Histol 15:88

Vogelzang NJ, Bosl GJ, Johnson K, Kennedy BJ (1981) Raynaud's phenomenon: a common toxicity after combination chemotherapy for testicular cancer. Ann Intern Med 95:288–292

Wakui S, Furusato M, Itoh T, et al. (1992) Tumor angiogenesis in prostatic carcinoma with and without bone marrow metastasis – a morphometric study. J Pathol 168:257–262

Waterman FM, Tupchong L, Nerlinger RE, Matthews J (1991) Blood flow in human tumors during local hyperthermia. Int J Radiat Oncol Biol Phys 20:1255

Weidner N (1995) Current pathologic methods for measuring intratumoral microvessel density within breast carcinoma and other solid tumors. Breast Cancer Res Treat 36:169–180

Weidner N, Folkman J, Pozza F, et al. (1992) Tumor angiogenesis – a new significant and independent prognostic indicator in early-stage breast carcinoma. J Natl Cancer Inst 84:1875–1887

Weidner N, Carroll PR, Flax J, Blumenfeld W, Folkman J (1993) Tumor angiogenesis correlates with metastasis in invasive prostate carcinoma. Am J Pathol 143:401–409

West CML (1993) Predictive assays in radiation therapy. Adv Radiat Biol, in press

White CM, Sondheimer HM, Crouch EC, Wilson H, Fan LL (1989) Treatment of pulmonary hemangiomatosis with recombinant interferon alfa-2a. N Engl J Med 320:1197

Wolff JEA, Guerin C, Laterra J, Bressler J, Indurti RR, Brem H, Goldstein GW (1993) Dexametasone reduces vascular density and plasminogen activator activity in 9L rat brain tumors. Brain Res 604:79

Wurschmidt F, Beck-Bornholdt HP, Vogler H (1990) Radiobiology of the rhabdomyosarcoma R1H of the rat: influence of the size of irradiation field on tumor response, tumor bed effect, and neovascularization kinetics. Int J Radiat Oncol Biol Phys 18:879

Yamaura H, Sato H (1974) Quantitative studies on the developing system of rat hepatoma. J Natl Cancer Inst 53:1229–1240

Yamaura HK, Yamada K, Matsuzawa T (1976) Radiation effect on the proliferating capillaries in rat transparent chambers. Int J Radiat Biol 30:179

Young JS, Lumsden CE, Stalker AL (1950) The significance of the "tissue pressure" of normal testicular and of neoplastic (Brown-Pearce carcinoma) tissue in the rabbit. J Pathol Bact 62:313–333

Zeintl H, Tompkins WR, Messmer K, Intaglietta M (1986) Static and dynamic microcirculatory video image analysis applied to clinical investigations. Prog Appl Microcirc 11:1–10

Zywietz F, Hahn LS, Lierse W (1994) Ultrastructural studies on tumor capillaries of a rat rhabdomyosarcoma during fractionated radiotherapy. Acta Anat (Basel) 150:80–85

4 Tumor Blood Flow

P. VAUPEL

4.1 Introduction

It is generally accepted that tumor blood flow, microcirculation, oxygen and nutrient supply, tissue pH distribution, and bioenergetic/metabolic status – factors that are usually closely linked and that define the so-called metabolic microenvironment – can markedly influence the therapeutic response of malignant tumors to conventional irradiation, chemotherapy, and other nonsurgical treatment modalities as well as cell proliferation activity and relevant biological characteristics of tumors (e.g., SUTHERLAND 1988; VAUPEL 1992, 1994; VAUPEL and JAIN 1991; VAUPEL et al. 1989). Information currently available on the blood flow of human tumors is presented in this chapter. According to these data, significant variations in this relevant factor are likely to occur between different locations within a tumor, and between tumors of the same grade and clinical stage. Therefore, evaluation of blood flow distribution of individual tumors before therapy and a corresponding "fine-tuning" of treatment protocols for individual patients may result in improved tumor response to treatment.

4.2 Tumor Microcirculation

Most, if not all, solid tumors originate as avascular aggregates of malignant cells. "Microscopic tumors" exchange nutrients and metabolic waste products with their surroundings by simple diffusion. Therefore, the growth of an avascular three-dimensional aggregate of tumor cells is self-limiting. The establishment of progressive expansion of malignant tumors is possible only if supply and drainage are initiated via blood flow through exchange vessels of the tumor microcirculatory bed. When considering the origin of blood vessels in a growing tumor, the existence of two different types of vessels must be taken into account:

- Preexisting normal host vessels incorporated into the tumor
- Newly formed tumor vessels arising from neovascularization stimulated by release of (one or more) angiogenesis factors; in general these vessels are void of smooth muscle cells and are not innervated

There is clear experimental evidence that the fraction of each of these vessel types depends on (a) tumor size, (b) whether the tumor is spontaneous (as in humans) or transplanted (as is the case in murine experimental systems), (c) tumor growth rate, (d) tumor growth site, and (e) the cell line investigated.

Newly formed microvessels in most solid tumors do not conform to the normal morphology of the host tissue vasculature (see Chap. 2). Microvessels in solid tumors exhibit a series of severe structural and functional abnormalities. They are often dilated, tortuous, elongated, and saccular. There is significant arteriovenous shunt perfusion accompanied by chaotic vascular organization without any regulation aimed at the metabolic demand or functional status of the tissue. Excessive branching is a common finding often coinciding with blind vascular endings. Incomplete or even missing endothelial

P. VAUPEL, Dr. med., Institute of Physiology and Pathophysiology, University of Mainz, Duesbergweg 6, D-55099 Mainz, Germany

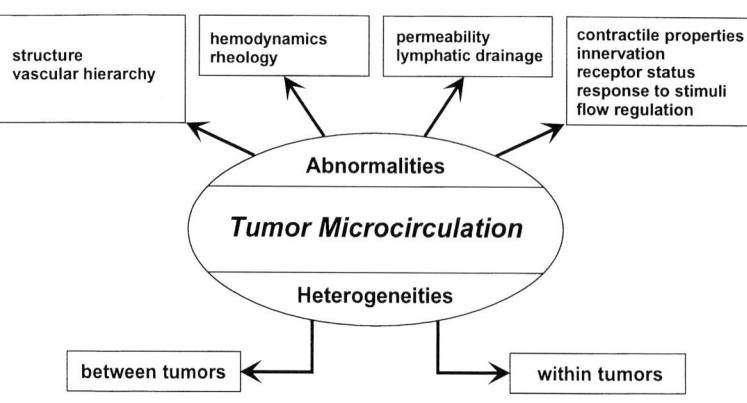

Fig. 4.1. Relevant structural and functional abnormalities of tumor neovasculature/microcirculation

lining and interrupted basement membranes result in increased vascular permeability with extravasation of red blood cells and blood plasma expanding the interstitial fluid space and drastically increasing hydrostatic pressure in the tumor interstitium, leading to a rise in *viscous resistance* to flow, mainly due to hemoconcentration.

Aberrant vascular morphology and a decrease in vessel density are responsible for an increase in *geometric resistance* to flow which, together with an enlargement of diffusion distances due to expansion of the extravascular space, can lead to perfusion with hypoxemic and nutrient-deprived blood. Substantial spatial heterogeneity in the distribution of tumor vessels and significant temporal heterogeneity in the microcirculation within a tumor may result in a considerably anisotropic distribution of tumor blood flow and flow-related factors. These are usually closely linked and define the so-called metabolic microenvironment (see Figs. 4.1, 4.2). Variations in these relevant parameters between tumors (even if

they are of the same grade and clinical stage) are often more pronounced than differences occurring between different locations or microareas within a tumor (for reviews see Vaupel et al. 1989; Vaupel 1990, 1992, 1994).

The morphological appearance of the tumor vascular bed does not necessarily allow direct judgements concerning functional aspects of tumor microcirculation or nutritive blood flow. In experimental tumor systems, only 20%–85% of the tumor vessels are perfused at a given time. The fraction depends on the tumor cell line investigated (Bernsen et al. 1995).

4.3
Tumor Blood Flow

A number of studies on blood flow through human tumors have been reported to date. Some of them are anecdotal reports rather than systematic investigations, and therefore definite conclusions cannot be drawn. Considering the data presently available, the following (preliminary) conclusions can be drawn when flow data derived from different reports are pooled (for reviews see Vaupel 1990, 1992, 1993, 1994, 1996; Vaupel and Jain 1991; Vaupel et al. 1989):

a) Blood flow can vary considerably despite similar histological classification and primary site (Fig. 4.3).

b) Tumors can have flow rates similar to those measured in organs with a high metabolic rate such as liver, heart, or brain.

c) Some tumors exhibit flow rates that are even lower than those of tissues with a low metabolic rate such as skin, resting skeletal muscle, or adipose tissue.

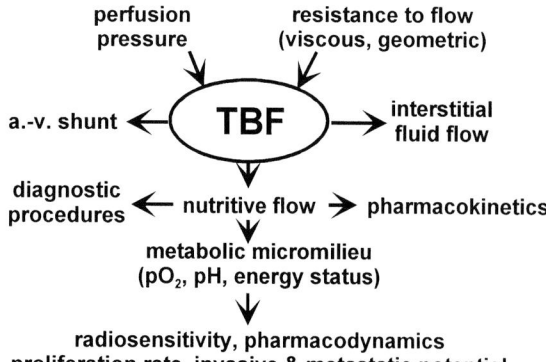

Fig. 4.2. Parameters determining tumor blood flow (perfusion pressure, resistance to flow) and possible consequences of flow efficacy on conventional treatment modalities (direct and indirect mechanisms). *TBF*, tumor blood flow

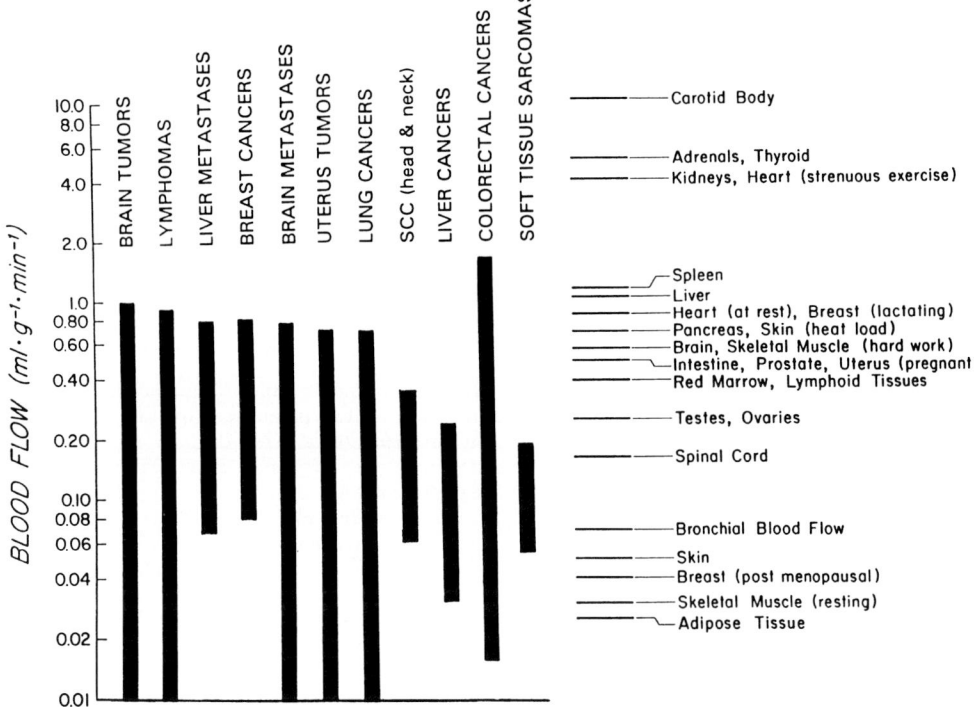

Fig. 4.3. Variability of blood flow in human malignancies (*black bars*) and mean flow values of normal tissues and organs. *SCC*, squamous cell carcinomas. (Pooled data, updated from VAUPEL 1997)

d) Blood flow in human tumors can be higher or lower than that of the tissue of origin, depending on the functional state of the latter tissue (e.g., average blood flow in breast cancers is substantially higher than that of the postmenopausal breast and significantly lower than flow data obtained in the lactating parenchymal breast).

e) The average perfusion rate of carcinomas does not deviate substantially from that of soft tissue sarcomas.

f) Metastatic lesions exhibit a blood supply that is comparable to that of the primary tumor, perhaps with the exception of those secondary lesions in the liver that are preferentially supplied by the portal system (due to lower perfusion pressure).

g) In some tumor entities, blood flow in the periphery is distinctly higher than in the center whereas in others, blood flow is significantly higher at the tumor center than at the tumor edge.

h) Flow data from multiple sites of measurement show marked heterogeneity within individual tumors. However, these flow differences are smaller than those seen between different tumors (i.e., tumor-to-tumor heterogeneity is more pronounced than intratumor heterogeneity).

i) There is substantial temporal flow heterogeneity on a microscopic level within human tumors, as shown by multichannel laser Doppler flowmetry (HILL et al. 1996; PIGOTT et al. 1996).

j) There is no association between tumor size and blood flow in most cancers. In bronchial carcinomas, blood flow through central tumor regions is inversely related to tumor volume (ROWELL et al. 1993). Only recently, an inverse relationship between flow and tumor volume has been described for colon carcinomas (HOLM et al. 1995) and a series of other tumor entities (HERING et al. 1995), a pattern well recognized in many experimental animal tumors.

k) Tumor blood flow (at least in ovarian and breast cancers) and vascularity seem to be positively correlated to vascular endothelial growth factor (VEGF) expression (D. MARMÉ, personal communication), the latter being dependent on the degree and duration of hypoxia. At the same time, there is evidence that increased VEGF expression coincides with malignant progression of the tumor (see Sect. 4.4).

l) Tumor blood flow is not regulated according to metabolic demand as is the case in normal tissues.

4.4
Role of Blood Flow and Flow-Related Microenvironmental Parameters in Chemo-, Immuno-, Radio-, and Thermotherapy

Blood flow and flow-related microenvironmental parameters (tissue oxygenation, pH distribution, bioenergetic status, and nutrient supply) are important factors known to be able to modulate the sensitivity of cancer cells to anticancer agents or cells, sparsely ionizing radiation, and heat treatment

Table 4.1. Possible role of microcirculation, blood flow, and flow-related microenvironmental parameters in the chemotherapy of solid tumors (i.e., in pharmacokinetics and pharmacodynamics)

Favorable aspects
- High vascular permeability (heterogeneity)
- High hydraulic conductivity within the interstitium
- Hypoxia (e.g., mitomycin C, misonidazole)
- $pH_e < 7$ (enhanced cellular uptake of weakly acid drugs)
- $pH_e < 7$ (e.g., mitomycin C, cisplatin)

Adverse aspects
- Low vascular density (heterogeneity)
- Large intercapillary distances (heterogeneity)
- Ischemic tissue areas (heterogeneity)
- Arteriovenous shunt perfusion
- Centrifugal bulk flow of free water
- Enlarged interstitial spaces
- High interstitial protein concentrations
- Hypoxia (O_2-dependent drugs)
- Cell cycle position (G_o cells)
- $pH_e < 7$ (impaired cellular uptake of weakly alkaline drugs)
- $pH_e < 7$ (e.g., doxorubicin, vinblastine)
- Formation of hypoxia-induced stress proteins
- Expression of genes responsible for drug resistance
- Selection of aggressive subclones

Table 4.2. Possible role of blood flow and flow-related parameters in humoral (e.g., monoclonal antibodies, cytokines) and cell-mediated (e.g., lymphokine-activated killer cells) immune mechanisms against cancer cells

Favorable aspects
- High vascular permeability
- High hydraulic conductivity within the interstitium

Adverse aspects
- Chaotic perfusion
- Large distances for transport
- Centrifugal bulk flow of free water
- High interstitial pressure (impaired extravasation)
- Arteriovenous shunt perfusion
- Anoxia (impaired cytokine action; e.g., TNF-α, IL 2)

Table 4.3. Possible impact of microcirculation, blood flow, and flow-related parameters on sparsely ionizing radiation

Favorable aspects
?

Adverse aspects
- Chaotic perfusion
- Hypoxia, anoxia (oxygen enhancement ratio ~3)
- $pH_e < 7$ (?)
- Cell cycle position (G_o cells)
- Selection of aggressive subclones

Table 4.4. Possible role of microcirculation, blood flow, and flow-related parameters on the efficacy of localized hyperthermia ($42°C < T < 45°C$)

Favorable aspects
- Low-flow areas (heterogeneity)
- Hypoxia, anoxia (heterogeneity)
- pH < 7
- Substrate and ATP depletion (heterogeneity)
- Lactate accumulation (heterogeneity)

Adverse aspects
- High-flow areas
- Cell cycle position (G_o cells)

Hyperthermia can act as a radio- and chemosensitizer at temperatures of 40° to 42°C.

(VAUPEL 1993). Besides *direct effects*, tumor blood flow and the flow-related parameters mentioned above can have more *indirect influences* on the efficacy of these treatment modalities (see Tables 4.1–4.4). Reviews covering the impact of blood flow and flow-related microenvironmental factors on treatment efficacy have been provided by DURAND (1991), GERWECK and SEETHARAMAN (1996), HALL (1994), HÖCKEL et al. (1996), JAIN (1993), SONG et al. (1993), TANNOCK and ROTIN (1989), TEICHER et al. (1990), VAUPEL (1993), VAUPEL et al. (1996) and WIKE-HOOLEY et al. (1984) (see also Chap. 13).

References

Bernsen HJJA, Rijken PFJW, Oostendorp T, van der Kogel AJ (1995) Vascularity and perfusion of human gliomas xenografted in the athymic nude mouse. Br J Cancer 71:721–726

Durand RE (1991) Keynote address: the influence of microenvironmental factors on the activity of radiation and drugs. Int J Radiat Oncol Biol Phys 20:253–258

Gerweck LE, Seetharaman K (1996) Cellular pH gradient in tumor versus normal tissue: potential exploitation for the treatment of cancer. Cancer Res 56:1194–1198

Hall EJ (1994) Radiobiology for the radiologist, 4th edn. Lippincott, Philadelphia

Hering ER, Blekkenhorst GH, Jones DTL (1995) Tumor blood flow measurements using coincidence counting on patients treated with neutrons. Int J Radiat Oncol Biol Phys 32:129–135

Hill SA, Pigott KH, Saunders MI, et al. (1996) Microregional blood flow in murine and human tumours assessed using laser Doppler microprobes. Br J Cancer Suppl 74:260–263

Höckel M, Schlenger KH, Aral B, Mitze M, Schäffer U, Vaupel P (1996) Association between tumor hypoxia and malignant progression in advanced cancer of the uterine cervix. Cancer Res 56:4509–4515

Holm E, Hagmüller E, Staedt U, et al. (1995) Substrate balances across colonic carcinomas in humans. Cancer Res 55:1373–1378

Jain RK (1993) Physiological resistance to the treatment of solid tumors. In: Teicher BA (ed) Drug resistance in oncology. Dekker, New York, pp 87–105

Pigott KH, Hill SA, Chaplin DJ, Saunders MI (1996) Microregional fluctuations in perfusion within human tumours detected using laser Doppler flowmetry. Radiother Oncol 40:45–50

Rowell NP, Flower MA, Cronin B, McCready VR (1993) Quantitative single-photon emission tomography for tumour blood flow measurement in bronchial carcinoma. Eur J Nucl Med 20:591–599

Song CW, Lyons JC, Luo Y (1993) Intra- and extracellular pH in solid tumors: influence on therapeutic response. In: Teicher BA (ed) Drug resistance in oncology. Dekker, New York, pp 25–51

Sutherland RM (1988) Cell and environment interactions in tumor microregions: the multicell spheroid model. Science 240:177–184

Tannock IF, Rotin D (1989) Acid pH in tumors and its potential for therapeutic exploitation. Cancer Res 49:4373–4384

Teicher BA, Holden SA, Al-Achi A, Herman TS (1990) Classification of antineoplastic treatments by their differential toxicity toward putative oxygenated and hypoxic tumor subpopulations in vivo in the FSaIIC murine fibrosarcoma. Cancer Res 50:3339–3344

Vaupel P (1990) Oxygenation of human tumors. Strahlenther Onkol 166:377–386

Vaupel P (1992) Physiological properties of malignant tumours. NMR Biomed 5:220–225

Vaupel PW (1993) Oxygenation of solid tumors. In: Teicher BA (ed) Drug resistance in oncology. Dekker, New York, pp 53–85

Vaupel PW (1994) Blood flow, oxygenation, tissue pH distribution, and bioenergetic status of tumors. Lecture 23. Ernst Schering Research Foundation, Berlin

Vaupel P (1996) Oxygen transport in tumors: characteristics and clinical implications. Adv Exp Med Biol 388:341–351

Vaupel P (1997) Blood flow and O_2-status of head and neck carcinomas. Adv Exp Med Biol, in press

Vaupel P, Jain RK (1991) Tumor blood supply and metabolic microenvironment. Characterization and implications for therapy. Fischer, Stuttgart

Vaupel P, Kallinowski F, Okunieff P (1989) Blood flow, oxygen and nutrient supply, and metabolic microenvironment of human tumors: a review. Cancer Res 49:6449–6465

Vaupel P, Thews O, Hoeckel M (1996) Tumor oxygenation: characterization and clinical implications. In: Smyth JF, Boogaerts MA, Ehmer BRM (eds) rhErythropoietin in cancer supportive treatment. Dekker, New York

Wike-Hooley JL, Haveman J, Reinhold HS (1984) The relevance of tumour pH to the treatment of malignant disease. Radiother Oncol 2:343–366

5 Clinical Investigations on Blood Perfusion in Human Tumors

H.J. Feldmann[1], M. Molls[2], and P. Vaupel[3]

5.1 Introduction

The responsiveness of tumors to radiotherapy, thermoradiotherapy, or treatment with chemical substances greatly depends on blood flow. With regard to the effectiveness of radiotherapy, the oxygen supply to the tumor is greatly influenced by the efficacy of blood flow. In general, approximately two to three times higher radiation doses are needed to kill hypoxic cells compared to well-oxygenated cells (GRAY et al. 1953; EVANS and NAYLOR 1963; VAUPEL and HÖCKEL, this volume). In chemotherapy, adequate vascular supply and blood flow are important for the delivery of cytotoxic agents to tumor cells. In the case of hyperthermia, sluggish tumor blood flow improves tumor heating since blood flow is the main route of heat dissipation. Therefore, the ability to heat tumors is markedly influenced by blood flow conditions. Experimental data suggest that poorly perfused areas are more sensitive to heat than highly perfused tissues that protect themselves by heat dissipation. Low blood flow during hyperthermia deprives tissues of nutrients and deteriorates the metabolic environment (hypoxia, acidosis, and energy depletion), thus increasing the cytotoxic effect of hyperthermia (STREFFER 1985; VAUPEL et al. 1989).

Several studies in humans underscore the importance of blood flow and address the heterogeneous perfusion within tumors as well as the different perfusion rates in tumor and normal tissues (LAMMERTSMÄ et al. 1985; SAMULSKI et al. 1987; MÄNTYLÄ et al. 1988; VAUPEL et al. 1989; FELDMANN et al. 1992). Investigations of blood flow have been performed using a variety of quantitative techniques (Table 5.1), e.g., positron emission tomography, isotope clearance techniques, thermal washout procedures, laser Doppler flowmetry, dynamic CT and MRI techniques, and color Doppler ultrasound (ACKER et al. 1990; BEANEY et al. 1984; ITO et al. 1982; LAGENDIJK et al. 1988; LAHTINEN et al. 1979; MÄNTYLÄ et al. 1982; VAUPEL et al. 1989; FELDMANN et al. 1992, 1993; SIEVERS et al. 1993; KEDAR et al. 1994; PIRHONEN et al. 1995). Noninvasive methods currently available for studying regional blood flow in patients are crude and provide only qualitative or semiquantitative information. Ultrasound Doppler measurements and new angiographic techniques assess flow velocities in arteries. Blood flow varies remarkably depending upon tumor type, volume, and site of growth. Clinical data show that there is a significant difference in relative flow rates between tumor centers and peripheries, which means that there is blood flow heterogeneity within the treatment volume (SAMULSKI et al. 1987; FELDMANN et al. 1992). This observation is in agreement with the differences in blood perfusion in most experimental tumors which have been found to be due to heterogeneity of the vascular density (JIRTLE 1988).

In this chapter investigations of regional tumor perfusion in different types of deep-seated tumors and normal tissues using methods available in the

[1]H.J. FELDMANN, Dr. med., and [2]M. MOLLS, Dr. med., Klinik und Poliklinik für Strahlentherapie und Radiologische Onkologie, Klinikum rechts der Isar der Technischen Universität München, Ismaninger Strasse 22, D-81675 München, Germany
[3]P. VAUPEL, Dr. med., Institute of Physiology and Pathophysiology, University of Mainz, Duesbergweg 6, D-55099 Mainz, Germany

Table 5.1. Techniques for measurement of tumor perfusion in superficial and deep-seated human malignancies

Technique	Tumor	Reference
PET [^{13}N]ammonia	Primary liver tumors	SHIBATA et al. 1988
PET ^{15}O inhalation technique	Primary cerebral tumors	ITO et al. 1982
PET ^{15}O inhalation technique	Primary and metastatic cerebral tumors	LAMMERTSMA et al. 1985
PET ^{15}O inhalation technique	Primary carcinoma of the breast Colorectal liver metastases	BEANEY et al. 1984 HOHENBERGER et al. 1993
Laser Doppler flowmetry	Lymph node metastases of head and neck cancer and breast cancer Chest wall recurrences of breast cancer Sarcomas of the extremities	ACKER et al. 1990; PIGOTT et al. 1996
Thermal clearance method	Advanced breast carcinoma Recurrent or metastatic colon carcinoma Recurrent pelvic tumors (soft tissue sarcoma, rectal cancer, cervical cancer)	LANGENDIJK et al. 1988 SAMULSKI et al. 1987 FELDMANN et al. 1992; MOLLS and FELDMANN 1991

^{133}Xe clearance method	Superficial tumor nodules of different
	    histologies (lymphomas, anaplastic and
	    differentiated carcinomas
	Carcinoma of the cervix and corpus uteri
		MÄNTYLÄ 1979
		
		NYSTRÖM et al. 1969

^{41}Ar, ^{85}Kr, ^{133}Xe clearance method	Superficial tumor nodules (e.g., breast cancer, head and neck cancer)	MÄNTYLÄ et al. 1988

Dynamic CT	Head and neck (primaries)
	Recurrent pelvic tumors (soft tissue sarcoma,
	    rectal cancer, cervical cancer, metastatic
	    hypernephroma)
		CLAUSSEN and LOCHNER 1983
		FELDMANN et al. 1992

Dynamic MRI	Recurrent pelvic tumors Recurrent rectal cancer Recurrent cervical cancer Meningeomas	FELDMANN et al. 1993 MÜLLER-SCHIMPFLE et al. 1993 HAWIGHORST et al. 1997 HAWIGHORST et al. 1996
Doppler ultrasound method	Ovarian cancer, cervical cancer Breast cancer Cervical cancer	SOHN et al. 1992 KEDAR et al. 1994 PIRHONEN et al. 1995

clinic will be presented and discussed briefly with regard to their relevance to therapy.

5.2
Methods for Measurement of Tumor Perfusion

5.2.1
Invasive Methods

5.2.1.1
Radioisotopic Clearance Techniques

Tumor blood flow is usually measured by radioisotope clearance techniques (^{85}Kr, ^{133}Xe, or ^{41}Ar) in animals and humans (e.g., SONG et al. 1972; VAUPEL et al. 1973; TANAKA 1974; MÄNTYLÄ 1979; MÄNTYLÄ et al. 1982). For the quantitative determination of

blood flow, the partition coefficient between the tissue and blood is required. The partition coefficients and the solubilities of krypton and xenon have been studied extensively in experiments in vitro and in animals (KITANI and WINKLER 1972; O'BRIEN and VEALL 1974). The values obtained have been used for blood flow calculations. For accurate determination of blood perfusion, the value of the partition coefficient should be measured in vivo at the same time as the blood flow rate. Therefore, MÄNTYLÄ et al. (1988) performed a systematic investigation in humans. They measured blood flow in human tumors using three diffusible radionuclides: argon, krypton, and xenon. To minimize the effects of various biological factors on the measurements, the radionuclides were injected into the tumor simultaneously. The injection was performed slowly with a fine needle inserted into the center of the tumor. The nuclides were dissolved in saline solution and a total amount of ap-

Table 5.2. Perfusion of different human tumors

Technique	Tissue	Perfusion units	Comments
PET ^{15}O inhalation technique			
Beaney et al. 1984	Breast carcinoma	0.08–0.31 ml·g^{-1}·min^{-1}	Regional blood flow higher
	Normal breast	0.01 ml·g^{-1}·min^{-1}	in nonnecrotic tumor tissue and normal breast
Ito et al. 1982	Tumor (nonenhancing hypodense center)	0.12 ml·g^{-1}·min^{-1}	Regional blood flow of the contrast-enhancing part was comparable to values obtained in white matter
Laser Doppler method			
Acker et al. 1990	Breast carcinoma	0.01–0.05 ml·g^{-1}·min^{-1}	In the majority of cases
	Soft tissue sarcoma (tumor core vs. tumor shell)	0.02–0.15 ml·g^{-1}·min^{-1}	flow in the periphery was higher than in the center
^{133}Xe clearance method			
Mäntylä 1979	Lymphomas	0.35 ± 0.21 ml·g^{-1}·min^{-1}	Lymphomas showed a
	Anaplastic carcinoma	0.15 ± 0.11 ml·g^{-1}·min^{-1}	statistically higher blood
	Differentiated carcinoma	0.23 ± 0.15 ml·g^{-1}·min^{-1}	flow than carcinomas
Nyström et al. 1969	Low-differentiated cervix carcinoma	0.20 ± 0.05 ml·g^{-1}·min^{-1}	Statistically significant
	High-differentiated cervix carcinoma	0.21 ± 0.04 ml·g^{-1}·min^{-1}	difference between poorly
	Low-differentiated corpus carcinoma	0.41 ± 0.09 ml·g^{-1}·min^{-1}	and well-differentiated
	High-differentiated corpus carcinoma	0.10 ± 0.04 ml·g^{-1}·min^{-1}	corpus carcinomas
Thermal washout			
Samulski et al. 1987	Tumor (hypodense CT)	0.06 ± 0.03 ml·g^{-1}·min^{-1}	Significant difference
	Tumor/normal interface	0.15 ± 0.07 ml·g^{-1}·min^{-1}	between normal and
	Tumor	0.09 ± 0.05 ml·g^{-1}·min^{-1}	tumor tissue
	Normal tissue	0.13 ± 0.07 ml·g^{-1}·min^{-1}	
Molls and Feldmann 1991	Tumor center	0.07 ± 0.04 ml·g^{-1}·min^{-1}	Significant difference
Feldmann et al. 1992	Tumor periphery	0.12 ± 0.05 ml·g^{-1}·min^{-1}	between tumor center and periphery
Dynamic CT			
Claussen and Lochner 1983	Head and neck tumors		
	Glomus tumor	>100 HU	
	Carcinoma	30–50 HU	
	Lymphoblastoma	<10 HU	
Feldmann et al. 1992	Tumor center	6.8 ± 0.8 HU	Significant difference
	Tumor periphery	13.9 ± 1.1 HU	between tumor center and tumor periphery
Dynamic MRI			
Feldmann et al. 1993	Tumor center	105 ± 14 SI	Significant difference
	Tumor periphery	225 ± 19 SI	between tumor center and tumor periphery

HU, Hounsfield Unit; SI, Signal Intensity.

proximately 1 ml of saline was injected. Radioactivity measurements were performed with scintillation detectors. Equations derived from the theoretical study of tumor blood flow performed by Perkkiö et al. (1986) were used to analyze the results. Measurements using simultaneous injection of different radioactive tracers yielded information on the relative partition coefficients of the nuclides used and estimated for the blood volume of the tissue studied.

This kind of invasive blood flow measurement was predominantly performed in superficial tumors. Mäntylä (1979) investigated 97 superficial tumor nodules from 80 patients including lymphomas and anaplastic or differentiated carcinomas of different histologies (Table 5.1). In general, blood flow in the lymphomas was greater than in anaplastic or differentiated carcinomas (Table 5.2). The mean blood flow was 0.35 ± 0.21 (SD) ml·g^{-1}·min^{-1} in the

Table 5.3. Treatment-related changes of tumor perfusion and vascularity

Technique	Flow changes	Specific point of flow change	Reference
Laser Doppler method	Flow changes under hyperthermia	Flow increase to a plateau No change in tumor blood flow Flow decrease Flow drop after an initial flow increase	ACKER et al. 1990
^{133}Xe clearance method	Radiation-induced changes in regional blood flow	Increase from 0.21 ± 0.18 to 0.31 ± 0.25 ml·g^{-1}·min^{-1} during first week Decrease during second week to 0.27 ± 0.19 ml·g^{-1}·min^{-1}	MÄNTYLÄ et al. 1982
Thermal clearance method	Hyperthermia-induced changes in regional blood flow	Increase and decrease of perfusion after fractionated hyperthermia treatments Tumor blood flow increased by amounts varying from 15% to 250% during the first 20–50 min of heating at 41°C–45°C No sharp reduction of blood flow	LAGENDIJK et al. 1988 WATERMAN et al. 1987
Thermodynamic method	Radiation-induced changes	Blood flow increases during the first few days following a single dose of 4 to 8 Gy Fractionated radiation causes a gradual increase in both normal and tumor tissue during treatment	JOHNSON 1976
Thermal washout	Hyperthermia-induced changes in regional blood flow Changes in regional blood flow during single hyperthermia treatments	Significant increase, decrease, or no change in tumor perfusion during fractionated thermoradiotherapy Slight increase in global tumor perfusion (not significant) during treatment Significant increase in normal tissue perfusion during treatment	FELDMANN (unpublished data) WUST et al. 1995
PET ^{15}O inhalation	Changes in regional blood flow in normal tissue during radiotherapy	Increase in regional blood flow in normal brain tissue	BEANEY et al. 1984
Color Doppler ultrasound	Changes of vascularity during medical treatment of breast cancer Changes of vascularity during radiotherapy of cervical cancer	Reduction of vascularity No change or progression of vascularity Decrease in tumor vascularity Increase in tumor vascularity	KEDAR et al. 1994 PIRHONEN et al. 1995

lymphomas, 0.15 ± 0.11 ml·g^{-1}·min^{-1} in the anaplastic carcinomas, and 0.22 ± 0.15 ml·g^{-1}·min^{-1} in the differentiated tumors. Blood flow was significantly greater in the lymphomas than in the anaplastic carcinomas, but the difference between the lymphomas and the differentiated carcinomas was not significant. As a conclusion of these investigations, the increased blood flow in lymphomas may partly explain the radiosensitivity of lymphomas, which generally is higher than that of solid tumors. Due to the difficulties concerning radionuclide injection into the tumor, only casuistic experience exists in deep-seated tumors. NYSTRÖM et al. (1969) evaluated blood flow

in tumors of the corpus uteri in 12 patients using the Xe clearance method. The injection into the fundus myometrium was performed through a special cannula allowing control of the depth of injection. It was possible to relate variations in blood flow in the fundus myometrium to the histologic differentiation. The mean blood flow was 0.10 ml·g^{-1}·min^{-1} with a range of 0.01 to 0.24 ml·g^{-1}·min^{-1} in six cases of highly differentiated tumors. Six cases of low tumor differentiation had a mean blood flow of 0.41 ml·g^{-1}·min^{-1} (range 0.11 to 0.69 ml·g^{-1}·min^{-1}). This investigation demonstrated that the xenon clearance method is simple to apply to the myometrium.

5.2.1.2
Laser Doppler Flowmetry

Perfusion measurements were performed interstitially with a commercially available microprocessor-based instrument using a solid-state laser diode as the source of coherent infrared light (780 ± 20 nm wavelength) which is delivered to tissues via a quartz optical fiber power of 2 mW (ACKER et al. 1990). The volume of measurement approximates a hemisphere with a radius of 1 mm (SONG et al. 1987; VAUPEL et al. 1988). Because of the small sampling volume, measurements usually reflect microcirculatory flow. The processed signals yield voltages that are proportional to the mean erythrocyte velocity multiplied by the number of moving erythrocytes within the sampling volume (red blood cell flux).

The measurement of ACKER et al. (1990) were performed in 17 patients with superficial tumors of the breast, lymph node metastases of head and neck cancers, lung cancers, or soft tissue sarcomas of the leg or trunk during hyperthermia treatment given in combination with conventional radiotherapy (Table 5.3). Blood flow data from multiple sites of measurement showed marked heterogeneity within individual tumors (up to 55-fold differences) and among different tumors (>100-fold differences). Measurements along a tumor radius, beginning at the center and advancing to the edge, were consistent with a two-compartment tumor model (shell and core). Responses during hyperthermia treatment sessions were also investigated. Four different temporal patterns of flow were observed, ranging from a steady state increase in flow to a plateau level to

a steady drop in flow during treatment. In general, these results suggest that the technique might be valuable in monitoring changes in flow between treatment sessions.

ZOGRAFOS et al. (1990) evaluated perfusion in human rectal tumors using a laser Doppler flowmeter. Flow in the center of the tumors was found to be higher than in normal mucosa. Two patients with ulcerated lesions had a much lower flow in the center of the tumor.

PIGOTT et al. (1996) investigated eight individual human tumors, two primary and one locally recurrent breast carcinoma, two metastatic skin deposits, and three metastatic lymph nodes, using a multichannel laser Doppler system to monitor microregional changes in flow. The results show that in 54% of regions investigated, there was a change in microregional blood flow with time of a factor of 1.5 or more. In conclusion, these findings demonstrate that microregional fluctuations in perfusion occur frequently in human tumors.

5.2.1.3
Thermal Clearance Method
(Effective Blood Perfusion)

As mentioned above, blood flow is a very important parameter in hyperthermia. Changes in blood flow induced during heating have a significant impact on temperature distribution. The thermal clearance method is clinically feasible in humans and has been applied to study blood flow in human tumors (WATERMAN et al. 1987; SAMULSKI et al. 1987;

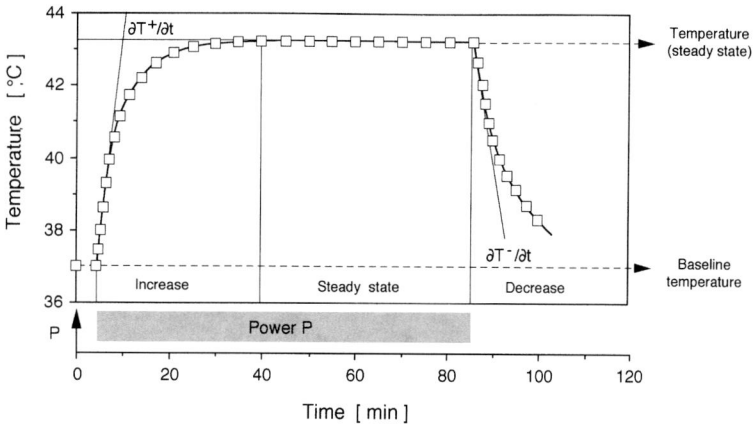

Fig. 5.1. Perfusion determination from thermal clearance curves ($\partial T^-/\partial t$)

LAGENDIJK et al. 1988; FELDMANN et al. 1992) as well as in animal tumors (MILLIGAN et al. 1983; LYONS et al. 1989; VAUPEL et al. 1989, 1992). Performing invasive thermometry, blood flow was determined at a specific position in the tissue from the rate of temperature decay when the power supply was turned off. In general, thermal decay is not caused by perfusion alone. The conduction of thermal energy through tissues also contributes to thermal decay. Therefore, the rate of temperature decay reflects the rate of energy removal by both conduction and perfusion. Consequently, studies utilizing the thermal clearance method did not actually measure blood flow, but rather a parameter related to blood flow, often referred to as the "effective blood flow" or the "local tissue cooling coefficient" (ROEMER 1990; NEWMAN et al. 1990). Blood flow can be determined from thermal clearance data if the rate of energy removal by conduction is negligible, as has been shown in regional hyperthermia of deep-seated tumors (FELDMANN et al. 1992; WUST 1992).

The thermal clearance method is illustrated in Fig. 5.1. The location of the temperature probes and washout measurement points were correlated with tissue type by reconstruction of the probe position by CT scans. Tissue categories in which thermal washout was measured were: tumor center, tumor periphery, and normal tissue (muscle, fat).

The mathematical method to assess perfusion by temperature decay curves is based on the well-known bioheat transfer equation for the temperature distribution T (ROEMER et al. 1985):

$$Q_a(t) - \varrho c \, dT(t)/dt = W_e \, c_b \varrho_b \left[T(t) - T_a(t) \right]$$
$$= Q_c(t) + Q_b(t) - Q_m(t)$$

where c, c_b = specific heat capacity of tissue, blood Ws/g°C); ϱ, ϱ_b = density of tissue, blood (g/cm³); T(t) = temperature at time t (°C); Q_a = specific absorption rate (mW/cm³); Q_m = metabolic heat production (mW/cm³); Q_c = heat loss by thermal conductivity; Q_b = heat loss by blood flow; W_e = effective perfusion rate (ml·100 g^{-1}·min^{-1}); and T_a(t) = arterial blood temperature at time t (°C).

Tissue constants are approximately set at c = 4 Ws/g°C and ϱ = 1 g/cm³. As an approximative approach, an effective perfusion can be introduced into the bioheat transfer equation which includes the conduction term with k (ROEMER 1990). In general, there are two special situations for calculating effective perfusion rates with regard to the bioheat trans-

fer equation (ROEMER et al. 1985). First, consider the situation for which power Q_a is applied to a subject and a steady state condition at temperature, T_0, is maintained. The effective perfusion rate can then be determined by:

$$W_e = Q_a / c_b \left(T_0 - T_a \right)$$

In the second situation the nonstationary bioheat transfer equation is reduced for the power-off state to:

$$\varrho c \, dT/dt + W_e \, c_b \left(T - T_a \right) = 0$$

If it is further assumed that W_e and T_a remain constant during the short observation period this has the commonly encountered exponential solution

$$T - T_a = \left(T_0 - T_a \right) e^{-\left(W_e c_b / \varrho c \right) t}$$

with T_0 as the initial probe temperature.

Several groups who treated patients with hyperthermia used the thermal clearance method and published blood flow data for superficial and deep-seated tumors (SAMULSKI et al. 1987; WATERMAN et al. 1987; LAGENDIJK et al. 1988; MOLLS and FELDMANN 1991; MOLLS et al. 1992; FELDMANN et al. 1992). In general, washout rates were significantly different among various tissue categories. Tissue categories such as tumor vs normal tissue, tumor/normal tissue interface vs hypodense tumor areas, and tumor center vs tumor periphery have been determined by computed tomography. The regions of interest on the CT scans were the sites where the temperature probes were located and temperature monitoring was performed. In general, it appears that blood flow is distinctly higher in the tumor periphery than in the tumor center. A more pronounced perfusion of the periphery has also been found in experimental tumors (rat fibroblasts, ras-transformed), as shown by KALLINOWSKI (1996). However, according to VAUPEL (1994), experimental tumors exist in which the blood flow in the center is significantly higher than that in the periphery.

5.2.2
Noninvasive Methods

5.2.2.1
Positron Emission Tomography

Positron emission tomography (PET) is a technique capable of providing absolute in-vivo measurements of regional blood flow, oxygen extraction ratio, oxygen utilization, and fractional blood volume. The PET scan measures regional tracer concentration and permits quantification of tissue perfusion as milliliters per gram and minute in the organ (HUANG et al. 1985). In clinical studies [^{13}N]ammonia and [^{15}O]water are available for the assessment of regional blood flow. [^{13}N]ammonia is extracted from plasma and trapped in tissue in proportion to blood flow. SHIBATA et al. (1988) examined the blood supply of hepatocellular carcinomas and metastatic liver tumors in 23 patients. They found a strong correlation between the demonstration of large tumor vessels by hepatic angiography and [^{13}N]ammonia uptake by the tumor. In seven liver tumors, accumulation of [^{13}N]ammonia was lower than in the normal liver throughout the scan. The results suggested that blood flow in liver tumors can be qualitatively assessed with dynamic PET.

[^{15}O]water is the most widely used inert flow tracer. It is extracted by tissue, with a first-pass extraction fraction close to 100%, and has been used for regional blood flow determination in brain studies. With the use of dynamic PET scanning, blood and tissue activity can be differentiated, and regional blood flow can be quantified with reasonable accuracy. Only few studies in oncology have been performed with [^{15}O] water to determine blood flow (Table 5.2). The greatest amount of experience was collected in patients with primary brain tumors and metastatic cerebral tumors (ITO et al. 1982; BEANEY et al. 1984). Tumor regional blood flow in these patients was found to be extremely variable, the mean value being close to that of contralateral white matter. In addition, perfusion of tumors and metastases can be defined by this method, as shown by HOHENBERGER et al. (1993) in colorectal liver metastases. More details concerning the PET method are given in Chap. 15.

possible to take the contrast agent as a physiological factor and gain information about function. The enhancement of contrast material (ΔHU) over baseline in the early phase after bolus injection is a reliable parameter for blood perfusion in the region of interest (CLAUSSEN and LOCHNER 1983). The ability of dynamic CT to provide quantitative functional information in addition to good anatomical detail is not well recognized. Only a certain amount of clinical data is available from CLAUSSEN and LOCHNER (1983), FELDMANN et al. (1992), and MOLLS and FELDMANN (1991). CLAUSSEN and LOCHNER investigated tumors of the head and neck region of different histologies. The enhancement of contrast material in the early phase in differentiated, and undifferentiated carcinomas and angiofibromas ranged from 30 to 50 Hounsfield units. FELDMANN et al. (1992) investigated recurrent pelvic tumors of different histologies. The enhancement of contrast material (ΔHU) was evaluated in 225 regions of interest. As in the case of the thermal washout procedure, there was a significant difference between tumor center (ΔHU 6.8 $\pm$ 0.8) and tumor periphery (ΔHU 13.9 $\pm$ 1.1).

Some methodological development has been undertaken for the quantification of tissue perfusion using dynamic CT. By applying a nuclear-medicine data processing technique to time-density data from a single-location dynamic CT sequence, tissue perfusion can be determined from the maximum gradient of the tissue time-density curve divided by the peak enhancement of the aorta. This technique is simple and can be applied to most tissues. In addition, a new application in CT imaging in which a quantifiable map of tissue perfusion is created and displayed by means of a color scale has been developed by Miles and colleagues (MILES 1991; MILES et al. 1991). The main advantage of this technique is that quantifiable information about perfusion is combined with good anatomical detail in a single image.

5.2.2.2
Dynamic Computed Tomography

Dynamic computed tomography (CT) describes a rapid sequence of images acquired following intravascular injection of contrast medium. If the sequence of scans is performed at the same slice location, it is possible to construct time-density curves by defining a region of interest over a particular structure. The curve may display the change of X-ray attenuation, and the change of iodine concentration within that structure with time. It is therefore

5.2.2.3
Xenon-Enhanced CT

Blood flow in brain tumors has been measured using stable-xenon computed tomography (Xe CT). To obtain the xenon wash-in data for this study, subjects inhaled a Xe/O$_2$ mixture. During inhalation time, multiple scans were taken sequentially at each scan level. Using a software package for calculation of cerebral blood flow from the Xe wash-in data, it was possible to display side by side both the anatomical CT scan and the corresponding xenon clearance cerebral blood flow scan of the same cross section at each scan level.

TATAGIBA et al. (1991) have used this approach to measure peritumor blood flow of intracranial meningeomas in 12 patients. They measured average values for the overall tumor, the peritumor edematous region, and the ipsilateral and contralateral hemispheric regions. Tumor perfusion values ranged from 0.08 to 1.23 $ml \cdot g^{-1} \cdot min^{-1}$.

TEGLIA et al. (1996) investigated 26 patients who suffered either from a glioblastoma multiforme or an anaplastic astrocytoma using stable-xenon computed tomography. Perfusion values were determined for each of the following anatomical regions: the low density tumor core, the enhancing active shell of the tumor, the low density peripheral region of edema, an ipsilateral region of normal brain adjacent to the tumor, and a region of remote normal tissue on the contralateral side of the brain.

The ipsilateral normal brain tissue adjacent to the tumor was found to have a perfusion rate of 0.84 $ml \cdot g^{-1} \cdot min^{-1}$, the edematous tissue had a perfusion rate of 0.52 $ml \cdot g^{-1} \cdot min^{-1}$, the active tumor 0.78 $ml \cdot g^{-1} \cdot min^{-1}$, and the core 0.39 $ml \cdot g^{-1} \cdot min^{-1}$. There was a significant negative correlation between tumor blood flow and tumor volume.

5.2.2.4
Dynamic Magnetic Resonance Imaging

Results of dynamic MRI have already been reported as a useful parameter in describing tissue perfusion (ERLEMANN et al. 1988, 1990; ERLEMANN 1993; FRANK et al. 1988; VAN NAEGELE et al. 1989; COHEN and WEISSKOPF 1991). The principles of using paramagnetic contrast agents to assess blood flow have been reviewed elsewhere (ROSEN et al. 1989). A concentrated bolus of a paramagnetic agent can produce loss of signal intensity in the tissue. The bolus distorts the local magnetic field homogeneity. This results in dephasing and signal loss. T2-related signal changes are probably primarily related to flow within microvessels rather than larger vessels. Gd-DTPA is distributed in both the interstitial and intravascular spaces. Therefore, changes in signal intensity of a tumor relate both to blood volume and to the volume and distribution of contrast agent that leaks into the tumor interstitium. Clinical investigations in brain tumors (EDELMAN et al. 1988) suggest that changes of signal intensity relate primarily to the intravascular – rather than interstitial – distribution of contrast agent. This method may therefore be useful in pro-

viding a dynamic assessment of tumor blood flow. Other approaches to dynamic enhanced studies have involved ultrafast gradient-echo sequences. These techniques generate T1 contrast similiar to that in conventional inverse-recovery imaging (HAASE et al. 1989; ATKINSON et al. 1990). In general, the fast inversion-recovery method produces T1-weighted images with lower spatial resolution and lower Gd-DTPA-induced ratios of change in signal intensity to noise than the T2 methods, but may still be useful for assessing flow dynamics.

Our own results (FELDMANN et al. 1993; SIEVERS et al. 1993) were obtained using relatively simple dynamic MR imaging evaluation methods (time-to-peak, slope, and maximum signal intensity at a given time). After bolus injection of Gd-DTPA, a biphasic course of signal intensity was evident. The first increase in signal intensity of 15–30 s after bolus injection was chosen for perfusion analysis. Flow is expressed as the difference of signal intensity (ΔSI). Although numerous methods using MRI in perfusion imaging are available, only a few studies have quantified tumor blood flow (FELDMANN et al. 1993; SIEVERS et al. 1993). In recurrent pelvic tumors, significant differences have been observed between various tumor regions (tumor center vs. tumor periphery) and different tumor entities (FELDMANN et al. 1993).

There is literature that describes methods yielding standardized tissue-specific data. These seem to be rather independent of the patient's hemodynamics and of the effects resulting from the mode of bolus injection. Using a two-compartment model, signal-time curves are characterized by three parameters: the amplitude describing the intensity of contrast medium enhancement, the distribution rate constant ($k21$) describing the exchange rate between the intravasal and extracellular space, and the elimination rate constant (kel; BRIX et al. 1991; MÜLLER-SCHIMPFLE et al. 1989, 1993; KNOPP et al. 1995). The amplitude depends on the properties of the tissue (native relaxation times, relaxivity, extracellular volume), on the sequence used, and on the dose of contrast medium. The parameter $k21$ is influenced by time resolution and the duration of infusion (MÜLLER-SCHIMPFLE et al. 1993; KNOPP et al. 1995).

In a prospective clinical study on 314 patients with undetermined breast lesions, signal-time curves revealed differences in the $k21$ parameter between benign lesions (0.56 ± 0.46) and malignant lesions (1.25 ± 0.80; $p \leq 0.0001$) as well as between different histological classifications. The authors (KNOPP et al. 1995) concluded that the MR technique used (FLASH

3D) allows better evaluation of vascular permeability with the MR contrast medium in tissue lesions. Distinct differences could be detected that might be due to evaluated expression of angiogenesis factors such as the vascular endothelial growth factor (VEGF).

HAWIGHORST et al. (1995, 1996) analyzed the diagnostic accuracy in suspected pelvic lesions in patients with treated cervical carcinoma. Twenty-one women with 24 suspected pelvic lesions and a history of treated cervical carcinoma were included in this study. Seventeen of 24 suspected lesions were histologically verified as tumor recurrences and seven were classified as benign masses. The pharmacokinetic analysis of dynamic MR images showed a significantly shorter and stronger contrast medium enhancement of malignant lesions than of benign lesions.

In a further study, the diagnostic accuracy in suspected local recurrences of rectal carcinomas was investigated. Analysis of the MR data showed greater and more rapid enhancement of malignant lesions ($n = 12$) than of benign or scarred lesions ($n = 7$). The mean amplitude was significantly higher ($p = 0.0038$) and the distribution time t21 significantly shorter ($p = 0.0018$) in the recurrent carcinomas (MÜLLER-SCHIMPFLE et al. 1993). In conclusion, pharmacokinetic mapping of dynamic MR imaging data has the potential to differentiate between benign and malignant pelvic lesions in patients with cervical and rectal carcinomas.

Other MR imaging approaches that do not employ contrast agents have been proposed for assessing blood flow (LE BIHAN et al. 1988). For instance, it may be possible to use velocity-dependent phase effects to distinguish between very slow molecular motion (diffusion) and faster capillary flow (perfusion). Further study is needed to determine the clinical viability of these methods.

5.2.2.5
Color Doppler Ultrasonography

Most tumors seem to develop a chaotic vascular network rather than a systematic vascular supply system with hierarchical vessel arrangement. Tumor microvessels are not uniform in either functional or morphological aspects. Besides capillary-like microvessels, sinusoidal or lacuna-like microvessels are obligatorily found, often showing a discontinuous endothelial cell lining.

During tumor growth, vascularity is dynamic with new vessels appearing rather suddenly. Speed and direction of blood flow in the tumor microvascular bed are unstable (KEDAR et al. 1994; MADJAR et al. 1994; PIRHONEN et al. 1995). These conditions lead to alterations in blood flow velocity which become manifest as flow acceleration in Doppler sonography. In addition, recent developments allow flow detection in small vessels. Moreover, tumor vessels often terminate in wide sinusoids, which in turn decrease peripheral resistance. This decrease supposedly becomes manifest as a loss of systolic-diastolic pressure (KEDAR et al. 1994; PIRHONEN et al. 1995). Numerous studies have therefore focused on evaluation of tumor vascularity, resistance to flow, and the complete sum of flow velocities in tumor vessels in carcinoma of the breast and cervical carcinoma (KEDAR et al. 1994; PIRHONEN et al. 1995). However, the data obtained by various research groups concerning differentiation of benign and malignant lesions vary considerably.

In 70% of patients with locally advanced breast cancer, a significant reduction of vascularity could be observed under chemo- or hormone therapy. In addition, PIRHONEN et al. (1995) observed a reduction of vascularity in patients with advanced cervical carcinomas under radiation therapy. An increase in vascularity was evident only in a minority of patients. These investigations demonstrate that it is possible to measure tumor vascularization using color Doppler ultrasonography.

5.3
Changes in Tumor Perfusion Under Therapy

A number of studies investigating changes in blood flow in human tumors following therapy have been reported (Table 5.3). Most of them are more or less case reports rather than systematic investigations.

WATERMAN et al. (1987) investigated the effect of heat on blood flow in superficial tumors during 1 h of local hyperthermia. Tumor blood flow increased by amounts varying from 15% to 250% during the first 20–50 min of heating at 41–45°C, after which flow remained relatively constant during the remainder of the treatment session. In contrast, perfusion analyses in advanced breast carcinoma during hyperthermia showed no change in blood flow during the stationary part of the individual hyperthermic sessions. Perfusion can both increase and decrease upon successful hyperthermia.

ACKER et al. (1990) investigated responses during hyperthermia treatment sessions using the laser

Doppler method. Several temporal patterns of flow were observed, ranging from a steady increase in flow up to a plateau level to a steady drop in flow during heating (see Sect. 5.2.1.2).

A statistical analysis of a relatively large sample of in vivo perfusion measurements (thermal washout procedure) by Wust et al. (1995) showed a slightly higher perfusion in tumors than in normal tissues in the pretherapeutic state. Under hyperthermic conditions, blood flow conditions invert, i.e., perfusion increases considerably in normal tissues (nearly double on average), but only moderately in tumors.

With regard to follow-up measurements of global tumor perfusion during fractionated thermoradiotherapy of about 4–6 weeks, our own data (thermal clearance method) show no clear direction of changes in global tumor perfusion in 38 patients. In most of the patients ($n = 18$), no change in global tumor perfusion ($\pm 20\%$ of the pretherapeutic value) was evident during therapy. In 12 patients a significant decrease ($\geq 20\%$ of the pretherapeutic value) of the perfusion values was seen, and in 8 patients a significant increase ($\geq 20\%$ of the pretherapeutic value) of global tumor perfusion was evident. One problem is that the thermal washout procedure is not a good tool for evaluation of minor flow changes (up to 20% of the initial value). It is remarkable that in nearly one third of the patients, flow changes of more than 20% could be demonstrated.

Mäntylä et al. (1982) investigated radiation-induced changes in regional blood flow in 48 superficial metastatic tumors with the ^{133}Xe washout method. Blood flow increased significantly during the first week of radiotherapy in the whole series from 0.20 ± 0.18 to $0.31 \pm 0.25\,\mathrm{ml \cdot g^{-1} \cdot min^{-1}}$. During the second week of radiotherapy, blood flow decreased to $0.27 \pm 0.19\,\mathrm{ml \cdot g^{-1} \cdot min^{-1}}$. The gradual decrease in blood flow seemed to continue in subsequent weeks.

In a study of the acute effects of a fractionated course of whole brain irradiation in patients with intracranial tumors by Beaney et al. (1984), changes in normal tissue blood flow were also investigated. In normal cortex, regional cerebral blood flow was seen to increase by the end of a fractionated course of radiotherapy. When these patients were rescanned 3–4 months later, a fall in regional cerebral blood flow was noted.

Kedar et al. (1994) performed measurements of tumor response to primary medical (chemo- or hormone) therapy with color Doppler flow imaging. Thirty-four women with a histologic or cytologic di-

agnosis of breast cancer whose tumors were treated medically before radiation therapy and/or surgery form the basis of this study. Changes in vascularity were assessed in every patient during the course of 126 treatment cycles. In 77% of the treatment cycles, changes in the vascularity were concordant with changes in the size of the tumor. Pirhonen et al. (1995) investigated the effects of external radiotherapy on uterine blood flow in patients with advanced cervical carcinoma as assessed by color Doppler ultrasonography. A decrease in tumor vascularity during treatment was associated with improved outcome.

5.4
Comparison of Methods

No systematic investigations comparing different methods of monitoring tumor perfusion in human tumors have been performed. Our own studies in recurrent pelvic tumors confirm the existence of a clear correlation between blood flow-related data obtained by the thermal washout procedure and those obtained by dynamic CT and/or MRI in comparable regions of interest (Feldmann et al. 1992, 1993). Scatter plots of 127 measurements in comparable regions of interest show a highly significant correlation between the thermal washout procedure and enhancement of contrast material ($r = 0.74$; $p < 0.001$) for all tissue categories (tumor center, tumor periphery, muscle, fat). An average enhancement of contrast material of about 20 ΔHU reflects a global perfusion rate (thermal washout) of $0.23\,\mathrm{ml \cdot g^{-1} \cdot min^{-1}}$. The good correlation between thermal and nonthermal methods of global tumor perfusion underlines the assumption that in deep-seated tumors, thermal conduction will be negligible in determining effective perfusion under regional hyperthermia.

In addition, upon analysis of flow values obtained in the same region of interest with dynamic CT and dynamic MRI, a highly significant correlation ($r = 0.81$; $p = 0.001$) was evident between the results of dynamic CT and MRI (Fig. 5.2).

Other investigations (Miles 1991; Miles et al. 1991) have shown that the value for spleen perfusion obtained by dynamic CT is close to the value obtained from intra-arterial xenon studies (Williams et al. 1968). The higher value obtained from CT may reflect the vasodilatory effect of the contrast medium. Although a nonionic contrast medium, which is less vasodilatory, was used in these CT studies, this

Fig. 5.2. Scatter plot of flow values obtained by dynamic CT (ΔHU) and flow values obtained by dynamic MRI (ΔSI) in comparable regions of interest

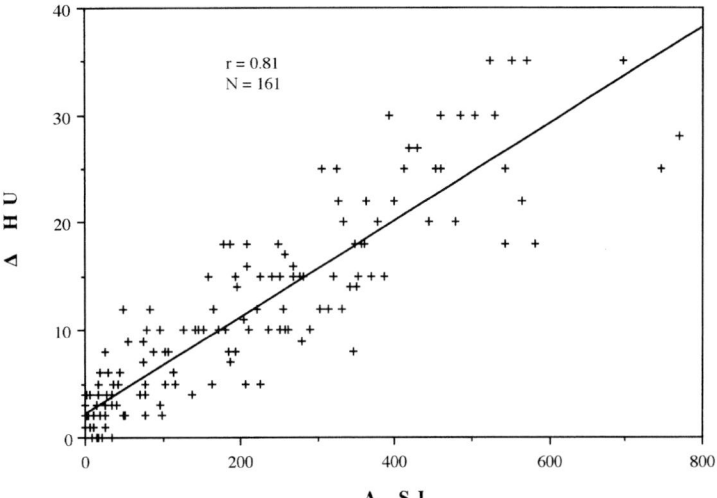

effect is likely to be significant. On the other hand, the CT-derived value for the perfusion of the renal cortex is significantly lower than those derived from studies using intra-arterial xenon (ROSEN 1968). Contrast agents are known to decrease renal blood flow, and this may account for the lower values obtained for the renal cortex.

5.5
Prognostic Relevance of Tumor Perfusion

There are several studies in humans that underscore the importance of blood flow for the quality of heat treatments combining radiotherapy and hyperthermia (SAMULSKI et al. 1987; MOLLS and FELDMANN 1991; FELDMANN et al. 1992; WUST 1992). In regional hyperthermia, the achieved steady state temperature above baseline (ΔT) depends predominantly on the relative perfusion rates. Scatter plots of 387 flow values (Fig. 5.3) which we obtained in our patients (60 tumors) by the thermal washout procedure and steady state temperature elevation (ΔT) demonstrate a significant inverse correlation ($r = 0.42$). High flow values correlated with low steady state temperatures (ΔT $\leq$ 3°C), and low flow values with high steady state temperatures (ΔT $\geq$ 3°C).

Several studies have shown a dependence of tumor growth on angiogenesis (FOLKMAN 1990) and a direct relationship between blood flow and tumor growth rate and bleeding at surgery, as well as the potential for metastasis (WEIDNER et al. 1991). Using color Doppler flow imaging, the vascularity of a tumor can be assessed quantitatively and follow-up examinations can be performed. KEDAR et al. (1994) investigated changes in vascularity in 43 patients with located breast carcinomas during chemo- or hormone therapy. Reduction in vascularity was seen in 77% of the treatment cycles and was concordant with response to therapy. In contrast, an increase in vascularity accompanies progression of disease.

PIRHONEN et al. (1995) investigated 14 patients with advanced cervical carcinomas during the course of external radiotherapy using color Doppler flow imaging. Decrease in tumor vascularity during treatment was associated with better outcome, whereas persistence of excessive vascularity at the end of radiation was associated with modest therapeutic response. These clinical results suggest that measurements of tumor vascularity and perfusion may be useful in early assessment of therapeutic response during medical (chemo- or hormone) therapy or radiotherapy.

In a pilot study of patients with meningeomas of the base of the skull, dynamic MR investigations were performed before, during, and after radiotherapy (HAWIGHORST et al. 1997). During treatment, the flow of contrast medium was delayed and the amplitude of enrichment increased. Those patients who responded to therapy showed a prolonged distribution of contrast medium in the tissue, whereas in the nonresponding patients the time period of distribution was shortened.

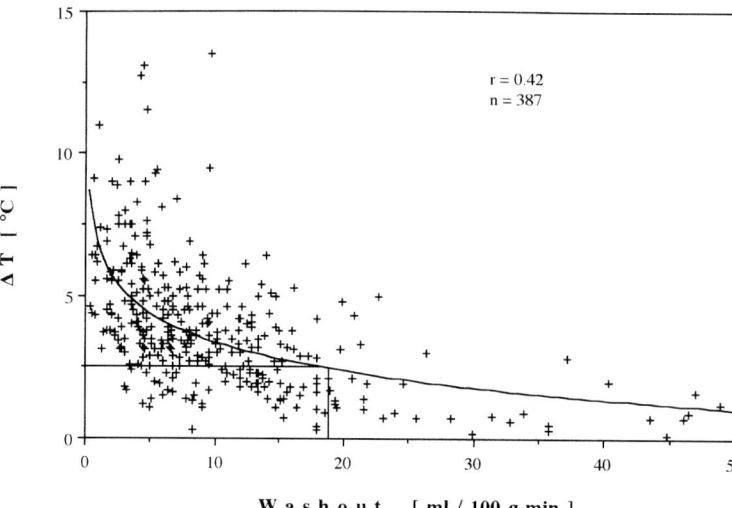

Fig. 5.3. Scatter plot of steady state temperature above baseline versus global tumor perfusion

5.6
Methodological Problems

In general, none of the clinical approaches to measurement of tumor perfusion provide information that explains events occurring at the microscopic level.

In principle, every thermal method for determination of effective perfusion fails to precisely differentiate between blood perfusion (i.e., convection) and conduction along thermal gradients. This problem has been discussed extensively by ROEMER (1990), SAMULSKI et al. (1987), WATERMAN et al. (1987), and GROEBE et al. (1987), who summarized the difficulties of blood flow determination from temperature decay curves and concluded that such analyses can provide nothing more than qualitative information about tissue/tumor perfusion. Significant errors for perfusion values using the thermal clearance method were found in association with enormous thermal gradients of up to 6°C (WONG et al. 1988). During regional hyperthermia, significantly lower thermal gradients are expected than those cited above. The conduction effect may play a role in tumors with poorly perfused or necrotic regions and well-cooled peripheries. The cooling process in the necrotic areas could be the result of heat conduction. However, despite these methodological problems, the thermal washout method of calculating effective perfusion values can be a very useful procedure, providing blood flow data in tumors, different tumor areas, and normal tissues upon regional hyperthermia.

In addition, some methodological problems exist concerning the noninvasive methods. Studies using radioisotopic clearance techniques suffer from uncertainties due to physical and geometrical factors. Using PET, LAMMERTSMA and JONES (1992) pointed out that the oxygen-15 steady state technique has only limited value in assessment of the pathophysiology of tumors. More accurate techniques, which are less sensitive to tissue heterogeneity, are required. A kinetic model for the measurement of tumor blood flow has been described which is far less sensitive to tissue heterogeneity (LAMMERTSMA and JONES 1992).

Dynamic MRI investigations are limited by the fact that Gd-DTPA is not a purely intravascular tracer. As it is distributed into both the interstitial and the intravascular spaces, one might expect that the change of signal intensity of a tumor relates both to blood volume and to the volume and distribution of the contrast agent that leaks into the tumor interstitium.

Despite some methodological problems, all methods may be useful in providing a dynamic assessment of tumor blood flow and/or vascularity. Comparison of different methods (FELDMANN et al. 1992, 1993) shows that there is good correlation between data obtained by thermal washout and noninvasive methods (dynamic CT/MRI). The approach of quantifying blood flow-related data by different methods is clinically relevant because tumor blood flow has a significant impact on the efficacy of different treatment modalities. In general, oxygen concentration and radiosensitivity depend on effective tumor perfusion. Furthermore, the cytotoxic effect of drugs (e.g., Adriamycin) is clearly related to the efficacy of tumor perfusion. With regard to the combination of radiotherapy or chemotherapy with hyperthermia, limitations exist for the heat treatment of high flow tumors. In contrast, poorly perfused tumors suffer from hypoxia, nutritional deprivation, and acidosis. Experimental investigations have proven a strong enhancement of the heat-induced cytotoxic effect under such environment conditions. Therefore, it makes sense to additionally apply hyperthermia under such conditions.

Endothelial cells are probably the most radiosensitive elements of mesenchymal tissues. The effect of irradiation on vessels includes endothelial cell degeneration, vasoconstriction, and thrombus formation (FAJARDO and BERTHRONG 1988). These changes affect the tumor's microcirculatory network which determines the oxygenation status of the tumor. The effect of oxygenation and the fraction of proliferating cells are well-established parameters affecting radiosensitivity. However, radiosensitivity of individual tumors or cells is the result of numerous exogenous and endogenous factors, including intrinsic radiosensitivity (STEEL 1989). The quantitative serial investigation of tumor vascularity using color Doppler ultrasound during radiotherapy or medical therapy appears to be a promising indicator in assessing the early response to therapy (KEDAR et al. 1994; PIRHONEN et al. 1995). These are very promising investigations that demonstrate the prognostic relevance of serial investigations in tumor blood flow and vascularity.

5.7
Modification of Tumor Blood Flow

Much effort has been put into investigating modifications of blood flow in experimental tumors by various vasoactive drugs. The changes in tumor blood flow caused by vasoactive compounds are believed to be mostly due to changes in blood flow in the surrounding normal tissue (JAIN and WARD-HARTLEY 1984; JIRTLE 1988). Agents used in this context include vasoactive and rheologically active agents, cytokines, and glucose. The responsiveness of tumor microvessels to pharmacological agents depends on the fraction of preexisting host vessels in the tumor mass that possess responsive smooth musculature. In addition, intratumor responsiveness can be influenced by steal and antisteal phenomena in conjunction with the vascular bed of the normal tissue adjacent to the tumor tissue. Similar effects of vasoactive agents on blood flow have to be expected if the tumor circulation is in series with the normal

tissue vascular bed, opposite effects if they are in parallel (JAIN and WARD-HARTLEY 1984; JIRTLE 1988; VAUPEL 1994). This probably is the main reason for the disappointing clinical results using vasoactive drugs in patients. DEWHIRST et al. (1990) investigated the use of hydralazine to manipulate tumor temperatures during hyperthermia in patients. In contrast to experimental data in canine subjects, in patients hydralazine resulted only in a slight reduction of blood pressure, which was ineffective in increasing tumor temperatures. Vasoconstriction in normal tissues can raise arterial blood pressure, leading to an increase in tumor blood flow. Such an increase in tumor blood flow caused by vasoconstrictors lasts for only a short period, since subsequent constriction of arteries feeding the tumors causes a decrease in tumor blood flow.

Improvement of blood flow by increasing the deformability of red and white blood cells by pentoxifylline or calcium channel blockers has been reported to increase tumor blood flow. However, the clinical relevance of these modifiers remains an open question (JIRTLE 1988).

5.8
Conclusion

Several different invasive and noninvasive methods are available for the quantification of tumor blood flow in human tumors in the clinic. At present, no clear advantage of either invasive methods (thermal clearance method, laser Doppler flowmetry, radioisotope methods) or non-invasive methods is evident. Furthermore, the thermal clearance method can only be used in a special treatment setup (local heat treatment). With regard to noninvasive methods, there is a lack of information about the pathophysiological mechanism of contrast material enhancement. Therefore, it is necessary to investigate the correlation between histopathologic and molecular biologic factors and the pharmacokinetic parameters of contrast medium enhancement more systematically in experimental and human tumors. Current studies underline the assumption that a correlation exists between angiogenetic and pharmacokinetic parameters. Despite some methodological problems, all of the methods may be useful in providing a dynamic assessment of tumor blood flow and/or vascularity. This was underlined by the comparison of our own blood flow data obtained by thermal washout and dynamic CT or MRI.

Considering the clinical data, and in accordance with VAUPEL et al. (1989), the following conclusions can be drawn: considerable intra- and interindividual heterogeneity of flow rates exists despite similar histologic classification and tumor site. Tumors may have flow rates similar to those of well-perfused organs such as liver, heart, and brain, or flow rates even lower than those of poorly perfused organs such as the skin, resting skeletal muscle, and adipose tissue. Most tumors contain both highly perfused areas and regions with compromised and sluggish perfusion, often associated with the development of necrosis. No significant difference exists between tumor blood flow and normal tissue blood flow. In general, due to a large heterogeneity of tumor perfusion within a given tumor, no clear relationship between tumor type and tumor perfusion is evident.

In future, the investigation of pretreatment blood flow rates of an individual tumor will be important in patient selection for treatment modalities. Furthermore, measurement of the extent of blood flow changes during therapy will play a major role in the optimization of individual treatments. In addition, the prognostic relevance of changes in tumor vascularity observed early during therapy may help to individualize and modify treatment schedules in order to achieve a better outcome in advanced or high-risk cases.

References

Acker JC, Dewhirst MW, Honore GM, Samulski TV, Tucker JA, Oleson JR (1990) Blood perfusion measurements in human tumors: evaluation of laser Doppler methods. Int J Hyperthermia 6:287–304

Atkinson DJ, Burstein D, Edelman RR (1990) First past cardiac perfusion: evaluation with ultrafast MR imaging. Radiology 174:757–762

Beaney RP, Lammertsma AA, Jones T, McKenzie CG, Halnan KE (1984) Positron emission tomography for in-vivo measurements of regional blood flow, oxygen utilisation, and blood volume in patients with breast carcinoma. Lancet 1:131–134

Brix G, Semmler W, Port R, Schad LR, Layer G, Lorenz W (1991) Pharmacokinetic parameters in CNS Gd-DTPA enhanced MR imaging. J Comput Assist Tomogr 15:620–628

Claussen C, Lochner B (1983) Dynamische Computertomographie. Springer, Berlin Heidelberg New York, pp 16–24

Cohen MS, Weisskopf RM (1991) Ultra-fast imaging. Magn Reson Imaging 9:1–37

Dewhirst MW, Prescott DM, Clegg S, et al. (1990) The use of hydralazine to manipulate tumor temperatures during hyperthermia. Int J Hyperthermia 6:971–983

Edelman RR, Mattle HD, Atkinson DJ, et al. (1988) Cerebral blood flow: assessment with dynamic contrast-enhanced

T2-weighted MR imaging at 1.5 T. Radiology 176:211–220

Erlemann R (1993) Dynamic gadolinium-enhanced MR imaging to monitor tumor response to chemotherapy. Radiology 186:904

Erlemann R, Reiser M, Peters PE (1988) Time-dependent changes in signal intensity in neoplastic and inflammatory lesions of the musculoskeletal system after i.v. administration of Gd-DTPA. Radiology 28:269–276

Erlemann R, Scuik T, Bosse A (1990) Assessment of response to preoperative chemotherapy in imaging osteosarcomas and Ewing's sarcomas with dynamic and static MR imaging and skeletal scintigraphy. Radiology 175:791–796

Evans NTS, Naylor PFD (1963) The effect of oxygen breathing and radiotherapy upon tissue oxygen tension of some human tumors. Br J Radiol 36:418–425

Fajardo LF, Berthrong M (1988) Vascular lesions following radiation. Pathol Annu 1:297–330

Feldmann HJ, Molls M, Hoederath A, Krümpelmann S, Sack H (1992) Blood flow and steady state temperatures in deep seated tumors and normal tissues. Int J Radiat Oncol Biol Phys 23:1003–1008

Feldmann HJ, Sievers K, Füller J, Molls M, Löhr E (1993) Evaluation of tumor blood perfusion by dynamic MRI and CT in patients undergoing thermoradiotherapy. Eur J Radiol 16:224–229

Folkman J (1990) What is the evidence that tumors are angiogenesis dependent? J Natl Cancer Inst 82:4–6

Frank JA, Choyke ME, Girton M, et al. (1988) Gd-DTPA-enhanced dynamic MRI: effect of diuretics on renal enhancement. Radiology 169:383

Gray LH, Conger AD, Ebert M, Hornsey S, Scott OCA (1953) The concentration of oxygen dissolved in tissue at the time of irradiation as a factor in radiotherapy. Br J Radiol 26:638–642

Groebe K, Kallinowski F, Vaupel P (1987) Is the division of heated tissue into temperature equivalent zones suitable for estimation of tumor blood flow from thermal clearance curves? Int J Radiat Oncol Biol Phys 13:917–920

Haase A, Matthaei D, Bartkowski R, Dühmke E, Leibnitz E (1989) Inversion recovery snapshot flash MR imaging. J Comput Assist Tomogr 13:1036–1040

Hawighorst H, Knapstein PG, Schaeffer U, Brix G, Essig M, Zuna I, Knopp MV (1995) Rezidivdiagnostik beim Zervixkarzinom mittels der dynamischen MRT: Korrelation von pharmakokinetischen Parametern und der Histopathologie. Radiologe 35:945–951

Hawighorst H, Knapstein PG, Schaeffer U, et al. (1996) Pelvic lesions in patients with treated cervical carcinoma: efficacy of pharmacokinetic analysis of dynamic MR images in distinguishing recurrent tumors from benign conditions. Am J Roentgenol 166:401–408

Hawighorst H, Engenhart R, Knopp MV, Brix G, Essig M, Miltner P, Zuna I, Fuß M, van Kaick G (1997) Intracranial meningiomas: time and dose dependent effects of irradiation on blood flow, vascular permeability and interstitial space monitored by dynamic MR imaging. Am J Roentgenol (in press)

Heß T, Müller-Schimple M, Brix G, et al. (1994) Dynamic Gd-DTPA enhanced MRI in the follow-up of cervical carcinoma after combined radio/chemotherapy. Adv MRI Contrast 2:85–93

Hohenberger P, Strauss LG, Lehner B, Frohmuller S, Dimitrakopoulou A, Schlag P (1993) Perfusion of colorectal liver metastases and uptake of fluorouracil assessed by H2 (15)O and [18F] uracil positron emission tomography (PET). Eur J Cancer 29:1682–1686

Huang SC, Schwaiger M, Carson RE, Phelps ME, Schelbert HR (1985) Quantitative measurement of myocardial blood flow with oxygen-15 water and positron computed tomography: an assessment of potential and problems. J Nucl Med 25:616–624

Ito M, Lammertsma AA, Wise RJS, et al. (1982) Measurement of regional cerebral blood flow and oxygen utilisation in patients with cerebral tumors using O-15 and positron emission tomography: analytical techniques and preliminary results. Neuroradiology 23:63–74

Jain RK, Ward-Hartley K (1984) Tumor blood flow – characterization, modification, and role in hyperthermia. IEEE Trans Sonics Ultrasonics 31:504–526

Jirtle RL (1988) Chemical modification of tumor blood flow. Int J Hyperthermia 4:355–371

Johnson R (1976) A thermodynamic method for investigation of radiation induced changes in the microcirculation of human tumors. Int J Radiat Oncol Biol Phys 1:659–670

Kallinowski F (1996) Charakterisierung des Mikromilieus maligner Tumoren – Ergebnisse an experimentellen Modellen und menschlichen Rektumkarzinomen. Habilitationsschrift, Medizinische Fakultät Heidelberg

Kedar RP, Cosgrove DO, Smith IE, Mansi JL, Bamber JC (1994) Breast carcinoma: measurements of tumor response to primary medical therapy with color Doppler flow imaging. Radiology 190:825–830

Kitani K, Winkler K (1972) In vitro determination of solubility of 137 Xenon and 85 Krypton in human liver tissue with varying triglyceride content. Scand J Clin Lab Invest 29:167–172

Knopp MV, Hoffmann U, Brix G, Hawighorst H, Junkermann HJ, van Kaick G (1995) Schnelle MR Kontrastmitteldynamik zur Charakterisierung von Tumoren. Radiologe 35:964–972

Lahtinen T, Karjalainen P, Alhava M (1979) Measurement of bone blood flow with a 133-Xe washout method. Eur J Nucl Med 4:435–439

Lammertsma AA, Jones T (1992) Low oxygen extraction fraction in tumours measured with the oxygen-15 steady state technique: effect of tissue heterogeneity. Br J Radiol 65:697–700

Lammertsma AA, Wise RJS, Cox TCS, Thomas GT, Jones T (1985) Measurement of blood flow, oxygen utilisation, oxygen extraction ratio, and fractional blood volume in human brain tumors and surrounding oedematous tissues. Br J Radiol 58:725–734

Lagendijk JJW, Hofmann P, Schipper J (1988) Perfusion analysis in advanced breast carcinoma. Int J Hyperthermia 5:479–495

Le Bihan D, Bretou E, Laltemand D, Aubin ML, Kignaud J, Laval-Jautet M (1988) Separation of diffusion and perfusion in intravoxel incoherent motion MR imaging. Radiology 168:497–505

Lyons BE, Samulski TV, Cox RS, Fessenden P (1989) Heat loss and blood flow during hyperthermia in normal canine brain. I. Empirical study and analysis. Int J Hyperthermia 5:225–248

Madjar H, Prömpeler H, Wohlfahrt R, Bauknecht T, Pfleiderer A (1994) Farbdopplerflußdaten von Mammatumoren. Ultraschall Med 15:69–76

Mäntylä MJ (1979) Regional blood flow in human tumors. Cancer Res 39:2304–2306

Mäntylä MJ, Toivanen JT, Pitkänen MA, Rekonen AH (1982) Radiation-induced changes in regional blood flow in human tumors. Int J Radiat Oncol Biol Phys 8:1711–1717

Mäntylä MJ, Heikkonen J, Licphil, Perkkiö J (1988) Regional blood flow in human tumors measured with argon, krypton and xenon. Br J Radiol 61:379–382

Miles KA (1991) Measurement of tissue perfusion by dynamic computed tomography. Br J Radiol 64:409–412

Miles KA, Hayball M, Dixon AK (1991) Colour perfusion imaging: a new application of computed tomography. Lancet 337:643–645

Milligan AJ, Conran PB, Ropar MA, McCulloch HA, Ahuja RK, Dobelbower RR (1983) Predictions of blood flow from thermal clearance during regional hyperthermia. Int J Radiat Oncol Biol Phys 9:1335–1343

Molls M, Feldmann HJ (1991) Clinical investigations on blood flow in malignant tumors of the pelvis and abdomen. In: Vaupel P, Jain RK (eds) Tumor blood supply and metabolic microenvironment: characterization and implications for therapy. Fischer, Stuttgart, pp 143–153

Molls M, Feldmann HJ, Sievers K (1992) Clinical investigations on blood perfusion in human malignancies of the pelvis and abdomen: significance for tumor therapy. In: Steiner R, Weisz PB, Langer R (eds) Angiogenesis: key principles – science – technology – medicine. Birkhäuser, Basel, pp 368–372

Müller-Schimpfle M, Rieber A, Kurz S, Stern W, Claussen CD (1989) Dynamische 3D-MR-Mammographie mit Hilfe einer schnellen Gradienten-Echo-Sequenz. Fortschr Röntgenstr 150:602–605

Müller-Schimpfle M, Brix G, Layer G, et al. (1993) Recurrent rectal cancer: diagnosis with dynamic MR imaging. Radiology 189:881–889

Naegele M, van Goetz AE, Gamarra F, et al. (1989) Control of tumor perfusion of shockwave treated tumors by MRI and Gd-DTPA. Fortschr Röntgenstr 150:602–605

Newman WH, Lele PP, Bowman HF (1990) Limitations and significance of thermal washout data obtained during microwave and ultrasound hyperthermia. Int J Hyperthermia 6:771–784

Nyström C, Forssman L, Roos B (1969) Myometrial blood flow studies in carcinoma of the corpus uteri. Acta Radiol Ther 8:193–198

O'Brien MD, Veall N (1974) Partition coefficients between various brain tumors and blood for 133-Xe. Phys Med Biol 19:472–475

Perkkiö J, Keskinen R, Heikkonen J, Mäntylä M (1986) Theoretical analysis of regional blood flow studies. Med Phys 13:229–232

Petersson HI (1991) Modification of tumor blood flow – a review. Int J Radiat Biol 60:201–210

Pigott KH, Hill SA, Chaplin DJ, Saunders MI (1996) Microregional fluctuations in perfusion within human tumor detected using laser Doppler flowmetry. Radiother Oncol 40:45–50

Pirhonen JP, Grenman SA, Bredback AB, Bohado-Singh R, Salmi TA (1995) Effects of external radiotherapy on uterine blood flow in patients with advanced cervical carcinoma assessed by color Doppler ultrasonography. Cancer 76:67–71

Roemer RB (1990) The local tissue cooling coefficient: a unified approach to thermal washout and steady state perfusion calculations. Int J Hyperthermia 6:421–430

Roemer RB, Fletcher AM, Cetas TC (1985) Obtaining local SAR and blood perfusion data from temperature measurements: steady state and transient techniques compared. Int J Radiat Oncol Biol Phys 11:1539–1550

Rosen BR, Belliveau JW, Chieu D (1989) Perfusion imaging by nuclear magnetic resonance. Magn Reson Imaging 5:263–281

Rosen SM (1968) Blood flow through organs and tissues. In: Bain WH, Harper AM (eds) Livingstone, Edinburgh, pp 458–465

Samulski TV, Fessenden P, Valdagni R, Kapp DS (1987) Correlations of thermal washout rate, steady state temperatures, and tissue type in deep seated recurrent or metastatic tumors. Int J Radiat Oncol Biol Phys 13:907–916

Shibata T, Yamamoto K, Hayashi N (1988) Dynamic positron emission tomography with 13N-ammonia in liver tumors. Eur J Nucl Med 14:607–611

Sievers KW, Feldmann HJ, Füller J, Molls M, Sack H (1993) Über die Wertigkeit der dynamischen MRT in der Perfusionsbeurteilung von Beckentumoren unter Hyperthermie. Fortschr Röntgenstr 159:245–250

Sohn C, Grischke EM, Wallwiener E, Kaufmann M, von Fournier D, Bastert G (1992) Die sonographische Durchblutungsdiagnostik gut- und bösartiger Tumoren. Geburtshilfe Frauenheilkd 52:397–403

Song CW, Payne JT, Levitt SH (1972) Vascularity and blood flow in x-irradiated Walker carcinoma 256 of rats. Radiology 104:693–697

Song CW, Rhee JG, Haumschild DJ (1987) Continuous and non-invasive quantification of heat induced changes in blood flow in the skin and RIF-1 tumor of mice by laser Doppler flowmetry. Int J Hyperthermia 3:71–77

Steel GG (1989) The 5Rs of radiobiology. Int J Radiat Biol 56:1045–1048

Streffer C (1985) Review: metabolic changes during and after hyperthermia. Int J Hyperthermia 1:305–309

Tanaka Y (1974) Regional tumor blood flow and radiosensitivity. In: Sugahara T, Revesz L, Scott O (eds) Fraction size in radiobiology and radiotherapy. Urban und Schwarzenberg, Munich, pp 13–26

Tatagiba M, Mirzal S, Samii M (1991) Peritumoral blood flow in intracranial meningeomas. Neurosurgery 28:400–404

Teglia A, Kittelson JM, Roemer RB, Hodak JA, Carter LP (1996) Cerebral blood flow in and around spontaneous malignant gliomas. Int J Hyperthermia 12:461–478

Vaupel P (1994) Blood flow, oxygenation status, tissue pH distribution and bioenergetic status of tumors. Lecture 23. Ernst Schering Research Foundation, Berlin

Vaupel PW, Höckel M (1995) Oxygenation status of human tumors: a reappraisal using computerized pO$_2$

histography. In: Vaupel PW, Kelleher DK, Günderoth M (eds) Tumor oxygenation. Fischer, Stuttgart, pp 219–232

Vaupel P, Braunbeck W, Schulz V, Günther H, Thews G (1973) Critical O$_2$ and glucose supply and microcirculation in tumor tissue. Bibl Anat 12:527–533

Vaupel P, Kluge M, Ambroz C (1988) Laser Doppler flowmetry in subepidermal tumors and in normal skin of rats during localized ultrasound hyperthermia. Int J Hyperthermia 4:307–321

Vaupel P, Kallinowski F, Okunieff P (1989) Blood flow, oxygen and nutrient supply and metabolic microenvironment of human tumors: a review. Cancer Res 49:6449–6465

Vaupel P, Schlenger K, Höckel M (1992) Blood flow and tissue oxygenation of human tumors: an update. Adv Exp Med Biol 317:139–151

Waterman FM, Nerlinger RE, Moylan DJ III, Leeper DB (1987) Response of human tumor blood flow to local hyperthermia. Int J Radiat Oncol Biol Phys 13:75–83

Waterman FM, Tupchong L, Liu CR (1991) Modified thermal clearance technique for determination of blood flow during local hyperthermia. Int J Hyperthermia 7:719–733

Weidner N, Semple JP, Welch WR, Folkman J (1991) Tumor angiogenesis and metastasis – correlation in invasive breast carcinoma. N Engl J Med 324:1–8

Williams R, Condon RE, Williams HS, Blendis LM, Kreel L (1968) Splenic blood flow in cirrhosis and portal hypertension. Clin Sci 34:441–452

Wong TZ, Mechling JA, Jones EL, Strohbehn JE (1988) Transient finite element analysis of thermal methods used to estimate SAR and blood flow in homogeneously and nonhomogeneously pefused tumor models. Int J Hyperthermia 4:571–592

Wust P (1992) Hyperthermie in der Tumortherapie: Methodische Entwicklung und klinische Evaluation. Habilitationsschrift, Freie Universität Berlin

Wust P, Stahl H, Löffel J, Seebass M, Riess H, Felix R (1995) Clinical, physiological and anatomical determinants for radiofrequency hyperthermia. Int J Hyperthermia 11:151–167

Zongrafos GC, Iffikar SY, Harrison J, Morris DC (1990) Evaluation of blood flow in human rectal tumors using a laser Doppler flowmeter. Eur J Surg Oncol 16:497–499

6 Oxygenation of Human Tumors

P. Vaupel[1] and M. Höckel[2]

CONTENTS

6.1 Introduction

Apart from inherent cell sensitivity, cell kinetics, and tumor microenvironmental factors (e.g., metabolic and energetic status, pH distribution), which are closely linked to the efficacy of tumor microcirculation and which – to a certain extent – are interdependent, tumor hypoxia is considered as one of the multifactorial causes of tumor treatment resistance. Experimental and clinical evidence suggests that the hypoxic fraction in solid tumors may influence their growth, may increase their malignant progression due to gene amplification and enhanced metastatic potential, and may reduce their sensitivity to conventional treatment modalities (e.g., sparsely ionizing radiation, certain chemotherapeutic agents; Höckel et al. 1996; Raleigh 1996).

6.2 Oxygen Status of Solid Tumors

Tumor oxygenation is dependent on the *cellular O_2 consumption rate* and on the *O_2 supply to the respiring cells*. The latter is preferentially determined by the convective transport via the blood and by the diffusional flux from microvessels to O_2 consuming sites. Peculiarities of tumor tissue oxygenation can therefore mainly be attributed to characteristic structural and functional abnormalities of tumor microcirculation (perfusion-limited O_2 delivery; Fig. 6.1) and to deterioration of diffusion geometry (diffusion-limited O_2 delivery). As a result of a compromised and anisotropic microcirculation, the O_2 availability to cancer cells shows great variability, and many human malignancies reveal hypoxic tissue areas that are heterogeneously distributed within the tumor mass and that may be located next to a well-perfused tumor area (intratumoral heterogeneity). As a rule, in most solid malignancies the tissue O_2 status is poorer than in normal tissue at the site of tumor growth. This has been shown for a series of solid tumors (Vaupel 1990, 1992, 1993, 1994, 1996; Vaupel et al. 1989, 1996).

6.3 Oxygenation of Human Tumors: Methods of Measurement

The oxygenation status of human tumors has been evaluated using a series of direct or indirect assays. Earlier studies have used a *cryospectrophotometric ex vivo microtechnique* that allows for the measurement of HbO_2 saturation of individual red blood cells in tumor microvessels (Mueller-Klieser et al. 1981; Wendling et al. 1984; Vaupel and Kallinowski 1987). As a rule, the mean oxyhemoglobin (HbO_2) values observed in cancers of the rectum were distinctly lower than those found in the normal mucosa at the site of tumor growth. The same holds true for squamous cell carcinomas of the

[1] P. Vaupel, Dr. med., Institute of Physiology and Pathophysiology, University of Mainz, Duesbergweg 6, D-55099 Mainz, Germany
[2] M. Höckel, Dr. med., Department of Obstetrics and Gynecology, University of Mainz, Langenbeckstrasse 1, D-55101 Mainz, Germany

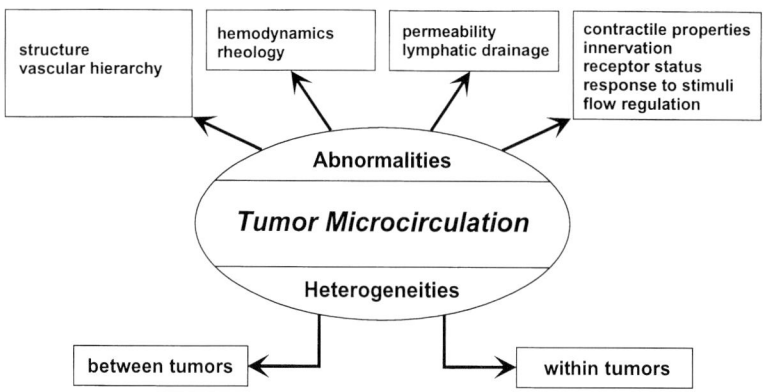

Fig. 6.1. Relevant structural and functional abnormalities of tumor neovasculature/microcirculation

oral cavity. The medians of HbO_2 frequency distributions of normal oral mucosa and of tumors at this site decreased from 80 sat. % to 49 sat. % and correlated well with changes in vascular density.

In various malignant tumors, considerable inter- and intraindividual differences were observed, even when tumors of the same clinical stage and grade were investigated. Although these investigations have provided a detailed insight into the oxygenation status of human malignancies, the experimental data provided only refer to the O_2 saturation status of red blood cells in microvessels. The O_2 profiles within the cellular and stromal compartments thus have to be extrapolated (calculated) from the O_2 loading of the "biological indicator" hemoglobin.

In addition to *magnetic resonance spectroscopy and imaging methods* for measuring tumor oxygenation and near infrared spectroscopy (that so far has only supplied very preliminary data on human solid tumors; e.g., McCoy et al. 1996; Steinberg et al. 1997), *nuclear medicine assays* for the detection of tumor hypoxia have been developed and applied to human tumors (Chapman 1991). With this latter technique, tumor oxygenation has been "indirectly" estimated using [^{123}I]-iodoazomycin arabinoside (IAZA; Groshar et al. 1993; Urtasun et al. 1996) or [^{18}F]-fluoromisonidazole (F-MISO; Koh et al. 1991, 1995; Valk et al. 1992; Rasey et al. 1996) as probes. Correlations between various methods of assessment of the oxygenation status of solid tumors in humans and their response to therapy have been evaluated a few years ago (Stone et al. 1993). As a "gold standard technology" for assessing the oxygenation of human malignancies, the polarographic O_2 electrode system has been suggested (Stone et al. 1993).

Oxygen partial pressure distributions for solid tumors in humans have been described over the past 40 years (e.g., Urbach 1956; Urbach and Noell 1958; Cater and Silver 1960; Evans and Naylor 1963; Cater 1964; Jamieson and van den Brenk 1965; Kolstad 1968; Badib and Webster 1969; Bergsjö and Evans 1971; Pappova et al. 1982; Gatenby et al. 1985, 1988). Most of these studies were anecdotal and/or case studies. Despite these earlier data, the clinical importance of tumor hypoxia remains uncertain since valid methods for the *routine measurement of intratumoral O_2 tensions* in patients have so far been lacking (Vaupel 1990, 1992, 1993, 1994, 1996; Vaupel et al. 1989, 1991a,b, 1992, 1995, 1996). During the past decade, a clinically applicable standardized procedure has been established which enables the determination of intratumoral O_2 tensions in human primary tumors and metastatic lesions by a computerized *polarographic needle electrode system* (pO_2 histography; Eppendorf, Hamburg, Germany). In the following sections, presently available pO_2 data from human tumors obtained with this reliable system, which is well tolerated by patients and has no effect on metastatic potential and proliferation characteristics (Lartigau et al. 1992a), are summarized.

6.4 Oxygenation Status of Primary Human Tumors

6.4.1 Breast Cancers

Extensive studies on the pretreatment tissue oxygenation of *breast cancers* have been performed by several groups (Vaupel et al. 1991b; Falk et al. 1992; Runkel et al. 1994). As a result of a compromised and anisotropic microcirculation (see Fig. 6.1), many breast cancers reveal hypoxic tissue areas that are heterogeneously distributed within the tumor mass. Mean and median O_2 tensions (pO_2) obtained from different pathological stages and histological grades are, on average, distinctly lower than in normal tissues (see Fig. 6.2; Vaupel et al. 1989, 1991b; Höckel et al. 1991). Oxygen tensions measured in normal breast revealed a mean (and median) pO_2 of 65 mmHg, whereas in cancers of the breast of stages pT1-4, the median pO_2 was 28 mmHg (Fig. 6.3). Thus far, one third of the breast cancers investigated exhibited pO_2 values between 0 and 2.5 mmHg, i.e., tissue areas with less than half-maximum radiosensitivity. In contrast, in the normal breast pO_2 values ≤12.5 mmHg could not be detected (Vaupel et al. 1991b). In all systematic studies on breast cancers, bimodal pO_2 distribution curves have been obtained (Vaupel et al. 1991b; Runkel et al. 1994), either indicating the coexistence of large normoxic and hypoxic tumor areas or a relevant

contribution of pO_2 readings in the (partially inflamed?) stromal compartment of breast cancers.

When pooled data for stages pT1 and pT2 and pT3 and pT4 breast cancers are compared, there is no evidence of statistically significant differences between the two groups (median pO_2 in pT1 and pT2 tumors: 28 mmHg; pT3 and pT4 tumors: 29 mmHg; Fig. 6.3). This implies that the oxygenation in breast cancers and the occurrence of hypoxia and/or anoxia do not correlate with the clinical stage (Vaupel et al. 1991b; Falk et al. 1992; Runkel et al. 1994; Füller et al. 1994). Similarly, there was no association between tumor size and blood flow (Grischke et al. 1994; Wilson et al. 1992). The proportion of pO_2 values between 0 and 2.5 mmHg was ~6% in pT1 and pT2, and ~7% in pT3 and pT4 tumors. In addition, there is substantial evidence that the oxygenation patterns do not correlate with histological grade (Vaupel et al. 1991b; Falk et al. 1992; Runkel et al. 1994), menopausal status, tumor histology (ductal vs. lobular), extent of necrosis or fibrosis, or with a series of other clinically relevant parameters (e.g., hormone receptor status, hemoglobin level, smoking habits; Vaupel et al. 1991b; Falk et al. 1992).

6.4.2
Cancers of the Uterine Cervix

Analysis of O_2 tensions measured in the normal cervix of nulliparous women resulted in oxygenation patterns characteristic of normal, adequately supplied tissues (median pO_2: 48 mmHg; Fig. 6.4). As a rule, the mean (and median) pO_2 values were distinctly lower in the normal cervix of parous women (most probably due to tissue changes following vaginal delivery). Here, the median pO_2 was 13 mmHg (with ~14% of the pO_2 readings in the 0–2.5 mmHg class). In *cancers of the uterine cervix* (stages FIGO I–IV), the median pO_2 is 10 mmHg before treatment (Fig. 6.4). To date, one third of the cervical cancers investigated exhibited pO_2 values between 0 and 2.5 mmHg. The relative number of pO_2 readings between 0 and 2.5 mmHg ranged from 1% (in a FIGO IV cancer) to 82% (in a FIGO III tumor; Höckel et al. 1991).

In FIGO I and II tumors, the median pO_2 is 11 mmHg, and in FIGO III and IV tumors it is 10 mmHg, with 6% and 18% of readings, respectively, in the lowest pO_2 class (0–2.5 mmHg). As was the case with breast cancers, the pretreatment oxygenation pattern in cervical cancers and the occurrence of hypoxia and/or anoxia did not correlate

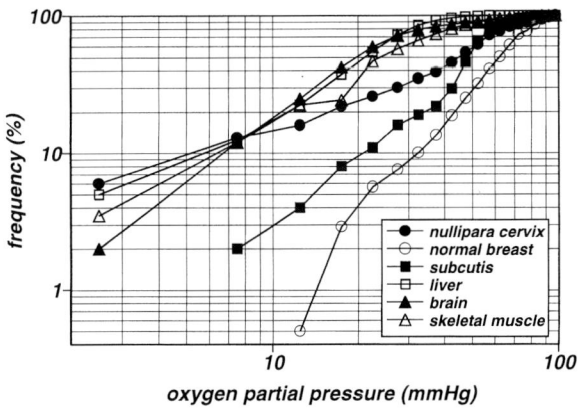

Fig. 6.2. Compilation of cumulative frequency distributions of measured pO_2 values in various normal tissues. (Data adapted from Vaupel et al. 1989)

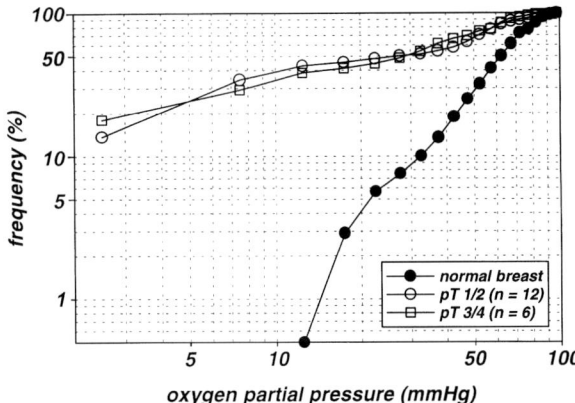

Fig. 6.3. Cumulative frequency distributions of measured tissue pO_2 values in normal breast and in breast cancers of stages pT1 and 2 and pT3 and 4

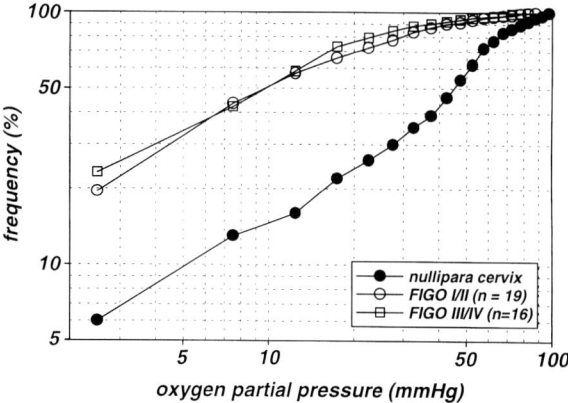

Fig. 6.4. Cumulative frequency distributions of measured tissue pO_2 values in the normal cervix of nulliparous women and in cervical cancers of stages FIGO I and II and FIGO III and IV

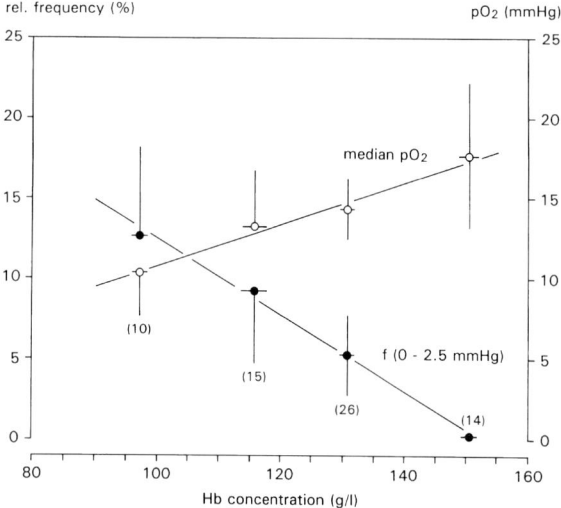

Fig. 6.5. Tumor median pO_2 and fraction of pO_2 values ≤ 2.5 mmHg as a function of hemoglobin concentration in cervical cancer patients. Values are means ± SEM, with the number of tumors investigated given in parentheses

with any of the above-mentioned clinically relevant parameters, with the exception of hemoglobin concentration.

From an earlier evaluation of oxygenation data of 65 cervical cancers there is an indication that severely anemic patients ([Hb] $\leq$ 100 g/l) tend to have lower pO_2 values in cervical cancers than patients with normal Hb concentrations ([Hb] > 140 g/l; p = 0.05). The fraction of "hypoxic" pO_2 values (0–2.5 mmHg) is significantly higher in the anemic group than in patients with normal Hb concentrations (p = 0.004; Fig. 6.5). Blood transfusion in a severely anemic patient ([Hb] $\leq$ 100 g/l) resulted in an increase in the median pO_2 value while, at the

same time, a significant reduction in the number of pO_2 readings in the 0–2.5 mmHg range occurred (VAUPEL 1994).

From our clinical studies on breast and cervical cancers there is clear indication that the oxygenation status of individual tumors before therapy cannot be predicted on the basis of tumor staging and/or grading. The lack of predictability is predominantly due to pronounced tumor-to-tumor variabilities even if tumors of the same clinical stage and histological grade are compared. Tumor-to-tumor variability in the oxygenation pattern is more pronounced than intratumor heterogeneity.

Oxygen tension measurements have also been performed in patients with untreated squamous cell carcinomas of the uterine cervix by LARTIGAU et al. (1992b). These authors observed slightly higher median pO_2 values of 21 mmHg, together with a higher proportion of pO_2 readings in the lowest class (0–2.5 mmHg). At the time of brachytherapy after external radiotherapy, the median pO_2 was $\approx$15 mmHg in cervical cancers (LARTIGAU et al. 1992c).

FYLES et al. (1996) reported a median pO_2 value of 5 mmHg in cervix cancers prior to radiation treatments. Eighteen of 31 patients had median pO_2 values of <10 mmHg and the median proportion of pO_2 values <5 mmHg was 52%, which is twice as high as data presented by HÖCKEL et al. (1993a,b). Recent communications by LYNG et al. (1996, 1997) also report low median pO_2 values in cancers of the uterine cervix ($\sim$2 mmHg) with a fraction of pO_2 readings below 5 mmHg of $\sim$68%. LYNG et al. (1996) have looked for a relationship between pretreatment oxygen tensions and vascular density in cervix cancers. In this latter study, tumor regions with a vascular density below 24 mm/mm^3 always showed low pO_2 values whereas tumor areas with a vascular density above 24 mm/mm^3 exhibited both high and low pO_2 values. These authors concluded that a low vascular density might be a useful predictor of tissue hypoxia in cervix carcinoma. High vascular density, on the other hand, can probably not be used to exclude hypoxic regions in cancers of the uterine cervix.

6.4.3
Head and Neck Cancers

Oxygenation data derived from primary head and neck tumors are sparse and thus preliminary (MUELLER-KLIESER et al. 1981; FLECKENSTEIN et al. 1993; SAUMWEBER et al. 1995). In Fig. 6.6, pooled oxygenation data obtained from primary head and

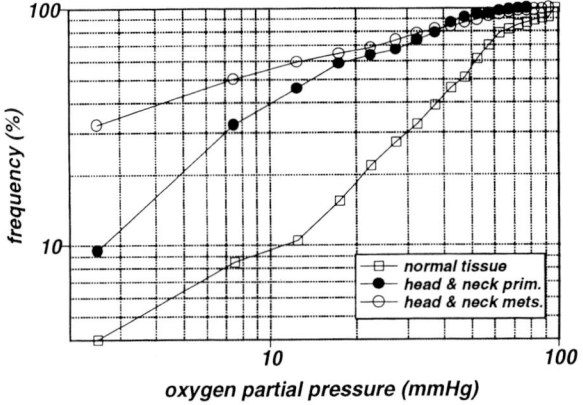

Fig. 6.6. Cumulative frequency distributions of pO₂ values in primary head and neck tumors compared to metastatic lesions of this tumor entity together with values obtained in the normal tissue adjacent to the lesions. (Adapted from VAUPEL et al. 1996)

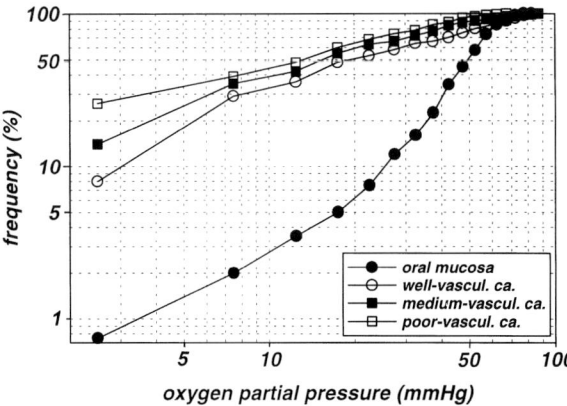

Fig. 6.7. Cumulative frequency distributions of pO₂ values in normal oral mucosa and in squamous cell carcinomas (*ca.*) of the oral cavity as a function of vascular density. Data have been calculated from oxyhemoglobin saturation values of individual red blood cells in tumor microvessels measured in cryobiopsies under ex vivo conditions. (From MUELLER-KLIESER et al. 1981)

neck tumors using the pO_2 histography system are plotted as a cumulative pO_2 distribution curve. Mean and median O_2 tensions obtained from different clinical stages and histological grades are, on average, distinctly lower than in normal tissue. Oxygen tensions measured in the subcutis of the head and neck region revealed a median pO_2 of approximately 50 mmHg, whereas in primary tumors the median pO_2 was 16 mmHg. Thus far, about 10% of the head and neck cancers investigated exhibited pO_2 values ≤5 mmHg.

The vascularity of primary head and neck tumors seems to have a significant impact on the oxygen-

ation status of these malignancies. This statement is supported by the data presented in Fig. 6.7: In squamous cell carcinomas of the oral cavity, HbO_2 saturation of individual red blood cells in tumor microvessels has been measured in cryobiopsies under ex vivo conditions (MUELLER-KLIESER et al. 1981). From these intravascular HbO_2 values, tissue oxygen tensions have been calculated and plotted together with the respective data of normal oral mucosa. There is a clear correlation between the oxygenation status and vascular density. Well-vascularized tumors showed a better oxygenation status and less hypoxia than poorly vascularized lesions. Tumors with medium-quality vascularization exhibited an oxygenation status between these two extreme patterns. The use of estimates of tumor vasculature alone, however, may not be a reliable (prognostic) indicator of hypoxia in all situations (HORSMAN 1993).

6.4.4
Soft Tissue Sarcomas

Oxygen tension measurements have been performed in patients with soft tissue sarcomas by BRIZEL et al. (1994, 1995), FELDMANN (1994), FÜLLER et al. (1994), HOHENBERGER and DRAGON (1995), and NORDSMARK et al. (1994, 1996a; see Table 6.1).

In a recent study by NORDSMARK et al. (1996a), a large interpatient variability was found with respect to the median pO_2 ≤ 5 mmHg, and the percentage of pO_2 ≤ 2.5 mmHg, respectively. The median pO_2 ranged from 1 mmHg to 58 mmHg, and the percentage of pO_2 ≤ 5 mmHg and ≤2.5 mmHg ranged from 0% to 60% and 0% to 54%, respectively. For all soft tissue sarcomas the overall median of the median pO_2 was 18 mmHg, and the overall median of the percentage of pO_2 ≤ 5 mmHg and ≤2.5 mmHg was 17% and 5%, respectively.

No correlation was found between oxygenation status and volume, histopathology, grade of malignancy, or hemoglobin levels. In this study the most strongly proliferating tumor cells were found in the most poorly oxygenated soft tissue sarcomas.

Relevant parameters of the pretherapeutic oxygenation status of various human malignancies are summarized in Table 6.1. In most tumor entities, no significant correlation with tumor size was found. Only in soft tissue sarcomas and in lymph node metastases of head and neck cancers was the hypoxic fraction found to increase with enlarging tumor volume.

Table 6.1. Pretherapeutic oxygenation status of human malignancies (assessed by computerized pO$_2$ histography)

Tumor	n	f (0–2.5 mmHg) [%]	f (0–5 mmHg) [%]	f (<10 mmHg) [%]	Mean pO$_2$ [mmHg]	Median pO$_2$ [mmHg]	Size dependency	References
Breast carcinomas	18	5.9	15.4	32.3	32	28	∅	Vaupel et al. (1991b)
	5	4.5		13.0		24	∅	Falk et al. (1992)
	18	7.5	15.5	26.0		23	∅	Runkel et al. (1994)
Cervical carcinomas	37	11.1	25.9	45.6	18	10	∅	Höckel et al. (1993a)
	6	15.0	17.0	21.0	29	21		Lartigau et al. (1992c)
	9		61.0		10	5		Brizel et al. (1995)
	31		52.0		17	5	∅	Fyles et al. (1996)
	36	45.0	70.0			2		Lyng et al. (1997)
Rectal cancers	5		35.0			25		Feldmann (1994)
	14					25		Kallinowski and Buhr (1995)
	15				26	19		Mattern et al. (1996)
Lung cancers	6	12.7		36.3		14		Falk et al. (1992)
Soft tissue sarcomas	4		17.0		21	· 20		Feldmann (1994)
	9	19.0	29.0	44.0	24	21	+ (Hypoxic fraction)	Brizel et al. (1994)
	18		10.0			23		Nordsmark et al. (1994)
	8		10.2		25			Füller et al. (1994)
	15		31.0		22	18		Brizel et al. (1995)
	15					27		Hohenberger and Dragon (1995)
	22	54.0	17.0			18		Nordsmark et al. (1996a)
Glioblastomas	10	38.0	45.5	61.0		7	∅	Rampling et al. (1994)
Head and neck cancers	7		9.8	33.0	22	19		Fleckenstein et al. (1993)
	16					26		Saumweber et al. (1995)
Lymph node metastases of head and neck cancers	31		22.8			12		Nordsmark et al. (1994)
	8		5.8		18			Füller et al. (1994)
	8		36.0		18	13		Brizel et al. (1995)
	20			47.7	20	16	+	Martin et al. (1993)
	15	22.5	34.5	54.0	21	9	+	Lartigau et al. (1993)
	6	9.0	19.0	35.0	14	22		Eble et al. (1995)
	35	21.6	34.6			15		Nordsmark et al. (1996b)
	16				27		+	Terris and Dunphy (1994)
	14	18.6	34.6	44.4	22	20		Strnad et al. (1996)
	5	4.2	7.0	14.5		25		Lyng et al. (1997)
Melanomas (metastases)	13	16.0	28.5	49.5	25	10		Guichard and Lartigau (personal communication)

n, number tumors investigated; f, fraction.

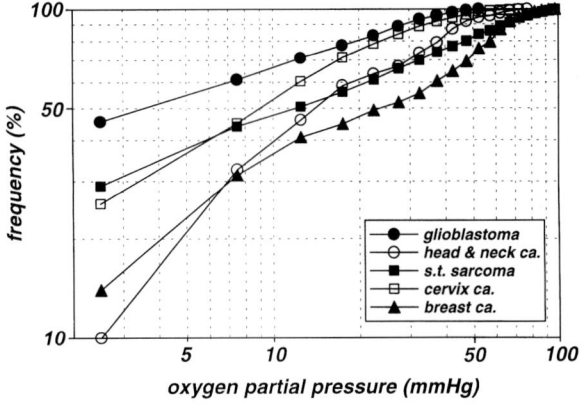

Fig. 6.8. Compilation of cumulative frequency distributions of measured pO$_2$ values in different primary tumors. *ca.*, Carcinoma; *s.t.*, soft tissue. Data have been adapted from published values. (From Höckel et al. 1991; Vaupel et al. 1991b; Fleckenstein et al. 1993; Brizel et al. 1994; Rampling et al. 1994)

In Fig. 6.8, relevant oxygenation data obtained from primary tumors using the pO$_2$ histography system are plotted as cumulative pO$_2$ distribution curves (Vaupel et al. 1991b; Höckel et al. 1991; Brizel et al. 1994; Fleckenstein et al. 1993; Rampling et al. 1994). From this compilation of pO$_2$ data there is evidence that different oxygenation patterns and different fractions of hypoxic tissue volumes have to be expected when considering different tumor entities. So far, breast cancers exhibit a significantly better oxygenation status than glioblastomas, with the other primary malignancies grouped between these tumors. Extensive hypoxia in glioblastomas correlates with poor therapeutic outcome and coincides with high vascular density.

6.5
Oxygenation Status of Metastatic Lesions

As was the case with the primary tumors, blood flow (Hill et al. 1996; Pigott et al. 1996) and oxygenation of metastatic lesions is generally anisotropic and compromised as compared to normal tissues at the site of metastatic growth. The median pO$_2$ values of the secondary tumors are lower than those recorded in the tumor surroundings (Fig. 6.6). This holds true for metastatic lesions of squamous cell carcinomas (Lartigau et al. 1994; Martin et al. 1993) as well as metastatic melanomas (Guichard and Lartigau, personal communication) and metastatic gastrointestinal tumors (Falk et al. 1994).

When comparing the oxygenation status of primary head and neck carcinomas (Fleckenstein et al. 1993; Mueller-Klieser et al. 1981; Saumweber et al. 1995) with metastatic lesions of this entity (Martin et al. 1993; Lartigau et al. 1993, 1994; Nordsmark et al. 1994, 1996b; Füller et al. 1994; Terris and Dunphy 1994; Brizel et al. 1995; Eble et al. 1995; Strnad et al. 1996), the fraction of pO$_2$ readings ≤5 mmHg is somewhat higher in the latter group (Fig. 6.6). In *breast and rectal cancers*, metastatic lesions also exhibited a poorer oxygenation status than the primaries (Füller et al. 1994; Kallinowski and Buhr 1995). Whether this pattern is characteristic of these three entities or a general biological phenomenon has to be elucidated in ongoing studies. *Local recurrences of breast cancers* (Füller et al. 1994) *and of cervix cancers* (see Table 6.2; Vorndran 1996) also seem to have a higher hypoxic fraction than the primary tumors.

In contrast to primary tumors investigated so far, the oxygenation status of metastatic lesions of the head and neck appears to be linked to tumor size

Table 6.2. Oxygenation status of primary cancers (pretherapeutic measurements) and of recurrent tumors of the uterine cervix (Vorndran 1996)

Stage	N	n	f (≤2.5 mmHg) [%]	f (≤5 mmHg) [%]	f (≤10 mmHg) [%]	Mean pO$_2$ [mmHg]	Median pO$_2$ [mmHg]
FIGO I	5	352	0	15	28	18	16
FIGO II	28	2510	6	24	45	20	13
FIGO III	23	1932	7	25	44	20	13
FIGO IV	5	344	2	31	52	13	9
FIGO I–IV	61	5138	6	23	43	19	13
Recurrent tumors	18	1449	20	36	48	19	11

N, number of tumors; *n*, number of pO$_2$ measurements; f, fraction.

with the lower median pO$_2$ values preferentially occurring in larger nodal sizes (LARTIGAU et al. 1993; TERRIS and DUNPHY 1994).

Recent studies confirmed earlier observations by GATENBY et al. (1988) suggesting that the presence of hypoxia corresponded with poor radiation response of metastatic head and neck tumors (NORDSMARK et al. 1996b). Furthermore, information is accumulating which indicates that pretreatment oxygenation in head and neck tumors can predict radiation response in advanced squamous cell carcinomas of the head and neck (NORDSMARK et al. 1996b). Similar observations were made earlier by HÖCKEL et al. (1996) for advanced stage cancers of the uterine cervix.

6.6
Conclusions

Using the pO$_2$ histography system for assessment of the oxygenation status of solid tumors, several investigations have clearly shown that:

1. Tumor oxygenation, as a rule, is anisotropic and compromised as compared to normal tissues.
2. Tumor oxygenation is not regulated according to metabolic demand as is the case in normal tissues.
3. On average, the median pO$_2$ values in primary, metastatic and recurrent tumors are lower than in normal tissue at the site of growth.
4. Many solid tumors contain hypoxic tissue areas (pO$_2 \leq 2.5$ mmHg).
5. Tumor-to-tumor variability in oxygenation is significantly greater than intratumor variability.
6. Tumor oxygenation is unpredictable considering clinical stage and grade.
7. Tumor oxygenation is independent of other known oncologic parameters.
8. Oxygenation status in some tumors is dependent on the vascularity.
9. Metastases seem to have a higher hypoxic fraction than primary tumors.
10. Tumor oxygenation appears to be a new, independent prognostic factor influencing overall survival and local control upon radiotherapy (± chemotherapy) in locally advanced cervical cancer (see Chap. 7) and in advanced metastatic lesions of the head and neck region.

As an optimistic outlook, tumor oxygenation status evaluated by computerized pO$_2$ histography may enable pretherapeutic selection of hypoxic tumors as candidates for modified treatment approaches.

References

Badib AO, Webster JH (1969) Changes in tumor oxygen tension during radiation therapy. Acta Radiol Ther Phys Biol 8:247–257

Bergsjö P, Evans JC (1971) Oxygen tension of cervical carcinomas during the early phase of external irradiation. Scand J Clin Lab Invest 27:71–82

Brizel DM, Rosner G, Harrelson J, Prosnitz LR, Dewhirst MW (1994) Pretreatment oxygenation profiles of human soft tissue sarcomas. Int J Radiat Oncol Biol Phys 30:635–642

Brizel DM, Rosner GL, Prosnitz LR, Dewhirst MW (1995) Patterns and variability of tumor oxygenation in human soft tissue sarcomas, cervical carcinomas, and lymph node metastases. Int J Radiat Oncol Biol Phys 32:1121–1125

Cater DB (1964) Oxygen tension in neoplastic tissues. Tumori 50:435–444

Cater DB, Silver IA (1960) Quantitative measurements of oxygen tensions in normal tissues and in the tumours of patients before and after radiotherapy. Acta Radiol 53:233–256

Chapman JD (1991) Measurement of tumor hypoxia by invasive and non-invasive procedures: a review of recent clinical studies. Radiother Oncol 20:13–19

Eble MJ, Lohr F, Wannenmacher M (1995) Oxygen tension distribution in head and neck carcinomas after peroral oxygen therapy. Onkologie 18:136–140

Evans NTS, Naylor PFD (1963) The effect of oxygen breathing and radiotherapy upon the tissue oxygen tension of some human tumours. Br J Radiol 36:418–425

Falk SJ, Ward R, Bleehan NM (1992) The influence of carbogen breathing on tumour tissue oxygenation in man evaluated by computerised pO$_2$ histography. Br J Cancer 66:919–924

Falk SJ, Ramsay JR, Ward R, Miles K, Dixon AK, Bleehan NM (1994) BW12C perturbs normal and tumour tissue oxygenation and blood flow in man. Radiother Oncol 32:210–217

Feldmann HJ (1994) Optimierungsansätze und Limitationen in der regionalen Thermoradiotherapie von Beckentumoren. Thesis, University of Essen, Germany

Fleckenstein W, Jungblut JR, Suckfüll M, Hoppe W, Weiss C (1993) Sauerstoffdruckverteilungen in Zentrum und Peripherie maligner Kopf-Hals-Tumoren. Dtsch Z Mund Kiefer Gesichts Chir 12:205–211

Füller J, Feldmann HJ, Molls M, Sack H (1994) Untersuchungen zum Sauerstoffpartialdruck im Tumorgewebe unter Radio- und Thermoradiotherapie. Strahlenther Onkol 170:453–460

Fyles A, Milosevic M, Kavanagh M-C, et al. (1997) Hypoxia measured with a polarographic electrode correlates with radiation response in cervix cancer. Radiother Oncol (in press)

Gatenby RA, Coia LR, Richter MP, et al. (1985) Oxygen tension in human tumours: in vivo mapping using CT-guided probes. Radiology 156:211–214

Gatenby RA, Kessler HB, Rosenblum JS, Coia LR, Moldofsky PJ, Hartz WH, Broder GJ (1988) Oxygen distribution in squamous cell carcinoma metastases and its relationship to outcome of radiation therapy. Int J Radiat Oncol Biol Phys 14:831–838

Grischke EM, Kaufmann M, Eberlein-Gonska M, Mattfeldt T, Sohn C, Bastert G (1994) Angiogenesis as a diagnostic factor in primary breast cancer: microvessel quantitation by stereological methods and correlation with color Doppler sonography. Onkologie 17:35–42

Groshar D, McEwan AJB, Parliament MB, et al. (1993) Imaging tumor hypoxia and tumor perfusion. J Nucl Med 34:885–888

Hill SA, Pigott KH, Saunders MI, et al. (1996) Microregional blood flow in murine and human tumours assessed using laser Doppler microprobes. Br J Cancer 74:260–263

Höckel M, Schlenger K, Knoop C, Vaupel P (1991) Oxygenation of carcinomas of the uterine cervix: evaluation by computerized O_2 tension measurements. Cancer Res 51:6098–6102

Höckel M, Knoop C, Schlenger K, Vorndran B, Mitze M, Knapstein PG, Vaupel P (1993a) Intratumoral pO_2 predicts survival in advanced cancer of the uterine cervix. Radiother Oncol 26:45–50

Höckel M, Vorndran B, Schlenger K, Baussmann E, Knapstein PG, Vaupel P (1993b) Tumor oxygenation: A new predictive parameter in locally advanced cancer of the uterine cervix. Gynecol Oncol 51:141–149

Höckel M, Schlenger K, Aral B, Mitze M, Schäffer U, Vaupel P (1996) Association between tumor hypoxia and malignant progression in advanced cancer of the uterine cervix. Cancer Res 56:4509–4515

Hohenberger P, Dragon S (1995) In situ oxygen partial pressure measurements in human soft tissue sarcomas. In: Vaupel PW, Kelleher DE, Günderoth M (eds) Tumor oxygenation. Fischer, Stuttgart, pp 327–333

Horsman MR (1993) Hypoxia in tumours: Its relevance, identification, and modification. In: Beck-Bornholdt HP (ed) Medical radiology. Current topics in clinical radiobiology of tumors. Springer, Berlin Heidelberg New York, pp 99–112

Jamieson D, van den Brenk HAS (1965) Oxygen tension in human malignant disease under hyperbaric conditions. Br J Cancer 19:139–150

Kallinowski F, Buhr HJ (1995) Tissue oxygenation of primary, metastatic and xenografted rectal cancers. In: Vaupel P, Kelleher DK, Günderoth M (eds) Tumor oxygenation. Fischer, Stuttgart, pp 205–209

Koh WJ, Rasey JS, Evans ML, et al. (1991) Imaging of hypoxia in human tumors with [^{18}F]Fluoromisonidazole. Int J Radiat Oncol Biol Phys 22:199–212

Koh WJ, Bergman KS, Rasey JS, Peterson LM, Evans ML, Graham MM, Grierson JR, Lindsley KL, Lewellen TK, Krohn KA, Griffin TW (1995) Evaluation of oxygenation status during fractionated radiotherapy in human nonsmall cell lung cancers using [F-18]Fluoromisonidazole positron emission tomography. Int J Radiat Oncol Biol Phys 33:391–398

Kolstad P (1968) Intercapillary distance, oxygen tension and local recurrence in cervix cancer. Scand J Clin Lab Invest Suppl 106:145–157

Lartigau E, Lespinasse F, Vitu L, Guichard M (1992a) Does the direct measurement of oxygen tension in tumors have any adverse effects? Int J Radiat Oncol Biol Phys 22:949–951

Lartigau E, Martin L, Lambin P, Haie-Meder C, Gerbaulet A, Eschwege F, Guichard M (1992b) Mesure de la pression partielle en oxygène dans des tumeurs du col utérin. Bull Cancer Radiother 79:199–206

Lartigau E, Vitu L, Haie-Meder C, et al. (1992c) Feasibility of measuring oxygen tension in uterine cervix carcinoma. Eur J Cancer 28A:1354–1357

Lartigau E, Le Ridant AM, Lambin P, Weeger P, Martin L, Sigal R, Lusinchi A, Luboinski B, Eschwege F, Guichard M (1993) Oxygenation of head and neck tumors. Cancer 71:2319–2325

Lartigau E, Randrianarivelo H, Martin L, et al. (1994) Oxygen tension measurements in human tumors: the Institut Gustave-Roussy experience. Radiat Oncol Invest 1:285–291

Lyng H, Sundfor K, Trope C, Rofstad EK (1996) Oxygen tension and vascular density in human cervix carcinoma. Br J Cancer 74:1559–1563

Lyng H, Sundfor K, Tanum G, Rofstad EK (1997) Oxygen tension in primary tumours of the uterine cervix and lymph node metastases of the head and neck. Adv Exp Med Biol, in press

Martin L, Lartigau E, Weeger P, et al. (1993) Changes in the oxygenation of head and neck tumours during carbogen breathing. Radiother Oncol 27:123–130

Mattern J, Kallinowski F, Herfarth C, Volm M (1996) Association of resistance-related protein expression with poor vascularization and low levels of oxygen in human rectal cancer. Int J Cancer 67:20–23

McCoy CL, McIntyre DJO, Robinson SP, Aboagye EO, Griffiths JR (1996) Magnetic resonance spectroscopy and imaging methods for measuring tumour and tissue oxygenation. Br J Cancer 74:226–231

Mueller-Klieser W, Vaupel P, Manz R, Schmidseder R (1981) Intracapillary oxyhemoglobin saturation of malignant tumors in humans. Int J Radiat Oncol Biol Phys 7:1397–1404

Nordsmark M, Bentzen SM, Overgaard J (1994) Measurement of human tumour oxygenation status by a polarographic needle electrode. Acta Oncol 33:383–389

Nordsmark M, Hoyer M, Keller J, Nielsen OS, Jensen OM, Overgaard J (1996a) The relationship between tumor oxygenation and cell proliferation in human soft tissue sarcomas. Int J Radiat Oncol Biol Phys 35:701–708

Nordsmark M, Overgaard M, Overgaard J (1996b) Pretreatment oxygenation predicts radiation response in advanced squamous cell carcinoma of the head and neck. Radiother Oncol 41:31–39

Pappova N, Siracka E, Vacek A, Durkovsky J (1982) Oxygen tension and prediction of the radiation response. Polarographic study in human breast cancer. Neoplasma 29:669–674

Pigott KH, Hill SA, Chaplin DJ, Saunders MI (1996) Microregional fluctuations in perfusion within human tumours detected using laser Doppler flowmetry. Radiother Oncol 40:45–50

Raleigh JA (1996) Hypoxia and its clinical significance. In: Tepper JE, Raleigh JA (eds) Seminars in radiation oncology, vol 6. Saunders, Orlando, Fla, pp 1–70

Rampling R, Cruickshank G, Lewis AD, Fitzsimmons SA, Workman P (1994) Direct measurement of pO_2 distribution and bioreductive enzymes in human malignant brain tumours. Int J Radiat Oncol Biol Phys 29:427–432

Rasey JS, Koh WJ, Evans ML, et al. (1996) Quantifying regional hypoxia in human tumors with positron emission tomography of [^{18}F]fluoromisonidazole: A pretherapy study of 37 patients. Int J Radiat Oncol Biol Phys 36:417–428

Runkel S, Wischnik A, Teubner J, Kaven E, Gaa J, Melchert F (1994) Oxygenation of mammary tumors as evaluated by ultrasound-guided computerized-pO_2-histography. Adv Exp Med Biol 345:451–458

Saumweber DM, Kau RJ, Arnold W (1995) Tumor tissue oxygenation in primary squamous cell carcinomas of the head and neck – preliminary results. In: Vaupel PW, Kelleher DK, Günderoth M (eds) Tumor oxygenation. Fischer, Stuttgart, pp 313–318

Steinberg F, Röhrborn HJ, Scheufler KM, et al. (1997) NIR reflexion measurements of hemoglobin and cytochrome aa₃ in healthy tissue and neoplasms. Correlations to oxygen consumption: preclinical and clinical data. Adv Exp Med Biol (in press)

Stone HB, Brown JM, Phillips TL, Sutherland RM (1993) Oxygen in human tumors: correlations between methods of measurement and response to therapy. Radiat Res 136: 422–434

Strnad V, Keilholz L, Kirschner M, Meyer M, Sauer R (1997) Sauerstoffdruckverteilung in Lymphknotenmetastasen und die Veränderungen während akuter respiratorischer Hypoxie. Strahlenther Onkol 173:267–271

Terris DJ, Dunphy EP (1994) Oxygen tension measurements of head and neck cancers. Arch Otolaryngol Head Neck Surg 120:283–287

Urbach F (1956) Pathophysiology of malignancy. I. Tissue oxygen tension of benign and malignant tumors of the skin. Proc Soc Exp Biol Med 92:644–649

Urbach F, Noell WK (1958) Effects of oxygen breathing on tumor oxygen measured polarographically. J Appl Physiol 13:61–65

Urtasun RC, Parliament MB, McEwan AJ, et al. (1996) Measurement of hypoxia in human tumours by non-invasive SPECT imaging of iodoazomycin arabinoside. Br J Cancer 74:S209–S212

Valk PE, Mathis CA, Prados MD, Gilbert JC, Budinger TF (1992) Hypoxia in human gliomas: demonstration by PET with fluorine-18-fluoromisonidazole. J Nucl Med 33:2133–2137

Vaupel P (1990) Oxygenation of human tumors. Strahlenther Onkol 166:377–386

Vaupel P (1992) Physiological properties of malignant tumours. NMR Biomed 5:220–225

Vaupel PW (1993) Oxygenation of solid tumors. In: Teicher BA (ed) Drug resistance in oncology. Dekker, New York, pp 53–85

Vaupel PW (1994) Blood flow, oxygenation, tissue pH distribution, and bioenergetic status of tumors. Lecture 23, Ernst Schering Research Foundation, Berlin

Vaupel P (1996) Oxygen transport in tumors: characteristics and clinical implications. Adv Exp Med Biol 388:341–351

Vaupel P, Kallinowski F (1987) Tissue oxygenation of primary and xenotransplanted human tumours. In: Fielden EM, Fowler JF, Hendry JH, Scott D (eds) Radiation research, vol 2. Taylor and Francis, London, pp 707–712

Vaupel P, Kallinowski F, Okunieff P (1989) Blood flow, oxygen and nutrient supply, and metabolic microenvironment of human tumors: a review. Cancer Res 49:6449–6465

Vaupel P, Schlenger K, Höckel M (1991a) Blood flow and tissue oxygenation of human tumors. Funktionsanal Biol Syst 20:165–185

Vaupel P, Schlenger K, Knoop C, Höckel M (1991b) Oxygenation of human tumors: evaluation of tissue oxygen distribution in breast cancers by computerized pO_2 tension measurements. Cancer Res 51:3316–3322

Vaupel P, Schlenger K, Höckel M (1992) Blood flow and tissue oxygenation of human tumors: an update. Adv Exp Med Biol 317:139–151

Vaupel PW, Kelleher DK, Günderoth M (1995) Tumor oxygenation. Fischer, Stuttgart

Vaupel P, Thews O, Höckel M (1996) Tumor oxygenation: characterization and clinical implications. In: Smyth JF, Boogaerts MA, Ehmer BRM (eds) rhErythropoietin in cancer supportive treatment. Marcel Dekker, New York, pp 205–239

Vorndran B (1996) Die intratumorale Oxygenierung als neuer und unabhängiger Prognosefaktor beim lokal fortgeschrittenen und rezidivierenden Karzinom der Cervix uteri. Thesis, Medical Faculty University of Mainz, Germany

Wendling P, Manz R, Thews G, Vaupel P (1984) Inhomogeneous oxygenation of rectal carcinomas in humans. A critical parameter for preoperative irradiation? Adv Exp Med Biol 180:293–300

Wilson CBJH, Lammertsma AA, McKenzie CG, Sikora K, Jones T (1992) Measurements of blood flow and exchanging water space in breast tumors using positron emission tomography: a rapid and noninvasive dynamic method. Cancer Res 52:1592–1597

7 The Prognostic Significance of Hypoxia in Cervical Cancer: A Radiobiological or Tumor Biological Phenomenon?

M. Höckel[1] and P. Vaupel[2]

CONTENTS

7.1 Introduction

Since Warburg's first observations and claims on anaerobiosis in cancer cells, hypoxia in malignant tumors has gained the continuous attention of researchers throughout this century. Mainly radiotherapists and radiobiologists became interested in this subject after it was found that the absence of molecular oxygen significantly reduces cell killing by sparsely ionizing radiation. In 1955 THOMLISON and GRAY presented evidence for hypoxic microregions within human tumors following a detailed histopathologic study of lung cancers. To our knowledge, URBACH (1956) was the first to measure low oxygen tension in malignant human tumors directly with pO$_2$ polarography. Over the following decades, numerous investigations were performed in vitro and in vivo with animal and human tumors to elucidate the role of tumor hypoxia, especially with respect to radiotherapy (for a review see VAUPEL et al. 1989). According to our present understanding, tumor hypoxia is a direct consequence of structural abnormalities of the microvasculature and functional abnormalities of the microcirculation in ma-

lignant tumors and results from either limited O$_2$ diffusion ("chronic hypoxia") or limited perfusion ("acute hypoxia", transient hypoxia, ischemic hypoxia). Although unequivocal evidence for the existence of hypoxic microregions in experimental and human tumors has been gathered, the clinical relevance of tumor oxygenation is unclear.

GATENBY et al. (1988) reported the results from polarographic oxygen tension measurements in lymph node metastases of head and neck cancers, showing a significant relationship between low mean intratumoral pO$_2$ values and failure to respond to fractionated radiotherapy, thus supporting older studies such as Kolstad's on cancer of the uterine cervix (KOLSTAD 1968). These investigations, however, did not receive widespread attention because of the small number of patients involved and methodological limitations (for a discussion see HÖCKEL et al. 1991).

In 1989 a novel computerized histography system was introduced allowing quick and reliable polarographic intratumoral pO$_2$ readings in the clinical setting (FLECKENSTEIN and WEISS 1986; KALLINOWSKI et al. 1990; VAUPEL et al. 1991). Applying this device according to the concepts of *systematic random sampling* for stereological analysis (WEIBEL 1969; ELIAS et al. 1971) to take into account intratumoral heterogeneity, we initiated a controlled prospective trial in patients with locally advanced cancer of the uterine cervix to evaluate the clinical relevance of tumor oxygenation for this tumor entity. After having presented preliminary data from the first analysis in 1992 demonstrating the powerful predictive value of tumor oxygenation on recurrence-free and overall survival for patients treated with radiation (HÖCKEL et al. 1993a,b), we have recently reported the results from a larger patient cohort which included a subgroup of patients who underwent radical surgical treatment (HÖCKEL et al. 1996). We show that the adverse outcome of patients with hypoxic cervical tumors is independent of standard treatment and apparently represents a tumor biological phenomenon.

[1] M. HÖCKEL, Prof. Dr. med., Dr. rer. nat., Department of Obstetrics and Gynecology, University of Mainz, Langenbeckstrasse 1, D-55101 Mainz, Germany
[2] P. VAUPEL, Prof. Dr. med., Institute of Physiology and Pathophysiology, University of Mainz, Duesbergweg 6, D-55099 Mainz, Germany

Supported by a grant from the Deutsche Krebshilfe (70-1920-Va2)

7.2
Methods

7.2.1
Patients and Treatment

All patients with cervical cancers of at least 3 cm in diameter as determined by clinical and CT or MRI investigation who had been referred to the Department of Obstetrics and Gynecology, University of Mainz Medical Center, were eligible for the prospective study. The study design was approved by the ethics committee. Patients were accrued once informed consent had been obtained.

Clinical staging was performed according to the FIGO classification. Patients treated with curative intent received either primary surgery or radiation with or without adjunctive chemotherapy. Surgical treatment consisted of radical hysterectomy or primary pelvic exenteration with pelvic lymph node dissection. If metastases were found in intraoperative frozen sections of pelvic lymph nodes, periaortic lymphadenectomy was performed as well.

Definitive radiation was administered as combined teletherapy and brachytherapy. External beam irradiation was delivered using 10-MV photons produced by a linear accelerator in the Division of Radiation Oncology. For the brachytherapy, a high-dose-rate Ir 192 afterloading machine in the Department of Obstetrics and Gynecology was used. The patients usually received 45 Gy whole-pelvis external beam radiation by the standard four-field box technique in 2 Gy/day fractions five times per week. Radiation treatment was completed by three to four endocavitary insertions with a Hentschke applicator delivering a total dose of 24–28 Gy to point A at a dose rate of 0.5–1.0 Gy/min in weekly intervals. Lateral tumor extension was treated with additional teletherapy boosts to the parametria of up to 15 Gy. Postoperative adjuvant whole-pelvic external beam radiation was given to a treatment field to include the upper half of the vagina caudally. The cephalad margin was L4/5, and the lateral borders were 1.5 cm beyond the linea terminalis.

Adjuvant systemic chemotherapy regimens were used as follows: carboplatinum and ifosfamide as postoperative treatment; *cis*-platinum, vincristine, and bleomycin as induction therapy followed by definitive radiation; *cis*-platinum or carboplatinum for concomitant chemoradiotherapy.

The majority of patients in this study were seen at regular follow-up intervals in the department. In the case of tumor recurrence, the site(s) of relapse were identified by clinical and radiological work-up. Complete follow-up reports were obtained from the patients whose after-care examinations were not performed at the University of Mainz. No patient was lost to follow-up.

7.2.2
Surgicopathological Evaluation

The surgical specimens obtained by radical hysterectomy were systematically dissected, fixed in formalin, and processed for routine histological examination. Tumor diameters, depth of cervical invasion, and bladder or rectal involvement were verified microscopically. The presence of tumor cell clusters within an endothelium-lined space, as seen in the H&E stained sections, was regarded as *lymphatic space involvement*. This feature was reported to be pronounced if it could be detected in nearly all of the high power fields at the invading tumor front.

The demonstration of a tumor cell embolus within a blood vessel identified by the tunica media in the H&E stained sections was interpreted as evidence of *vascular invasion*. Parametrial tumor infiltration was investigated in the axial sections of the lateral parts of the surgical specimen. Four sections from each lymph node removed were checked for tumor metastases.

7.2.3
Intratumoral pO$_2$ Histography

Tumor oxygenation was measured pretherapeutically with the Eppendorf histograph system adhering strictly to the standard procedure as developed and validated earlier (Höckel et al. 1993a,b). Immediately after calibration of the ethylene-oxide-sterilized needle electrode, the conscious patient was placed in a defined lithotomy position. pO$_2$ readings were performed along linear tracks, first in the normal fatty tissue of the mons pubis, followed by measurements at the 12 o'clock and 6 o'clock sites of the central cervical tumor avoiding macroscopically necrotic areas. Twenty-five to 35 pO$_2$ measurements were taken on each tumor track (50–70 readings in total) starting at a tissue depth of 5 mm. The measuring points were placed 0.7 mm apart from each other, resulting in an overall measuring track length of approximately 2.0 to 2.5 cm. Via an on-line computing system, the pO$_2$ data of each track were displayed (i) as absolute values of oxygen partial pressure related to the location of the measuring point along the track and (ii) as relative frequencies within a pO_2 histogram ranging from 0 to 100 mmHg with a class width of 2.5 mmHg.

After the pO_2 measurements, cylindrical punch biopsies of 2 mm in diameter and 2 cm in length were taken from those tumor areas where pO_2 determination had been performed using a Biopty device (Radiplast, Uppsala, Sweden). The biopsies were fixed in buffered formalin and processed for histology to assure that the pO_2 measurements had been carried out in the tumor tissue and not in the cervical stroma. Concomitantly with the pO_2 determinations, intravaginal temperature, heart rate, arterial blood pressure, hemoglobin concentration, and hematocrit were monitored. The pO_2 measurements were usually performed 1–5 days prior to oncologic treatment.

7.2.4
Data Analysis

The oxygenation status of each tumor was represented by the *median pO_2* and the *low pO_2 fraction* (corresponding to the relative frequency of pO_2 readings at 0–2.5 mmHg or 0–5 mmHg) derived from the pooled histograms.

Endpoints of the study were recurrence-free survival and overall survival. For comparisons the following statistical methods were applied: determination of correlation coefficients, Fisher's exact test, and Mann-Whitney-Wilcoxon test (U test). Survival and recurrence-free survival were calculated with the Kaplan-Meier life table method. Differences between survival curves were analyzed with the log-rank test. Variables influencing survival and recurrence-free survival were evaluated with the univariate and multivariate Cox proportional hazards model.

7.3
Results

From June of 1989 until June of 1995, 103 patients with advanced cancers of the uterine cervix, FIGO stages Ib bulky ($n = 13$), IIa,b ($n = 51$), IIIa,b ($n = 34$), and IVa,b ($n = 5$), entered the study. Fifty percent of the tumors had median $pO_2 < 10$ mmHg designated as *hypoxic tumors*; in 78 patients pO_2 readings ≤ 5 mmHg and in 57 patients pO_2 readings

≤ 2.5 mmHg were recorded. No correlation could be detected between age, menopausal status, parity of the patients, and the oxygenation pattern of their tumors. Likewise, no influence of clinical tumor stage and tumor size, histologic type, and differentiation on tumor oxygenation has been found. However, based on the histopathologic evaluation of the surgical specimens of 47 patients treated with radical hysterectomy or exenteration and complete pelvic lymph node dissection, *hypoxic tumors* more frequently exhibited pronounced *lymphatic space involvement* ($p = 0.03$) and (occult) parametrial extension ($p = 0.03$).

Eighty-nine out of 103 patients in whom intratumoral pO_2 measurements had been carried out were treated with curative intent. Forty-two patients received radiation therapy with or without preceding or concomitant chemotherapy, and 47 pa-

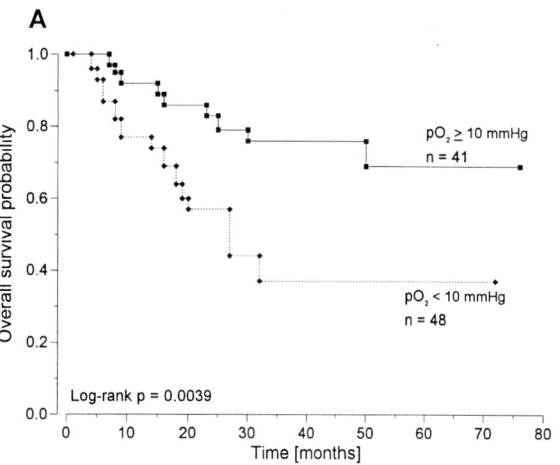

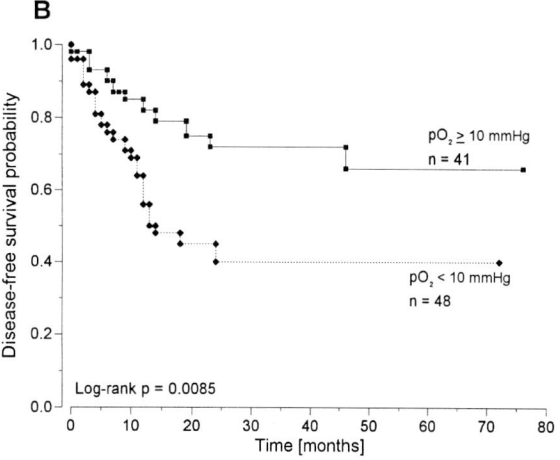

Fig. 7.1A,B. Overall (**A**) and disease-free (**B**) survival probabilities calculated with the Kaplan-Meier method for 89 patients treated with curative intent stratified for tumor oxygenation. (From HÖCKEL et al. 1996)

tients underwent radical surgery with or without induction or adjuvant chemotherapy. The entire group of treated patients and the radiation and surgery subgroups were analyzed for outcome. Within a median follow-up of 28 months (range 3–76 months), 35 patients overall, 15 of them in the radiotherapy group and 20 in the surgery group, progressed or relapsed. All but six patients failed locoregionally with or without simultaneous distant metastases. Patients with *hypoxic tumors* (median pO$_2$ < 10 mmHg) had a significantly worse disease-free and overall survival probabilities compared to better oxygenated tumors (pO$_2 \geq$ 10 mmHg; Fig. 7.1). Similar survival differences were found in the radiotherapy as well as surgery subgroups (Figs. 7.2, 7.3).

Cox regression analysis revealed tumor oxygenation as the strongest independent prognostic factor in the whole group of treated patients followed by

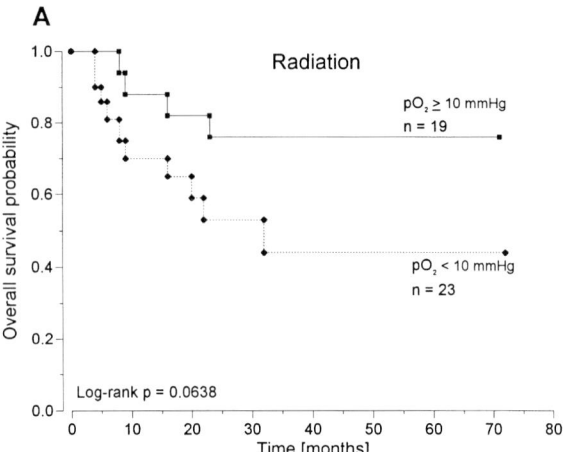

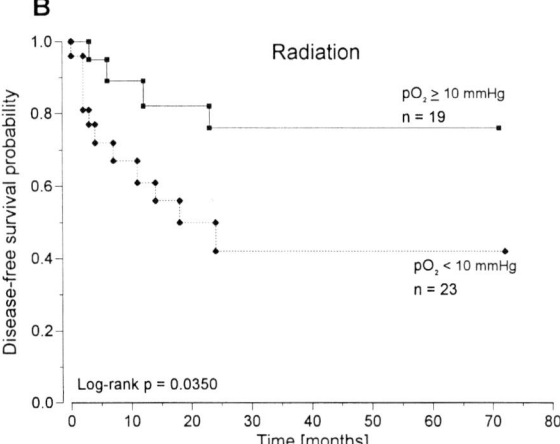

Fig. 7.3A,B. Overall (**A**) and disease-free (**B**) survival probabilities calculated with the Kaplan-Meier method for 42 patients treated with primary radiation stratified for tumor oxygenation. (From HÖCKEL et al. 1996)

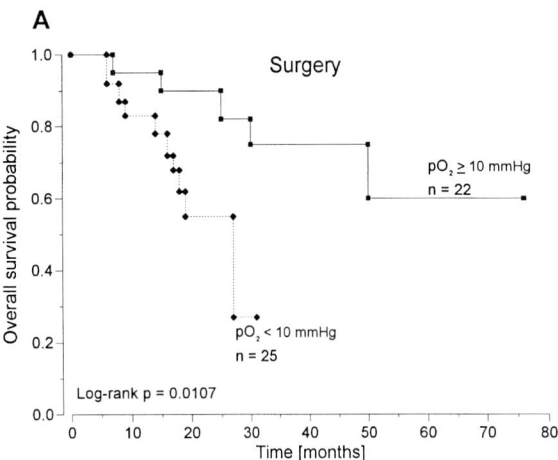

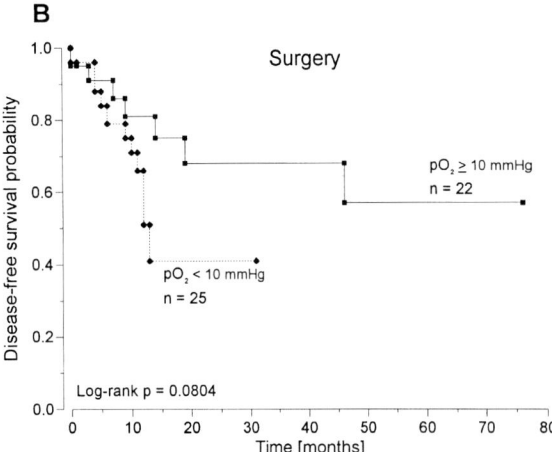

Fig. 7.2A,B. Overall (**A**) and disease-free (**B**) survival probabilities calculated with the Kaplan-Meier method for 47 patients treated with primary surgery stratified for tumor oxygenation. (From HÖCKEL et al. 1996)

FIGO stage. The potential of tumor oxygenation in predicting clinical aggressiveness in cervical cancer increased with tumor size: the difference in outcome between hypoxic and well-oxygenated tumors was significant in patients with tumors of more than 5 cm in clinical size. It did not reach statistical significance for smaller tumors (Fig. 7.4).

7.4
Discussion

The study of tumor oxygenation in more than 100 patients with locally advanced cervical cancer confirmed our earlier findings that this parameter is independent of clinical stage, clinical tumor size, and histological type and grading. However, contrary to

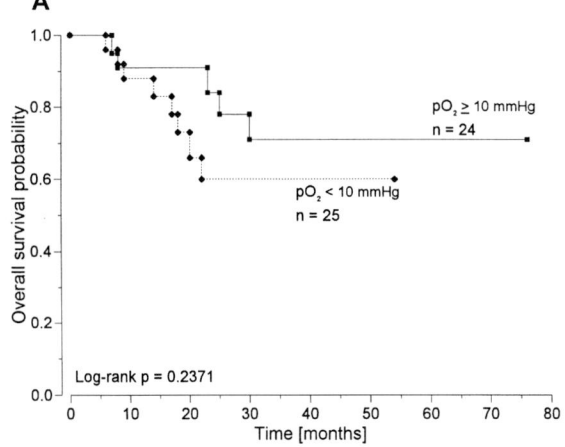

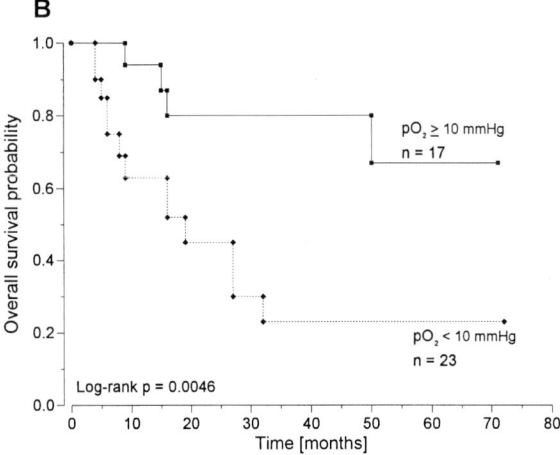

Fig. 7.4A,B. Overall survival probabilities calculated with the Kaplan-Meier method for 49 patients with cervical cancer up to 5 cm in maximum clinical diameter (**A**) and for 40 patients with tumors of more than 5 cm in diameter (**B**) stratified for tumor oxygenation. (From HöCKEL et al. 1996)

the pretherapeutic clinical findings, histopathological evaluation of the complete surgical specimens from 47 patients who underwent radical surgery revealed significantly more frequent (occult) parametrial infiltration as well as pronounced lymphatic space involvement at the tumor front in hypoxic cervical cancers as compared to the better oxygenated tumors. Clearly, tumor oxygenation measurements identified histologically more aggressive tumor phenotypes which were not adequately staged by clinical evaluation.

Five-year overall as well as disease-free survival probabilities calculated for 89 patients who underwent standard primary treatment for cure were significantly lower for hypoxic tumors than for well-oxygenated tumors of similar clinical stages and

sizes. Multivariate Cox regression analysis revealed tumor oxygenation as the most powerful pretreatment prognosticator in our study cohort. Most interestingly, the disadvantage in outcome for hypoxic tumors was independent of the mode of primary treatment – radiation or radical surgery.

Why do hypoxic cervical cancers have a poorer outcome? Since the majority of the patients who were not disease-free relapsed in the pelvis with or without simultaneous distant metastases, radioresistance and tumor spread beyond the resected tissue must be considered as primary causes of failure. One explanation could be that tumor hypoxia is a mere marker of the malignant phenotype. Alternatively, hypoxia *per se* may have an impact on the clinical aggressiveness of solid malignancies. This view is strongly supported by several mechanisms established from in vitro studies and animal tumor models:

(i) The radiosensitizing effect of molecular oxygen for photon irradiation should be diminished in low pO_2 tumors which may contain significant levels of radiobiologically hypoxic fractions (GRAY et al. 1953; HORSMAN et at. 1993; NORDSMARK et al. 1995). However, direct measurements of tumor oxygenation may not be representative of the radiobiological hypoxic fraction if a substantial proportion of cells within a tumor is nonclonogenic (FENTON et al. 1995). Likewise, the impact of pretreatment tumor hypoxia on *fractionated* radiotherapy with claimed reoxygenation is unknown at present (ZYWIETZ et al. 1995).

(ii) When starved in a hypoxic microenvironment, tumor cells, like normal cells, can respond with the expression of a variety of genes coding for *oxygen-regulated proteins* (HEACOCK and SUTHERLAND 1986; SUTHERLAND et al. 1996). Several potential mechanisms have been suggested to explain how some of these oxygen-regulated proteins, including p53 and vascular endothelial growth factor, influence, tumor aggressiveness in terms of malignant progression and decreased responsiveness to therapy (TAKAHASHI et al 1995; GIACCIA 1996; GRAEBER et al. 1996).

(iii) Finally, there is strong evidence that hypoxia enhances genetic instability and heterogeneity, leading to the evolution of metastatic and intrinsically resistant tumor cell variants (RICE and SCHIMKE 1986; YOUNG et al. 1988; RUSSO et al. 1995). Hypoxia-reoxygenation of tumor cells might be regarded as microregional counter-

Table 7.1. New treatment concepts for bulky hypoxic cervical cancers

Surgery	Extended radical hysterectomy Complete pelvic and periaortic lymph node dissection Reconstruction of ureteral, bladder, rectal functions
Radiation	In combination with: Hypoxic modification (e.g., hyperbaric oxygen, carbogen, nicotinamide, nitric oxide, hypoxic radiosensitizers) Hypoxia-specific cytotoxins (e.g., tirapazamine) Hyperthermia

parts of the general ischemia-reperfusion syndrome with the production of oxygen-derived free radicals (e.g., by the xanthine oxidase pathway) as the key event (MCCORD 1985). Hydroxyl radicals, thought to be causative for tumor-initiating DNA mutations in various solid neoplasms, might also play an essential role in malignant tumor progression (MALINS et al. 1993). Whether these mechanisms are relevant for the association observed between tumor oxygenation and therapeutic outcome in cancer of the uterine cervix remains to be established. Our observation that the hypoxia effect is much more pronounced in larger cervical cancers than in smaller tumors may support the hypothesis that the presence of a low pO_2 microenvironment indeed contributes to malignant progression instead of merely being a marker of biological aggressiveness.

Based on the results of this study and the putative underlying mechanisms of the hypoxia effect in solid neoplasms, new treatment concepts for surgical as well as for radiation therapy could improve the outcome of bulky hypoxic cervical cancers and should be tested in clinical trials (Table 7.1).

Although none of these approaches is active in preventing distant metastasis by direct systemic action, a great body of evidence suggests a major breakthrough in long-term survival if locoregional control could be improved in advanced cervical cancer (SUIT and WESTGATE, 1986; PONTÉN et al. 1995).

References

Elias H, Henning A, Schwartz DE (1971) Stereology: applications to biomedical research. Histol Rev 51:158–200

Fenton BM, Mohammad FK, Siemann DW (1995) Should direct measurements of tumor oxygenation relate to the radiobiological hypoxic fraction of a tumor? Int J Radiat Oncol Biol Phys 33:365–373

Fleckenstein W, Weiss C (1986) Local tissue pO_2 measured with "thick" needle probes. Funktionsanal Biol Syst 15:155–166

Gatenby RA, Kessler HB, Rosenblum JS, Coia LR, Moldofsky PJ, Hartz WH, Broder GJ (1988) Oxygen distribution in squamous cell carcinoma metastases and its relationship to outcome of radiation therapy. Int J Radiat Oncol Biol Phys 14:831–838

Giaccia AJ (1996) Hypoxic stress proteins: survival of the fittest. Semin Radiat Oncol 6:46–58

Graeber TG, Osmaninan C, Jacks T, Housman DE, Koch CK, Lowe SW, Giaccia AJ (1996) Hypoxia-mediated selection of cells with diminished apoptic potential in solid tumors. Nature 379:88–91

Gray LH, Conger AD, Ebert M, Hornsey S, Scott OCA (1953) The concentration of oxygen dissolved in tissues at the time of irradiation as a factor in radiotherapy. Br J Radiol 26:638–648

Heacock CS, Sutherland RM (1986) Induction characteristics of oxygen regulated proteins. Int J Radiat Oncol Biol Phys 12:1287–1290

Höckel M, Schlenger K, Knoop C, Vaupel P (1991) Oxygenation of carcinomas of the uterine cervix: evaluation by computerized O_2 tension measurements. Cancer Res 51:6098–6102

Höckel M, Knoop C, Schlenger K, et al. (1993a) Intratumoral pO_2 predicts survival in advanced cancer of the uterine cervix. Radiother Oncol 26:45–50

Höckel M, Vorndran B, Schlenger K, Baussmann E, Knapstein PG (1993b) Tumor oxygenation: a new predictive parameter in locally advanced cancer of the uterine cervix. Gynecol Oncol 51:141–149

Höckel M, Schlenger K, Aral B, Mitze M, Schäffer U, Vaupel P (1996) Association between tumor hypoxia and malignant progression in advanced cancer of the uterine cervix. Cancer Res 56:4509–4515

Horsman MR, Khalil A, Nordsmark M, Grau C, Overgaard J (1993) Relationship between radiobiological hypoxia and direct estimates of tumor oxygenation in a mouse tumor model. Radiother Oncol 28:69–71

Kallinowski F, Zander R, Höckel M, Vaupel P (1990) Tumor tissue oxygenation as evaluated by computerized-pO_2-histography. Int J Radiat Oncol Biol Phys 19:953–961

Kolstad P (1968) Intercapillary distance, oxygen tension, and local recurrence in cervix cancer. Scand J Clin Lab Invest 106:145–157

Malins DC, Holmes EH, Polissar NL, Gunselman SJ (1993) The etiology of breast cancer: characteristic alterations in hydroxyl radical-induced DNA base lesions during oncogenesis with potential for evaluating incidence risk. Cancer 71:3036–3043

McCord JM (1985) Oxygen-derived free radicals in postischemic tissue injury. N Engl J Med 312:159

Nordsmark M, Grau C, Horsman MR, Jörgensen HS, Overgaard J (1995) Relationship between tumour oxygenation, bioenergetic status and radiobiological hypoxia in an experimental model. Acta Oncol 34:329–334

Pontén J, Adami H-O, Bergström R, et al. (1995) Strategies for global control of cervical cancer. Int J Cancer 60:1–26

Rice GC, Hoy C, Schimke RT (1986) Transient hypoxia enhances the frequency of dihydrofolate reductase gene amplification in chinese hamster ovary cells. Proc Natl Acad Sci USA 83:5978–5982

Russo CA, Weber TK, Volpe CM, et al. (1995) An anoxia inducible endonuclease and enhanced DNA breakage as contributors to genomic instability in cancer. Cancer Res 55:1122–1128

Suit HD, Westgate SJ (1986) Impact of improved local control on survival. Int J Radiat Oncol Biol Phys 12:453–458

Sutherland RM, Ausserer W, Murphy B, Laderoute K (1996) Tumor hypoxia and heterogeneity: challenges and opportunities for the future. Semin Radiat Oncol 6:59–70

Takahashi Y, Kitadai Y, Bucana CD, Cleary KR, Ellis LM (1995) Expression of vascular endothelial growth factor and its receptor, KDR, correlates with vascularity, metastasis, and proliferation of human colon cancer. Cancer Res 55:3964–3968

Thomlinson RH, Gray LH (1955) The histologic structure of some human lung cancers and the possible implications for radiotherapy. Br J Cancer 9:537–549

Urbach F (1956) Pathophysiology of malignancy. I. Tissue oxygenation of benign and malignant tumors of the skin. Proc Soc Exp Biol Med 92:644–649

Vaupel P, Kallinowski F, Okunieff P (1989) Blood flow, oxygen and nutrient supply, and metabolic microenvironment of human tumors: a review. Cancer Res 49:6449–6465

Vaupel P, Schlenger K, Knoop C, Höckel M (1991) Oxygenation of human tumors: evaluation of tissue oxygen distribution in breast cancers by computerized O_2 tension measurements. Cancer Res 51:3316–3322

Weibel ER (1969) Stereological principles for morphometry in electron microscopic cytology. Int Rev Cytol 26:235–302

Young SD, Marshall RS, Hill RP (1988) Hypoxia induces DNA overreplication and enhances metastatic potential of murine tumor cells. Proc Natl Acad Sci USA 85:9533–9537

Zywietz F, Reeker W, Kochs E (1995) Tumor oxygenation in a transplanted rat rhabdomyosarcoma during fractionated irradiation. Int J Radiat Oncol Biol Phys 32:1391–1400

8 Changes in Tumor Oxygenation During Radiation Therapy

M. Molls, H.J. Feldmann, P. Stadler, and R. Jund

CONTENTS

8.1
Introduction

It is well known that radiobiologically hypoxic cells are present in rodent and xenografted human tumors (Moulder and Rockwell 1984; Suit et al. 1990; Guichard et al. 1983). Hypoxia is regarded as an important factor for radioresistance; two to three times higher radiation doses are needed to kill hypoxic cells than well-oxygenated cells (Gray et al. 1953; Evans and Naylor 1963). Thus the presence of hypoxic clonogenic cells could be a major factor influencing the response of tumors to radiation therapy and to some cytotoxic drugs (Sakata et al. 1991; Dische 1985; Overgaard 1989).

Because of the potential role of tumor oxygenation for the effectiveness of radiation therapy and also chemotherapy, there has been interest in the measurement of oxygen supply to human tumors since the 1950s, when significant hypoxia in human tumor tissue was demonstrated using polarographic techniques (Cater and Silver 1960; Evans and Naylor 1963). However, the technical limitations of the equipment used (e.g. compression effects, large needle diameter) made the results of these studies difficult to interpret (Cater and Silver 1960; Kolstad 1968). Therefore, since valid methods for the routine measurement of intratumoral tensions in patients are so far lacking (Vaupel et al. 1989;

Vaupel 1995), the clinical importance of tumor hypoxia has remained uncertain. During the past few years a clinically applicable standardized procedure has been established. It allows the determination of intratumoral O_2 tensions in primary tumors and metastatic lesions of patients using the Eppendorf computerized histograph, which causes minimal damage to the evaluated tissue (Schramm et al. 1992).

At present, investigations are focused on the comparison of the oxygenation pattern in normal tissue, benign and malignant tumors (Fleckenstein et al. 1990; Füller et al. 1994; Höckel et al. 1991; Lartigau et al. 1993; Saumweber et al. 1995; Vaupel et al. 1991, 1995). Furthermore, the evaluation of the biological and clinical significance of low pO_2 values in human tumors as a new predictive parameter indicating treatment outcome and survival has become a research topic of high interest (Gatenby et al. 1988; Höckel et al. 1993; Nordsmark et al. 1996).

With regard to follow-up measurements, there are only clinical case reports about the changes of the oxygenation status of tumors during radiotherapy alone, thermoradiotherapy, or radiochemotherapy in a variety of entities (Feldmann et al. 1994; Füller et al. 1994; Molls et al. 1995; Jund et al. 1996; Lartigau et al. 1995). Therefore, interpretation of the results is difficult. A well-designed experimental study on the influence of fractionated radiotherapy in rat rhabdomyosarcoma yielded interesting results (Zywietz et al. 1995). This chapter briefly reviews the published data. In addition, we describe the course of oxygenation status in primary tumors and metastatic neck lymph nodes of advanced head and neck carcinomas. The pO_2 was measured in 31 patients who were treated according to a well-defined radiochemotherapy schedule of our institution.

M. Molls, Prof. Dr. med., Klinik und Poliklinik für Strahlentherapie und Radiologische Onkologie, Klinikum rechts der Isar der Technischen Universität München, Ismaninger Strasse 22, D-81675 München, Germany

8.2
Methodological Problems

A reliable electrode system has recently become available for the measurement of tumor oxygenation in patients under clinical conditions (Sigma pO_2 histograph). This system uses readily available, mechanically stable needle electrodes. The succession of forward and backward motions of the electrode through the tissue minimizes the problem of compression artifacts. Since the electrode tip moves to a new tissue location for each measurement, an erroneous elevation of tissue O_2 tensions due to hemorrhages at the electrode tip is avoided. More technical details concerning the pO_2 measurements are given in Chap. 6.

Several groups have critically evaluated the electrode performance in vivo. Due to intratumor heterogeneity a certain number of measurements in different electrode tracks through a tumor are required in order to obtain sufficient information on the oxygenation status of an individual tumor. HÖCKEL et al. (1991) compared the pO_2 histograms obtained from two standard electrode tracks with those evaluated from multiple tracks. No significant differences were found when more than 40 pO_2 values were recorded per track. For the lowest pO_2 class (0–2.5 mmHg), there were no differences when multiple and single electrode tracks were compared. Additionally, similiar pO_2 histograms were usually obtained when the same tumor region was measured repeatedly during a 1-day observation period using different O_2-sensitive electrodes (HÖCKEL et al. 1991).

MOLLS et al. (1994) performed two measurements within 24 h in each of eight patients with different tumors before commencing radiotherapy. Pooled data of the first and second measurements demonstrated that there is good congruence between the two measurements. Although differences in the mean pO_2 values could be observed in seven of the eight patients, the differences in the pO_2 readings below 2.5 mmHg were minimal and statistically not significant.

NORDSMARK et al. (1995) investigated in lymph node metastases of head and neck cancer whether O_2 electrode measurements were able to resolve significant tumor-to-tumor variability in oxygenation status. Their analysis of variance clearly showed that it is possible to discriminate intertumor heterogeneity from intratumor heterogeneity using polarographic electrodes. However, the analysis also demonstrated that conclusions should not be based on a single

electrode track due to substantial variability between tracks.

8.3
Follow-up Measurements

There are only a few reports in the literature describing pO_2 measurements in irradiated experimental and human tumors. An improvement in tumor tissue oxygenation in a mouse mammary carcinoma 72–74 h after a single dose of 60 Gy, resulting in a marked reduction of pO_2 values below 2.5 mmHg, was noted by VAUPEL et al. (1984). An increase in the tissue oxygenation 3–4 days after irradiation of mouse mammary carcinomas with single doses of 32 and 65 Gy and a decrease in the frequency of pO_2 values below 2.5 mmHg was observed by KOUTCHER et al. (1992). These studies involved single large radiation doses.

ZYWIETZ et al. (1995) investigated rhabdomyosarcomas (R1H) of the rat with fractionated radiotherapy; a total dose of 60 Gy was given in 20 fractions. Tumor oxygenation did not change significantly during the first 3 weeks of irradiation (up to 45 Gy). After the 4th week of treatment, and following a total dose of 60 Gy, a significant decrease in tumor oxygenation was observed. The median pO_2 was 8 ± 2 mmHg, and 35% of the values recorded were between 0 and 5 mmHg. Thus, with increasing radiation dose (60 Gy) a significant decrease in tumor oxygenation was found.

In addition, ZYWIETZ et al. (1997) investigated changes in R1H tumor oxygenation during a fractionated irradiation in combination with local hyperthermia. In general, the pO_2 values decreased continuously from the start of the combined treatment with increasing radiation dose and number of heat fractions. These findings suggest that radiotherapy with adjuvant hyperthermia induces greater changes in tumor oxygenation than radiation alone.

With regard to the clinical situation, there are only case reports about changes in tissue oxygenation during radiotherapy or thermoradiotherapy (BADIB and WEBSTER 1969; BERGSJÖ and EVANS 1971; FELDMANN et al. 1994; FLECKENSTEIN et al. 1990; FÜLLER et al. 1994; MOLLS et al. 1995; JUND et al. 1996). Quantitative pO_2 measurements during conventional radiotherapy were performed for the first time in various tumors by BADIB and WEBSTER (1969). At weekly intervals, a progressive increase in tumor oxygenation during irradiation was observed which reached its maximum at the end of radiation.

FLECKENSTEIN et al. (1990) repeatedly measured pO_2 in a metastasis of a squamous cell carcinoma during the course of radiation therapy. Two weeks after the start of therapy, the mean pO_2 value within the entire tumor mass was lower than before treatment. After several weeks of treatment (70 Gy), the mean pO_2 was slightly increased and locally steep pO_2 gradients in the center and periphery of the tumor were observed.

FELDMANN et al. (1994) investigated changes in oxygenation patterns of recurrent tumors during thermoradiotherapy. Most patients suffered from chest wall recurrences of breast cancer and lymph node metastases of head and neck cancer. In 9 of 10 patients a significant decrease in the mean pO_2 was observed 24h after the first hyperthermic treatment. The observed decrease might have been the result of increased interstitial pressure due to hyperthermia-induced edema.

FÜLLER et al. (1994) performed follow-up measurements in 14 patients with different tumors (recurrent rectal cancer, soft tissue sarcoma, breast cancer, melanoma) during the course of radiotherapy or thermoradiotherapy. Most patients suffered from recurrent disease. A total of two to six measurements could be performed. In 4 of the 14 patients there was a decrease in median pO_2, in 4 no change, and in 6 an increase was observed between the first and the last measurement. In 7 patients there was an increase, in 6 no change, and in 1 a decrease of the hypoxic fraction (% values ≤5 mmHg) (Fig. 8.1). Due to the great heterogeneity of the treated tumors and the different treatment schedules and total radiation doses, no firm conclusions can be drawn from these results. In addition, in 50% of cases the last measurement was performed in the middle of the radiation course and not at the end.

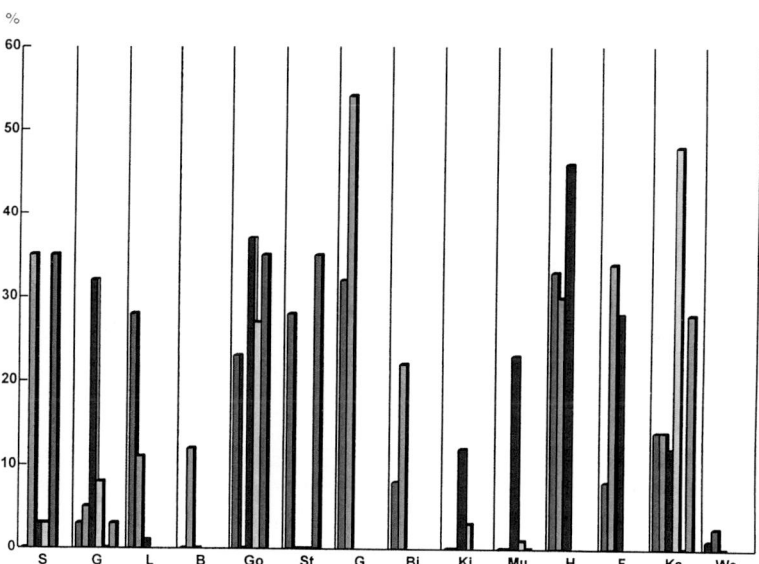

Patient	Tumor	Initial hypoxic fraction (%)
S	Melanoma	0
G	Chordoma	3
L	Rectal carcinoma-relapse	28
B	Neck node-relapse	0
Go	Rectal carcinoma-relapse	23
St	Rectal carcinoma-relapse	28
G	Rectal carcinoma	32
Bi	Neck node	8
Ki	Breast carcinoma-relapse	0
Mu	Breast carcinoma-relapse	0
H	Breast carcinoma-relapse	33
F	Sarcoma	8
Ks	Chondrosarcoma	14
W	Breast carcinoma-relapse	1

Fig. 8.1. Hypoxic fraction (percentage of pO_2 values ≤5 mmHg) in 14 human tumors. Measurements during thermoradiotherapy

Measurements have also been performed during the definitive treatment of cervical cancer combining percutaneous irradiation and brachytherapy. The patients ($n = 6$) were investigated at the beginning of the treatment and at the time of brachytherapy after external radiotherapy. A decrease of the median pO_2 from 21 to 15 mmHg was observed under these conditions by LARTIGAU et al. (1992).

In recent studies we determined the pO_2 status during split-course radiochemotherapy. In 31 patients (age 40–92 years, mean 58 years) with locally advanced head and neck cancer (T2–4 N1–3 M0) a total number of 25 metastatic neck nodes and 6 primaries were investigated pretherapeutically and during a defined course of radiochemotherapy. The tumors were classified according to the TNM staging system. All patients underwent primary radiochemotherapy (AUBERGER et al. 1995). The patients usually received 50 Gy in a standard three-field technique in fractions of 2 Gy/day, five fractions per week, followed by an additional 20 Gy in the same fractionation to the primary tumor including involved nodes by shrinking field technique. The treat-

ment schedule was a split-course treatment with a 2-week break after 30 Gy. In week 1 and week 6 (first week after break) the patients received chemotherapy: 10 mg/m² mitomycin C on day 1 of the course as bolus injection and 750–1000 mg/m² 5-fluorouracil on days 1–5 of the course as continuous infusion. The times for the follow-up measurements of the pO_2 are given in Table 8.1.

Figure 8.2 demonstrates changes in tumor oxygenation (median pO_2) during split-course radiochemotherapy. The reduction of the number of tumors investigated during therapy is due to the fact that some patients refused further measurements. In addition, at the end of the treatment shrinkage of the tumors came into play; nodes with a volume smaller than 0.7 cm³ could not be investigated.

There was a slight reduction in median pO_2 from 18.3 to 14.4 mmHg after the first cycle of radiochemotherapy (30 Gy). A clear increase to 23 mmHg was observed at the end of the 2-week break (30 Gy), whereas a further decrease in median pO_2 was evident at the end of the combined treatment (70 Gy). Further analysis of the pO_2 distribution showed that during the first 3 weeks of treatment (30 Gy) only a slow increase of the frequencies of pO_2 values below 5 mmHg from 31% to 32.3% could be observed, whereas a marked decrease was evident after the pause (24.3%). At the end of the treatment (70 Gy) a strong increase of the hypoxic fraction to 46% was found (Fig. 8.3).

In general, tumors undergoing radiochemotherapy shrank (Fig. 8.4). This was ascertained by ultrasonography. The investigations were performed at the same times as the pO_2 measurements.

Table 8.1. Times of pO_2 measurement

Measurement	Time
1	Before commencement of therapy
2	At end of first course of radiochemotherapy (30 Gy)
3	At end of 2-week break
4	At end of second course of radiochemotherapy (70 Gy)

median pO₂ (mm Hg)

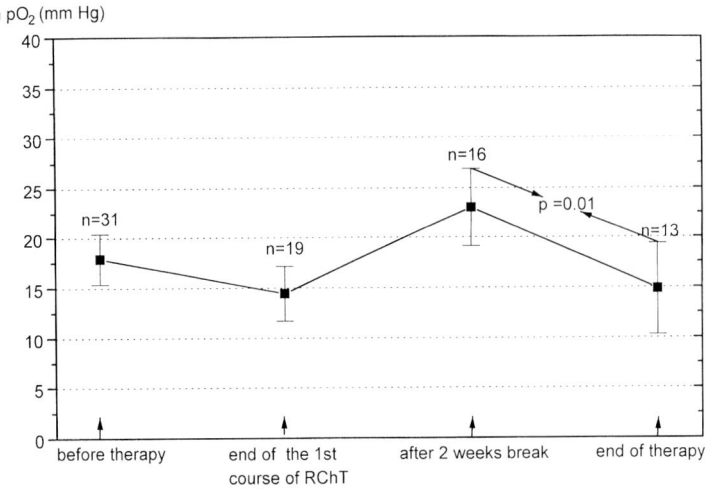

Fig. 8.2. Median O_2 tension (± standard error) in head and neck carcinomas during radiochemotherapy (split course). Terminal decrease is significant (Wilcoxon test $p = 0.01$)

Fig. 8.3. Hypoxic fraction (percentage of pO$_2$ values ≤5 mmHg ± standard error) in head and neck carcinomas during radio-chemotherapy (split course)

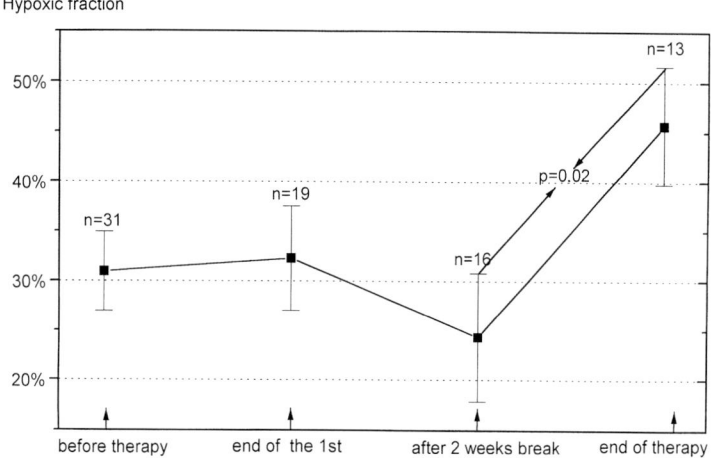

Fig. 8.4. Reduction in tumor volume of neck nodes (*n* = 10) during radiochemotherapy (measurements by ultrasonography; mean and standard error)

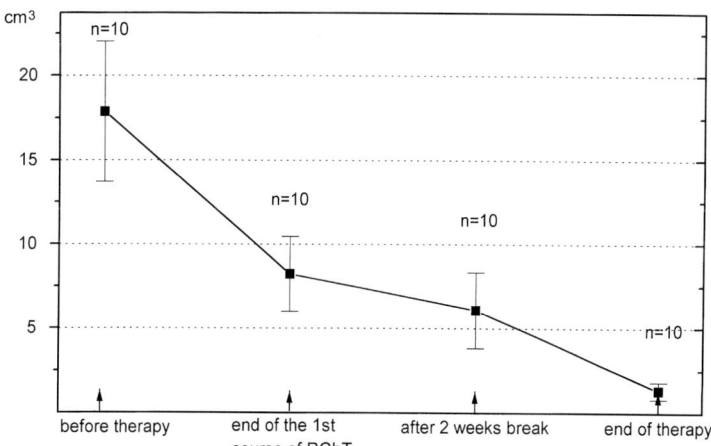

8.4
Conclusions and Future Directions

The published data show no clear direction in changes of tumor oxygenation during fractionated radiotherapy or thermoradiotherapy. Increase, decrease and no change in the oxygen supply are all reported. Only a trend to a lower median pO$_2$ or a higher hypoxic fraction at the end of a treatment course is evident. A significant decrease in median pO$_2$ could be observed under extreme conditions (e.g. 24 h after a cytotoxic hyperthermic treatment).

However, our data on changes of oxygenation in neck nodes and primaries in patients with locally advanced head and neck tumors during radiochemotherapy are in agreement with the findings of

experimental studies. ZYWIETZ et al. (1995) treated rhabdomyosarcomas in rats with clinically relevant doses and fractionation. Our own data and those of ZYWIETZ et al. (1995) suggest that oxygenation status is dose-dependent. After radiation doses in the range of 45–70 Gy the number of hypoxic tumor sites increased considerably. Despite possible differences in tumor biology, especially in vascularization, between human and transplanted rodent tumors, this result is interesting.

Unfortunately, the radiobiological significance of this phenomenon remains unclear. The following questions arise: Does the decrease in tumor oxygenation after comparatively high radiation doses reflect exclusively the radiation-induced necrosis of the tumor tissue? Does hypoxia during the late treatment period indicate that hypoxic clonogenic cells are still interspersed among living tumor cells? If the latter were true this would have consequences for the treatment design. Thus, the application of modalities such as additional hyperthermia, additional cytostatic drugs that are highly efficient under hypoxia and neutrons could be useful in this late phase of the treatment. It has also been pointed out that the presence of hypoxic cells might be exploited to the advantage of the cancer patient by combining radiotherapy with newly developed hypoxic cytotoxins (BROWN and GIACCIA 1994).

A further interesting observation in the clinical split-course radiochemotherapy was the trend towards better tumor oxygenation after the 2-week break than before the break. Again, no definitive interpretation of this finding is possible. However, if there were reoxygenation in the true radiobiological sense, then after the break the tumors should have become more sensitive towards radiation and also chemotherapy. As our clinical results (AUBERGER et al. 1995) are comparable in terms of local tumor control and patient survival with those obtained after non-split-course schedules, one can speculate that the advantage of reoxygenation after the break has counteracted the disadvantage of assumed repopulation during the break.

Regarding the oxygenation status during therapy, one of the most important problems which should be investigated in more detail is the question to what degree the intended effect of radiation-induced tumor necrosis causes low pO_2 readings in pO_2 measurements. Therefore, in future we should try to characterize the tumor morphology such that we know more accurately whether the pO_2 measurements are performed in necrotic or viable tumor areas. The tools by which necrotic areas can be detected macroscopically and their volume characterized to a certain extent are ultrasonography and computer tomography. A microscopic method would be the determination of necrotic zones in histological sections. The sections can be performed in those tumor regions in which the tracks of the pO_2 electrodes are found. We are aware that such approaches are relatively imprecise. However, we think it is worthwhile to start elucidating this problem.

With regard to future research activities, we would like to encourage radiooncologists to start investigations on the tumor microenvironment in patients. We need much more clinical information to help us arrive at a comprehensive understanding of the true significance of factors such as pH, blood supply, and pO_2 for radiooncological treatment outcome. This understanding is not only important in the sense of classical radiobiology; we also need it in the context of new research directions in radiobiology, namely signal transduction, gene regulation, activities of gene products, etc. It is evident that these molecular and cell biological processes do not act independently. The whole system of the tumor, including its physiological milieu, represents a structure in which there are interdependences between the different levels of biological organization, e.g. molecular, cellular, histomorphological, and physiological. Recent experimental findings favor this view of the malignant tumor as a highly complex biological entity in which systemic conditions may influence important molecular functions. Thus, it has been pointed out that a normal p53 protein may function abnormally in a disturbed microenvironment (HAINAUT and MILNER 1993). Activators of protein kinase C mediated cellular cytotoxicity to hypoxic cells but not normoxic cells (KOONG et al. 1994). According to GRAEBER et al. (1996), hypoxia can dramatically alter cellular phenotype and may be important in tumor progression.

References

Auberger T, Burzin M, Molls M, Hölzer-Müller L, Kau RJ, Herzog M, Hanauske AR (1995) Concomitant radiochemotherapy with mitomycin C and 5-FU in advanced head and neck cancer. In: Kogelnik HD (ed) Progress in radio-oncology V. Monduzzi Bologna, pp 67–72

Badib AO, Webster JH (1969) Changes in tumor oxygen tension during radiation therapy. Acta Radiol 8:247–257

Bergsjö P, Evans JC (1971) Oxygen tension of cervical carcinoma during the early phase of external irradiation. Scand J Clin Lab Invest 27:71–82

Brown JM, Giaccia AJ (1994) Tumor hypoxia: the picture has changed in the 1990s. Int J Radiat Biol 65:95–102

Cater DB, Silver IA (1960) Quantitative measurements of oxygen tension in normal tissue and in the tumors of patients before and after radiotherapy. Acat Radiol 53:233–256

Dische S (1985) Chemical sensitizers for hypoxic cells: a decade of experience in clinical radiotherapy. Radiother Oncol 3:97–115

Evans NTS, Naylor PFD (1963) The effect of oxygen breathing and radiotherapy upon tissue oxygen tension of some human tumors. Br J Radiol 36:418–425

Feldmann HJ, Molls M, Füller J, Stuben G, Sack H (1994) Changes in oxygenation patterns of locally advanced recurrent tumors under thermoradiotherapy. In: Vaupel P et al. (eds) Oxygen transport to tissue XV. Plenum, New York, pp 479–483

Fleckenstein W, Jungblut JR, Suckfall M (1990) Distribution of oxygen pressure in the periphery and centre of malignant head and neck tumors. In: Ehrly AM, Fleckenstein W, Hauss J, Huch R (eds) Clinical oxygen pressure measurement II. Blackwell Ueberreuter, Berlin, pp 81–90

Füller J, Feldmann HJ, Molls M, Sack H (1994) Untersuchungen zum Sauerstoffpartialdruck im Tumorgewebe unter Radio- und Thermoradiotherapie. Strahlenther Onkol 170:453–460

Gatenby RA, Kessler HB, Rosenblum JS, Coia LR, Moldofsky PJ, Hartz WH (1988) Oxygen distribution in squamous cell carcinoma metastases and its relationship to outcome of radiation therapy. Int J Radiat Oncol Biol Phys 14:831–838

Graeber TG, Osmanian C, Jacks T et al. (1996) Hypoxia-mediated selection of cells with diminished apoptotic potential in solid tumors. Nature 379:88–91

Gray LH, Conger AD, Ebert M, Hornsey S, Scott OCA (1953) The concentration of oxygen dissolved in tissues at the time of irradiation as a factor in radiotherapy. Br J Radiol 26:638–648

Guichard M, Dertinger H, Malaise EP (1983) Radiosensitivity of four human tumor xenografts: influence of hypoxia and cell-cell contact. Radiat Res 95:602–609

Hainaut P, Milner J (1993) Redox modulation of p53 conformation and sequence-specific DNA binding in vitro. Cancer Res 53:4469–4473

Höckel M, Schlenger K, Knoop C, Vaupel P (1991) Oxygenation of carcinomas of the uterine cervix: evaluation by computerized O_2 tension measurements. Cancer Res 51:6098–6102

Höckel M, Knoop C, Schlenger K, Vorndran B, Mitzl M, Knapstein PG, Vaupel P (1993) Intratumoral pO_2 predicts survival in advanced cancer of the uterine cervix. Radiother Oncol 26:45–50

Jund R, Feldmann HJ, Molls M (1996) Der Sauerstoffpartialdruck im Gewebe menschlicher Kopf-, Halskarzinome während primärer Radio-Chemotherapie. Laryngo Rhino Otol 75:43–47

Kolstad P (1968) Intercapillary distance, oxygen tension and local recurrence in cervix cancer. Scand J Clin Lab Invest 106[Suppl]:145–157

Koong AC, Chen EY, Kim CY, Giaccia AJ (1994) Activators of protein kinase C selectively mediate cellular cytotoxicityto hypoxic cells and not aerobic cells. Int J Radiat Oncol Biol Phys 29:259–265

Koutcher JA, Alfieri AA, Devitt ML, Rhee JG, Kornblith AG, Mahmood U, Merchant TE, Cowburn D (1992) Quantitative changes in tumor metabolism, partial pressure of oxygen, and radiobiological oxygenation status postradiation. Cancer Res 52:4620–4627

Lartigau E, Vitu L, Haie-Medar C, Cosset MF, Delapierre M, Gerbaulet A, Eschwege F, Guichard M (1992) Feasibility of measuring oxygen tension in uterine cervix carcinoma. Eur J Cancer 28A:1354–1357

Lartigau E, Le Ridant AM, Lambin P, Weeger P, Martin L, Sigal R, Lusinchi A, Luboinski B, Eschwege F, Guichard M (1993) Oxygenation of head and neck tumors. Cancer 71:2319–2325

Lartigau E, Lusinchi A, Randrianarivelo H, Weeger P, Wibault P, Luboinski B, Eschwege F, Guichard M (1995) Oxygen tension distribution before and during accelerated radiotherapy and carbogen breathing: preliminary results. In: Vaupel P, Kelleher DK, Günderoth M (eds) Tumor oxygenation. Fischer, Stuttgart, pp 305–311 (Funktionsanalyse biologischer Systeme, vol 24)

Molls M, Feldmann HJ, Füller J (1994) Oxygenation of locally advanced recurrent rectal cancer, soft tissue sarcoma and breast cancer. In: Vaupel P et al. (eds) Oxygen transport to tissue XV. Plenum Press, New York, pp 459–463

Molls M, Feldmann HJ, Auberger T (1995) Blood flow and oxygenation status under radiotherapy: clinical investigations. In: Kogelnik HD (ed) Progress in radio-oncology V. Monduzzi, Bologna, pp 123–127

Moulder JE, Rockwell S (1984) Hypoxic fractions of solid tumors: experimental techniques, methods of analysis and a survey of existing data. Int J Radiat Oncol Biol Phys 10:695–712

Nordsmark M, Bentzen SM, Overgaard J (1995) Inter- and intra-tumor variability in the oxygenation status of lymph node metastases from head and neck cancer. In: Vaupel P, Kelleher DK, Günderoth M (eds) Tumor oxygenation. Fischer, Stuttgart, pp 259–267 (Funktionsanalyse biologischer Systeme, vol 24)

Nordsmark M, Overgaard M, Overgaard J (1996) Pretreatment oxygenation predicts radiation response in advanced squamous cell carcinoma of the head and neck. Radiother Oncol 41:31–39

Overgaard J (1980) Sensitization of hypoxic tumor cells – clinical experience. Int J Radiat Biol Phys 56:801–811

Sakata K, Tak Kwok T, Murphy BJ, Laderoute KR, Gordon GR, Sutherland RM (1991) Hypoxia-induced drug resistance: comparison to P-glycoprotein-associated drug resistance. Br J Cancer 64:809–814

Saumweber DM, Kau RJ, Arnold W (1995) Tumor tissue oxygenation in primary squamous cell carcinomas of the head and neck – preliminary results. In: Vaupel P, Kelleher DK, Günderoth M (eds) Tumor oxygenation. Fischer, Stuttgart, pp 313–317 (Funktionsanalyse biologischer Systeme, vol 24)

Schramm U, Fleckenstein W, Weber C (1990) Morphological assessment of skeletal muscular injury caused by pO_2 measurements with hypodermic needle probes. In: Ehrly AM, Fleckenstein W, Hauss J, Huch R (eds) Clinical oxygen pressure measurement II. Blackwell Ueberreuter, Berlin, pp 38–50

Suit HD, Zeitman A, Tomkinson K, Ramsay J, Gerweck L, Sedlacek R (1990) Radiation response of xenografts of a human squamous cell carcinoma and a glioblastoma multiforme: a progress report. Int J Radiat Biol Phys 18:365–373

Vaupel P, Frinak S, O'Hara M (1984) Direct measurement of reoxygenation in malignant mammary tumors after a single large dose of irradiation. Adv Exp Med Biol 180:773–782

Vaupel P, Kallinowski F, Okunieff P (1989) Blood flow, oxygen and nutrient supply, and metabolic microenvironment of human tumors: a review. Cancer Res 49:6449–6465

Vaupel P, Schlenger K, Knoop C, Höckel M (1991) Oxygenation of human tumors: evaluation of tissue oxygen distribution in breast cancers by computerized O_2 tension measurements. Cancer Res 51:3316–3322

Vaupel P, Kelleher DK, Günderoth M (eds) (1995) Tumor oxygenation. Fischer, Stuttgart (Funktionsanalyse biologischer Systeme, vol 24)

Zywietz F, Reeker W, Kochs E (1995) Tumor oxygenation in a transplanted rat rhabdomyosarcoma during fractionated irradiation. Int J Radiat Oncol Biol Phys 32:1391–1400

Zywietz F, Reeker W, Kochs E (1997) Changes in tumor oxygenation during a combined treatment with fractionated irradiation and hyperthermia: an experimental study. Int J Radiat Oncol Biol Phys 37:155–166

9 Misonidazole Labeling as a Marker of Cellular Hypoxia

M. Parliament[1] and R. Urtasun[2]

CONTENTS

9.1
Introduction

Techniques which could directly or indirectly measure oxygen tension are of great value in understanding the natural history of tumor oxygenation. However, the opposite approach, that of using probes to specifically identify *hypoxic* cells within a tumor population, has generated interest. Ideally, consumption or binding of such hypoxia probes would be maximal in radiobiologically hypoxic cells, and minimal in normoxic normal tissues or tumor cells. Such probes should be detectable by standard laboratory or imaging techniques, so that pretreatment identification of patients whose tumors contain an appreciable number of positively labeling hypoxic cells may be performed.

[1,2] M. Parliament MD and R. Urtasun, MD, Department of Radiation Oncology, Cross Cancer Institute, 11560 University Avenue, Edmonton, Alberta, Canada, T6G 1Z2

The hypoxic radiosensitizer drug misonidazole (Miso) has shown promise as a hypoxic cell labeling probe. This chapter describes the nature of the labeling technique using Miso as a prototype, and discusses the applications and limitations of this technique particularly with respect to clinical trials using Miso or its analogues.

9.2
Mechanisms of Sensitizer-Adduct Formation in Hypoxic Cells

Miso was recognized to bind selectively to hypoxic cells (Chapman et al. 1979; Varghese and Whitmore 1980). Using autoradiography, [^{14}C]Miso was shown to label zones of spheroids and tumors in locations that would be predicted to be hypoxic, based on oxygen diffusion and consumption by intervening cell layers (Franko et al. 1982). The difference in the binding rates of Miso to anoxic versus maximally normoxic cells is 55-fold for EMT6 cells (Miller et al. 1982); this differential binding rate has subsequently been shown to vary considerably between cell lines. Bioreduction of Miso requires an activation energy of 33.5 kcal/mol, consistent with an enzymatic process. The half-life of [^{14}C]Miso adducts to the macromolecular (i.e., acid-insoluble) fraction of EMT-6 cells is 50–55 h (Chapman et al. 1983), whereas the serum elimination half-life is approximately 10 h. Based upon these properties, it has been suggested that hypoxic radiosensitizers could be developed as markers for hypoxic cells in tumors.

The bioreduction of 2-nitroimidazoles and other nitroheterocycles under hypoxic cellular conditions to generate reactive species has been reviewed (Franko 1986). In either normoxic or hypoxic cells, the nitro group can be reduced to the radical anion by cellular electron donors and enzymes (Mason 1982). However, in the presence of oxygen, the radical anion is quickly converted back to the intact nitro group and a superoxide ion is generated (Mason and HOLTZMAN 1975). In the absence of oxygen, further reduction steps can occur (RAUTH 1984). It is postulated that the nitroso and hydroxylamine moieties are reactive. The reactive species are available for binding nonprotein sulfhydryls (SMITH and BORNE 1984) or protein thiols (RALEIGH and KOCH 1990). Nitroreduction does not appear to be the exclusive domain of one particular enzyme or enzyme system. Rather, nitroreduction of these compounds can be achieved by NAD(P)H: cytochrome P450 oxidoreductase, NAD(P)H: quinone oxidoreductase (JOSEPH et al. 1994) and xanthine oxidase (PREKEGES et al. 1991). As discussed below, variability in the levels of different nitroreductases between cell lines and tissues is likely to result in variation of both normoxic and hypoxic binding rates.

In summary, the relative stability of sensitizer-adducts, once formed in hypoxic cells, raises the possibilities of tagging the nitroimidazole with a marker isotope or fluorescent molecule, to be detected as a marker of intracellular oxygen concentration. It has become clear over the past 15 years that this concept is an oversimplification, and uncritical application of this technique, without due consideration to possible modifying factors, is cautioned against.

9.3
Correlation Between Sensitizer-Adduct Formation and Other Factors Modulated by Tumor Hypoxia

9.3.1
Radiation Response

Under conditions in which the hypoxic fraction of EMT6 spheroids could be varied, FRANKO (1985) showed that the thickness of the zone heavily labeled with [^{14}C]Miso correlated well with hypoxic fraction. Contributions from noncellular material, non-viable cells and normoxic cells became substantial for hypoxic fractions of less than 20%. A correlation between the surviving fraction in EMT6 tumors in vivo after 20 Gy and [^{14}C]Miso retention was reported (CHAPMAN 1984). However, there is substantial scatter in the data, and the slope is greater than 1 on a log–log plot. HIRST et al. (1985) observed a similar problem in EMT6 and RIF-1 tumors. It was postulated that a slope greater than 1 could result from nonspecific binding to normoxic cells and necrosis; once this was subtracted, however, a direct correlation between Miso binding and hypoxic fraction could be obtained. Given these results, extrapo-

lation of absolute radiobiological hypoxic fraction from the quantity of 2-nitroimidazole bound is difficult to make with precision. However, elevated marker binding may be sufficient to answer the clinically relevant question of whether or not any given tumor may exhibit hypoxic radioresistance.

9.3.2
Correlation with pO$_2$ Histography

A microelectrode technique to measure the oxygenation in spheroids has been performed in EMT6 spheroids simultaneously incubated with Miso (MUELLER-KLIESER et al. 1991). In spheroids which the central pO$_2$ is lower than 10 mmHg, a marked increase in label density can be detected next to necrosis, confirming the expected inverse relationship between simultaneous pO$_2$ measurement and misonidazole labeling. Calibration of this relationship for a variety of cell types could prove useful.

9.3.3
Correlation with ^{31}P Magnetic Resonance Spectroscopy

In the Dunning R3327-AT rat tumor model, CHAPMAN (1991) used photodynamic therapy with variable light dose to investigate the relationship between Miso binding and ^{31}P magnetic resonance spectroscopy (MRS) signal. Using this technique, hypoxic fractions of various sizes can be induced therapeutically because this treatment causes reduction of tumor perfusion and diminished O$_2$ supply. The measure of tumor energy status, P$_i$/beta ATP, increases with greater [^{3}H]Miso uptake. Although there is substantial scatter in these data and the relationship is not linear, this study did suggest that ^{31}P MR spectral changes could be related to Miso binding when the hypoxic fraction is altered acutely.

9.4
Factors That May Modify Sensitizer-Adduct Formation

The kinetics of bioreduction of nitroimidazoles have been established for Miso. The binding rate depends on Miso concentration, but this is not true for all nitroimidazoles. For Miso, the binding rate is half-order in anoxic or hypoxic conditions, and first-order in normoxic conditions (CHAPMAN et al.

1983). In EMT6 cells, the oxygen concentration corresponding to half-maximal radiosensitivity, and that for half-maximal Miso binding, are almost identical (CHAPMAN et al. 1989). KOCH et al. (1984) showed that the O_2 level for half-maximal binding varied with Miso concentration but occurred at around 0.2% O_2 for V79 cells using 50 µM Miso. Finally, at high Miso concentrations, binding to EMT6 cells is inhibited by glucose depletion (LING and SUTHERLAND 1986).

9.5
Binding to Normal Tissues

Because of the marked differential between the rates of normoxic and anoxic binding of Miso, it was initially hoped that cellular oxygen tension alone would be the primary influence on sensitizer-adduct formation. GARRECHT and CHAPMAN (1983) showed retention of [^{14}C]Miso activity in EMT6 tumors and in a variety of normal tissues in BALB/c mice. These tissues include liver, nasopharyngeal and oral cavities, and tissues along the lumen and within the contents of the intestinal tract. It was initially postulated that normoxic liver cells bound activated Miso products due to a high concentration of drug detoxification enzymes. The oropharyngeal and gastrointestinal uptake were postulated to be due to an association with the bacterial flora of these organs. Subsequent studies have shown that binding to normal tissues varies enormously (COBB et al. 1989), with markedly elevated label density in some unexpected tissues.

9.5.1
Binding to Liver

Murine hepatocytes and mouse hepatoma 129P cells, when made hypoxic, bind [^{14}C]Miso. However, the binding rate is not uniquely high in comparison to other cell lines, suggesting that increased binding to liver in vivo is due to physiologic hypoxia (VAN OS-CORBY and CHAPMAN 1986; VAN OS-CORBY et al. 1987). Regional variations of autoradiographic grain patterns in murine liver have been noted on a microscopic level.

The lowest grain counts are adjacent to the portal tract (oxygenated blood, zone 1), and the highest grain count is adjacent to hepatic vein (deoxygenated blood, zone 3). Grain counts midway between the portal tract and hepatic vein are intermediate to

these extremes (MAXWELL et al. 1989). Induction of hypobaric hypoxia resulted in 2.5-fold higher Miso binding in liver as well as in kidney and spleen (MACMANUS et al. 1989). This pattern of binding can be explained by the oxygen tension gradient expected in vivo over the functional hepatic subunit.

9.5.2
Binding to Apparently Normoxic Tissues

Detailed microscopic assessment of Miso retention in a number of normal tissues in the mouse has been reported (COBB and NOLAN 1989; COBB et al. 1989, 1990b). It was demonstrated that significant Miso retention existed in esophagus and airway epithelium, liver, footpad, eyelid (Meibomian gland), sebaceous glands, stomach, and parotid gland. It is unlikely that all of these tissues are hypoxic, particularly the airway epithelium (COBB et al. 1990a). Two potential mechanisms were postulated to lead to the observed binding of 2-nitroimidazoles in normally oxygenated tissues: (1) nitroreductase activity may be sufficiently elevated in these tissues to cause an elevated rate of marker binding, despite competition from oxygen; (2) nitroreduction might occur due to an enzyme such as DT-diaphorase, which is a quinone reductase known to be an obligate 2-electron reductase. If Miso were a substrate for this enzyme, then the initial reduction step would not be reversible by oxygen (COBB et al. 1990a). In esophageal epithelium, elevated Miso binding was shown to be oxygen sensitive. The possibility of oxygen insensitive bioreduction was investigated using DT-diaphorase inhibitor, dicoumarol. In vitro, dicoumarol either had no effect or very slight inhibition on the binding of [^{3}H]Miso to esophagus; in vivo, dicoumarol had variable or little effect (PARLIAMENT et al. 1992a). In order to determine the relative effectiveness of specific reductases for the binding of 2-nitroimidazoles, JOSEPH et al. (1994) transfected cells to cause overexpression of P450 reductase and DT-diaphorase. An 80-fold overexpression of P450 reductase resulted in a five- to sevenfold increase in hypoxic marker binding, compared to a 1.5-fold increase in hypoxic binding associated with a 1000-fold overexpression of DT-diaphorase. Therefore, from these studies it is suggested that P450 reductase is the most important cellular enzyme for bioreduction of 2-nitroimidazoles.

In summary, the normal tissue binding of nitroimidazoles is oxygen-sensitive, with the most important enzyme activity being P450 reductase, which utilizes a 1-electron reduction step which can be inhibited by oxygen. Binding of 2-nitroimidazoles by normoxic normal tissues is not expected to pose a problem in the non-invasive detection of hypoxia markers, assuming that tumors arising from the same tissue contain a similar complement of nitrore-ductases. Further study is required to clarify this.

9.6
Sensitizer-Adduct Techniques

9.6.1
[^{3}H],[^{14}C]Misonidazole Autoradiography

A pilot study initiated early in 1985 investigated the use of [^{3}H]misonidazole as a hypoxic marker for human tumors in situ (URTASUN et al. 1986). The eligibility criteria included patients with accessible, subcutaneous, soft tissue or lymph node metastases. Lesions were at least 3 cm in diameter and in some cases were fixed to the skin or underlying structures. [^{3}H]Miso was given intravenously 20–24 h prior to tumor excision. Specimens were then fixed and embedded. To allow the development of grains identifying the location of [^{3}H]Miso adducts 4-μm sections were dipped in photographic emulsion and exposed for 4–8 weeks. In order to establish a threshold grain density level corresponding to radiobiological levels of hypoxia, in vitro calibration of the oxygen dependence of binding in the same tumor was determined concurrently. Fragments of the tumor were incubated with [^{14}C]Miso in degassed chambers equilibrated with specific O$_2$ concentrations for 3 h. Tumor fragments were sectioned and coated to prevent beta particles from ^{3}H exposing the emulsion. In this fashion, a standard calibration curve of relative grain density versus oxygen concentration could be derived (URTASUN et al. 1986). Patients were accrued with small cell carcinoma of the lung, melanoma, soft tissue sarcoma, mesothelioma, osteosarcoma and squamous cell carcinoma of the tongue. Half-maximal binding was observed at O$_2$ concentrations between 0.1%–0.3% in small cell lung cancer. This roughly corresponded to a grain density of greater than 10 grains/100 μm^2. Based on this, an estimate of the hypoxic area fraction in each tumor is possible. Results varied from <1% hypoxic area fraction in most soft tissue sarcomas, to 20%–40% in a small-cell lung carcinoma metastasis. Results of this study are shown in Table 9.1, along with the tumor/plasma ratio of [^{3}H]Miso activity. These results suggest intra- and intertumoral heterogeneity of Miso binding, with the conclusion that hypoxia in situ is by no

Table 9.1. Detection of hypoxia in situ in human tumors using [^{3}H]Miso autoradiography

Histology	Fraction positive	Percentage
Small-cell lung cancer	8/12	67
Melanoma	3/3	100
Soft tissue sarcoma	1/10	10
Squamous cell carcinoma of tongue	0/2	0

Positive result is defined as a zone of intense label >1% of sectional area. For methods, see text. (Reproduced with permission from URTASUN 1992.)

means universal but is likely related to intrinsic growth and metabolic conditions of each tumor histology. The autoradiographic approach has the disadvantage of requiring excision of the tumor. It is also rather time-consuming, requiring a minimum of 4 weeks for autoradiographic development. Radiation protection safeguards mandated hospital confinement because of the prolonged elimination of [^{3}H]Miso. Finally, the technique could not practically be repeated on the same tumor during a course of fractionated radiotherapy (RT).

These findings were an important demonstration of the utility of sensitizer-adduct techniques in the detection of hypoxia in situ in human tumors. Current interest in histological sensitizer-adduct techniques have utilized evolving immunohistochemical methods (Sect. 9.6.4).

9.6.2
Noninvasive Detection of Sensitizer-Adducts

The use of a noninvasive procedure to measure the oxygenation status of individual human tumors is an attractive notion. Potential advantages of this strategy include assessment of deep-seated and/or inaccessible tumors, the ability to repeat measurements during fractionated RT, lack of perturbation of vasculature, and the avoidance of sampling error. Potential disadvantages include the loss of detailed microscopic information, due to volume averaging.

9.6.2.1
[^{18}F]Fluoromisonidazole/PET

The Miso analogue fluoromisonidazole (F-Miso) was labeled with ^{3}H and ^{18}F in preclinical studies aimed towards the detection of sensitizer-adducts by positron emission tomography (PET). Cellular uptake studies in V-79 and RIF-1 cells suggest that bioreduction in anoxic conditions occurs via the usual mechanism (Rasey et al. 1990). Murine biodistribution studies suggest both hepatobiliary and renal modes of excretion with some evidence for specific retention in the KHT tumor in vivo. In vitro studies of [³H]F-Miso show that half-maximal binding occurs over a range of oxygen levels similar to Miso and is cell lineage dependent. In canine osteosarcomas chosen for PET imaging with [¹⁸F]F-Miso, intratumoral uptake is heterogeneous, with tumor/plasma ratios ranging from 0.3 to 2.0 at 3–4 h. The same group has reported its clinical experience with hypoxia imaging in human tumors with [¹⁸F]F-Miso (Koh et al. 1992; also summarized in Stone et al. 1993). [¹⁸F]F-Miso was administered intravenously. Immediate and delayed image acquisitions were performed, with concurrent arterial or venous blood sampling. From the plasma counts, derivation of tissue/plasma and tumor/plasma ratios of [¹⁸F]F-Miso were obtained. Preclinical data suggests that in normoxic tissues, 90% of tissue/blood ratios determined to date are <1.3. It was thus proposed that a tumor/blood ratio of ≥1.4 at 2 or more hours after injection of the tracer was indicative of hypoxia within the region of interest (although the relationship of this ratio to radiobiological hypoxic fraction is not currently known). The fractional hypoxic volume was defined as the fraction of pixels in the entire tumor volume with a tumor/blood ratio ≥1.4. Among 28 patients, fractional hypoxic volumes range from 0% to 91%, with a median of 10%–19% (Stone et al. 1993). These included eight non-small-cell lung carcinomas, eight head and neck carcinomas, four prostate carcinomas and three other tumors. Repeat [¹⁸F]F-Miso imaging during fractionated RT was proposed as a means of non-invasively assessing tumor reoxygenation. Seven of eight patients 2 weeks or more after the start of RT showed a decrease in fractional hypoxic volume, consistent with reoxygenation (Stone et al. 1993). The prognostic value of F-Miso hypoxic imaging in patients undergoing conventional RT is somewhat difficult to determine from the current data; many of the patients were treated with neutrons. Access to PET imaging remains uncommon in North America, and interest in hypoxia imaging using conventional nuclear medicine techniques has been growing.

9.6.2.2
Iodinated Nitroimidazoles

Radiohalogen-labeled nitroimidazoles (see also Sect. 9.6.2.3) have proven easier to synthesize than other gamma-emitter labeled nitroimidazoles. Misonidazole analogues radiolabelled with ¹²³I have been developed for detection of tumor hypoxia using conventional gamma camera techniques available in most nuclear medicine departments. Substantial clinical experience has developed with iodinated nitroimidazoles.

A series of radioiodinated azomycin nucleosides has been synthesized and characterized. Iodoazomycin riboside (IAZR; Jette et al. 1986; Wiebe et al. 1986) showed a more rapid hypoxic binding rate than Miso in vitro, but was subject to metabolic deiodination in vivo. Agents which were hypoxia-selective and more stable included 1-(6-iodo-6-deoxy-beta-D-galactopyranosyl)-2-nitroimidazole (IAZG), 1-(4-iodo-4-deoxy-beta-L-xylopyranosyl)-2-nitroimidazole (IAZP), and 1-(5-iodo-5-deoxy-beta-D-arabinofuranosyl)-2-nitroimidazole (IAZA, Fig. 9.1; Mercer et al. 1990; Mannan et al. 1991, 1992). These compounds show hypoxic cytotoxicity, and are hypoxic radiosensitizers in vitro. In BALB/c mice bearing EMT6 tumors, ¹²⁵I-IAZA showed a tumor/blood ratio of 8.7 at 8 h. The biodistribution data for IAZA show preferential retention in EMT6 tumors with significant uptake in kidney, liver, intestine and stomach from 0 to 4 h, consistent with combined hepatoliliary and renal excretion of the drug and/or metabolites. Finally, in EMT6 spheroids it was shown that [¹²⁵I]IAZA binds to the inner rim of intact cells adjacent to necrosis in a manner analogous to [³H]Miso (G.G. Miller, unpublished data). Despite higher tumor/blood ratios with IAZP, IAZA was chosen for initial clinical investigations because difficulties were encountered with exchange radiolabelling of IAZP. [¹²³I]Iodoazomycin arabinoside has been studied in patients by our group using planar and single-photon emission computed tomography (SPECT) techniques (Fig. 9.2; Parliament et al. 1992b; Groshar et al. 1993). [¹²³I]IAZA (176–370 MBq) was given intravenously by bolus injection. Immediate (1 h) and delayed (16–24 h) imaging was

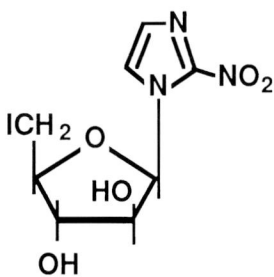

Fig. 9.1. Iodoazomycin arabinoside (IAZA). (Reproduced with permission from Parliament et al. 1992b)

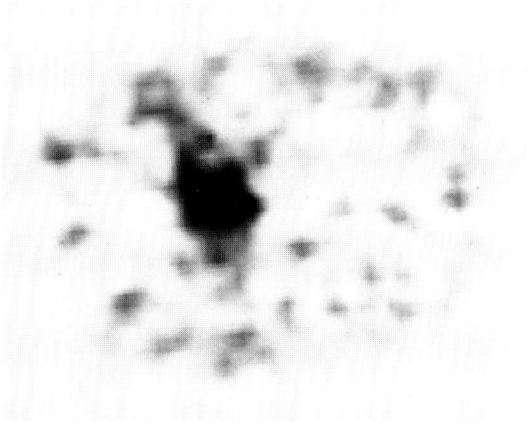

Fig. 9.2. [^{123}I]IAZA-avidity in a patient with small-cell lung carcinoma limited to the right hilum and mediastinum (axial reconstruction from SPECT). (Reproduced with permission from PARLIAMENT et al. 1992b)

Table 9.2. Clinical experience with [^{123}I]IAZA as a hypoxia imaging agent

Diagnosis	Lesion avidity
Soft tissue sarcoma	3/5
Brain metastases (various)	3/4
Glioblastoma	0/11[a]
Small cell carcinoma lung	6/11
SCC neck node metastases	5/15
Adenocarcinoma prostate	0/1
Melanoma	0/1

A total of 50 patients have been accrued; results are given for those evaluable ($n = 48$). Two patients with glioblastoma were not evaluable (failure to complete scanning protocol). Avid lesions are defined in the text.
SCC, squamous cell carcinoma.
[a] One patient with glioblastoma had qualitatively equivocal uptake.

performed. Delayed imaging is necessary to allow clearance of unbound [^{123}I]IAZA in plasma and extracellular fluid. Quantitative image analysis was performed using region of interest techniques to compare central tumor activity with activity in adjacent normal tissue. Tumor/normal tissue ratios which increase by a factor greater than 1.1 (comparing early and later images) were proposed as evidence for preferential hypoxic binding of IAZA (PARLIAMENT et al. 1992b). Efforts were made to image patients prior to any therapy; however, early in the study some patients had already commenced RT and/or chemotherapy. At the time of writing, 48 patients with a variety of histologies have undergone imaging (Table 9.2).

Regional tissue perfusion has been demonstrated noninvasively with several radiopharmaceuticals including ^{99m}Tc-labeled hexamethylpropylenamine oxime (HMPAO; SUGA et al. 1991; MOORE et al. 1992; ROWELL et al. 1989). A qualitative comparison of ^{99m}Tc-labeled HMPAO and [^{123}I]IAZA distribution in tumors has been performed (GROSHAR et al. 1993). Eight of 13 tumors (62%) with decreased HMPAO uptake showed increased IAZA uptake, whereas only five of 14 (36%) with normal or increased HMPAO uptake showed increased IAZA uptake. The absence of uptake of [^{123}I]IAZA in patients with glioblastoma multiforme was striking, despite evidence for reduced perfusion. Three of four metastatic brain lesions showed increased IAZA uptake, suggesting that blood/brain transfer is not impaired with IAZA. One explanation could be that cell death and inactivation occur in gliomas at unexpectedly high oxygen

tensions; appropriate nitroreductases for metabolism do appear to be present in gliomas (RAMPLING et al. 1994). If one excludes the glioblastoma data, it becomes clear that eight of nine extracranial tumors (89%) with decreased HMPAO uptake show increased IAZA uptake, compared to only five out of 14 tumors (36%) with normally increased HMPAO uptake showing increased IAZA uptake ($p < 0.05$). Thus, in extracranial tumors, an inverse correlation between tumor perfusion and hypoxic marker affinity can be demonstrated using volume averaging techniques. Yet the relationship between these parameters is complex and oversimplification is hazardous. The above data are in conflict with the finding of extensive hypoxia in situ in gliomas using pO_2 measurements with needle electrodes on anesthetized patients (RAMPLING et al. 1994). ^{78}F-Miso imaging data in glioma are limited; however, VALK et al. (1992) arguably showed increased F-Miso uptake in one of three patients. Our group has developed an animal model to investigate the apparently anomalous hypoxia imaging results in gliomas.

Preliminary data of the prognostic importance of [^{123}I]IAZA uptake in tumors prior to RT is currently being investigated by our group in patients with squamous cell carcinoma neck node metastases being treated with radical RT (Table 9.3). Patients are evaluated clinically and radiographically for their response at 3 months post-RT. Among IAZA-non-avid neck nodes, five out of nine showed a complete response, whereas among IAZA-avid neck nodes, none of four showed a complete response (M. PARLIAMENT, unpublished results). The study continues

Table 9.3. Preliminary results of [^{123}I]IAZA avidity as a prognosticator in squamous cell carcinoma neck node metastases treated with primary radiotherapy

	Complete response
IAZA – Avid	0/4
IAZA – Nonavid	5/9

Radiotherapy (RT) was given once daily, 5 days weekly. Dose fractionation schemes were: 66 Gy/33 fractions (f) (continuous), $n = 7$; 70 Gy/35 f (continuous), $n = 3$; 30 Gy/10 f + 30 Gy/10 f (split), $n = 2$; and 25 Gy/10 f + 25 Gy/10 f + 15 Gy/ 10 f (split), $n = 1$. Response evaluated clinically and radiologically at 3 months post-RT.

to accrue patients in hopes of establishing the independent prognostic importance of IAZA uptake, particularly in relation to nodal size. In a parallel study, patients with small cell lung cancer prior to any treatment are also being evaluated.

Finally, an (iodovinyl)-misonidazole has been synthesized and characterized as a hypoxia imaging agent (BISKUPIAK et al. 1991; MARTIN et al. 1993). This drug was designed to resist in vivo iodination. Further data are awaited to assess its clinical usefulness.

9.6.2.3
Brominated Nitroimidazoles

4-[^{82}Br]Bromomisonidazole was synthesized and judged to be too toxic as a clinical radiosensitizer. It was assessed as a possible agent for hypoxia imaging (RASEY et al. 1982, 1985). Spheroid data were deemed encouraging, but this compound has not been brought to clinical studies.

9.6.2.4
Technetium Nitroimidazoles

The high cost and limited availability of the cyclotron-produced radionuclides ^{18}F and ^{123}I has spurred efforts to design a 2-nitroimidazole containing ^{99m}Tc, the most commonly used radionuclide isotope in clinical nuclear medicine. The challenge has been to produce a chemically stable radioligand which retains the ability to enter intracellular compartments. However, successful synthesis of a technetium-labeled nitroimidazole would enable hypoxia imaging to be performed widely in clinical nuclear medicine departments, for oncologic, car-

diac, and neurologic applications. A number of ^{99m}technetium complexes have been synthesized for the detection of hypoxic tissue during physiological imaging (LINDER et al. 1993, 1994). The most promising of these is the ^{99m}Tc complex of a 2-nitroimidazole-derivatized propylene amine oxime designated BMS-181321. This has been shown to be selectively retained in a rat stroke model (DI ROCCO et al. 1993) and in an isolated rat heart ischemia model (RUMSEY et al. 1993). KUSUOKA et al. (1994) independently showed that BMS-181321, when injected before ischemia, is selectively retained in myocardium, but not in nonischemic controls. Preclinical tumor hypoxia marking studies with this compound are underway.

9.6.2.5
Fluorinated Nitroimidazoles Detected by ^{19}FMRS

^{19}F-labeled nitroimidazoles have the theoretical advantage of high sensitivity and low endogenous background. The two fluorinated nitroimidazoles which have proven useful for MRS detection of sensitizer-adducts are hexafluoromisonidazole (CCI-103F; JIN et al. 1990; MAXWELL et al. 1989; RALEIGH et al. 1989, 1991; KWOCK et al. 1992) and the monofluorinated nitroimidazole Ro 07-0741 (MAXWELL et al. 1989). Measurements of CCI-103F retention in SCC VII tumors of C3H/Km mice were reported by JIN et al. (1990). This technique uses proton MRS as an internal standard with which to calibrate absolute CCI-103F concentration from the ^{19}FMRS measurements. CCI-103F retention was longest in liver, then in experimental tumors, and then in brain. Using EMT6 tumors in BALB/c mice, MAXWELL et al. (1989) similarly found a progressive reduction in signal intensity in normal tissues, but tumor levels remained constant or declined more slowly. This is consistent with the detection of bound sensitizer-adducts. Higher tumor/brain ratios were seen at 6–7 h with CCI-103F (4.2) than with Ro 07-0741 (2.9). Relatively few centers have explored these techniques to date.

9.6.3
Fluorescent Nitroheterocycles

A variety of fluorescent-labeled nitroheterocycles have been found to undergo bioreductive activation and hypoxic binding to cellular macromolecules under hypoxic conditions (HODGKISS et al. 1991a, 1992;

OLIVE and DURAND 1981; OLIVE et al. 1986). These techniques have only been used in animal models to date. Given the number of 2-nitroimidazole drugs for which the hypoxic labeling efficiency and biodistribution is known, it seems likely that research interests will turn towards immunohistochemical methods of detecting adducts of these molecules in tumors, perhaps by fluorescent techniques.

9.6.4
Immunohistochemical Methods

9.6.4.1
Hexafluoromisonidazole (CCI-103F)

As noted above, the advantage of histologic detection of sensitizer-adducts rests in the ability of a microscopic technique to detect the exact location of adducts with respect to functional vessels and necrosis. This level of detailed information is necessarily absent in volume-averaging or imaging techniques. However, histologic detection of adducts requires administration of a nitroimidazole prior to biopsy (usually a second biopsy, after malignancy has been confirmed) and is potentially subject to sampling error.

Nonetheless, there are many accessible tumors (e.g., cervical carcinoma) which lend themselves to this approach. A nonradioactive nitroimidazole would have the additional advantage of safety and ease of administration. Earlier studies showed that the binding pattern of hexafluoromisonidazole (CCI-103F) was indistinguishable from Miso (RALEIGH et al. 1987). On histologic sections, antibodies to the protein adduct of CCI-103F may be applied (CLINE et al. 1990). A second antibody conjugated to a fluorescent molecule or enzyme is then applied in order to visualize the sensitizer-adduct–antibody complex. MILLER et al. (1989) reported quantitative image analysis of fluorescent-labeled CCI-103F adducts in spheroids. In spontaneous canine tumors, mean hypoxic area fractions of 12.2% ± 16.7% (range 0%–35%) have been reported (CLINE et al. 1994). The pattern of labeling qualitatively observed was felt to be consistent with detection of hypoxic cells arising from limited oxygen diffusion. An enzyme-linked immunosorbent assay analysis (ELISA) has been developed for the rapid detection of sensitizer-adducts in tumor biopsy samples as well (RALEIGH et al. 1992, 1994). This technique has been used to assess the natural history of nitroimidazole binding

before and after RT in treated spontaneous canine tumors. In ten of 11 tumors, the binding pattern remained unchanged after RT, consistent with a failure of reoxygenation to occur. The relationship between proliferating cells (measured by proliferating cell nuclear antigen immunostaining) and hypoxia (CCI-103F adduct formation) has been reported in spontaneous canine tumors (ZEMAN et al. 1993). These two parameters are related in a complex way with much overlap in the populations of cells staining positive for either proliferation or hypoxia. However, the finding of cellular proliferation despite moderate hypoxia is of fundamental biological interest. These are valuable preclinical studies. CCI-103F has not yet been approved for use in humans and no clinical data is available.

9.6.4.2
Other Nitroimidazoles

Several other nitroimidazoles have been developed as hypoxia markers. A polyclonal-antibody immunohistochemical technique for the detection of pimonidazole adducts has been developed by Raleigh's group (STONE et al. 1993). In histologic sections of biopsies from two patients given pimonidazole for cervical carcinoma, there was a diffuse pattern of binding. This is of clinical interest as the toxicity of pimonidazole in humans is known, and the concentrations required for immunodiagnosis would be several fold lower than those required for radiosensitization. LORD et al. (1993) have reported the immunohistochemical detection of a pentafluorinated derivative of etanidazole. This technique uses a monoclonal antibody to recognize adducts. Hypoxic regions have been demonstrated in EMT6 mouse and Morris 7777 hepatoma rat tumors. HODGKISS et al. (1991b) synthesized a new 2-nitroimidazole covalently bound to the hapten theophylline, denoted 7-(4′-(2-nitroimidazole-1-yl)-butyl)-theophylline (NITP). This nitroimidazole, which is delivered intraperitoneally in a peanut-oil DMSO vehicle shows half-maximal binding just above 10^3 ppm O_2. CaNT tumors labeled with NITP may be disaggregated and sorted using flow cytometry to distinguish the fraction of heavily labeled cells. Such techniques have demonstrated a reduction in sensitizer-adduct formation to tumor cells, following normobaric oxygen and carbogen breathing by tumor-bearing mice, consistent with improved tumor oxygenation (ROJAS et al. 1992).

9.7
Summary

The bioreductive formation of sensitizer-adducts by 2-nitroimidazoles in hypoxic cells can be utilized to positively detect cells whose intracellular pO_2 would be low enough to confer radioresistance. Correlations between sensitizer-adduct levels and other measures of tumor oxygenation are sparse, yet the available evidence suggests that elevated nitroimidazole binding should predict an inferior radiotherapeutic outcome. Since the absolute binding rate of 2-nitroimidazoles can be modified by a number of different variables, attempts to oversimplify the relationship between marker binding and radiobiological hypoxic fraction are potentially prone to pitfalls. As a result, the likely role of these techniques will be to rank tumors of the same type into two or three prognostic categories, e.g., well oxygenated, borderline oxygenation, and poorly oxygenated. Also, these assays are more relevant for the detection of chronically hypoxic cells than intermittently hypoxic cells since bioreductive binding occurs over several hours. The relevance of intermittently hypoxic cells to the radioresistance of human tumors remains speculative.

Noninvasive techniques (hypoxia imaging) have the potential advantage of assessing cellular hypoxia in deep-seated, unperturbed tumors. However, the volume averaging approach cannot provide detailed morphological information. Microscopic techniques provide detailed information about hypoxia at the cellular level; however, they are invasive and subject to sampling error. Development of both techniques in clinical trials and comparative assessment of oxygen tension with pO_2 histography will allow a better understanding of the relative importance of these issues. Data continues to be accrued to establish the independent prognostic significance of 2-nitroimidazole binding using non-invasive techniques.

Acknowledgements. Supported by the Alberta Cancer Board Research Initiatives Program Grant #RI-14(3). The authors would also like to thank Drs. J.D. Chapman (for leadership and encouragement), A.J.B. McEwan, J.R. Mercer, R.H. Mannan and L.I. Wiebe for their collaboration.

References

Biskupiak JE, Grierson JR, Rasey JS, Martin GV, Krohn KA (1991) Synthesis of an (iodovinyl) misonidazole derivative for hypoxia imaging. J Med Chem 34:2165–2168

Chapman JD (1984) The cellular basis of radiotherapeutic response. Radiat Phys Chem 24:283–291

Chapman JD (1991) Measurement of tumor hypoxia by invasive and non-invasive procedures: a review of recent clinical studies. Radiother Oncol Suppl 20:13–19

Chapman JD, Raleigh JA, Pedersen JE, et al. (1979) Potentially three distinct roles for hypoxic cell sensitizers in the clinic. In: Okada S, Imamura M, Terasima T, Yamaguchi H (eds) Radiation research. Japanese Association of Radiation Research, Tokyo, pp 885–892

Chapman JD, Franko AJ, Sharplin J (1981) A marker for hypoxic cells in tumors with potential clinical applicability. Br J Cancer 43:546–550

Chapman JD, Baer K, Lee J (1983) Characteristics of the metabolism-induced binding of misonidazole to hypoxic mammalian cells. Cancer Res 43:1523–1528

Chapman JD, Urtasun RC, Franko AJ, Raleigh JA, Meeker BE, McKinnon SA (1989) The measurement of oxygenation status of individual tumors. In: Paliwal BR, Fowler JF, Herbert DE, Kinsella RJ, Orton CG (eds) Prediction of response in radiation therapy. 1. The physical and biological basis. American Institute of Physics, New York, pp 49–60

Cline JM, Thrall DE, Page RL, Franko AJ, Raleigh JA (1990) Immunohistochemical detection of a hypoxic marker in spontaneous canine tumors. Br J Cancer 62:925–931

Cline JM, Thrall DE, Rosner GL, Raleigh JA (1994) Distribution of the hypoxia marker CCI-103F in canine tumors. Int J Radiat Oncol Biol Phys 28:921–933

Cobb LM, Nolan J (1989) Autoradiographic study of tritium-labelled misonidazole in the mouse. Int J Radiat Oncol Biol Phys 16:953–956

Cobb LM, Nolan J, O'Neill P (1989) Microscopic distribution of misonidazole in mouse tissues. Br J Cancer 59:12–16

Cobb LM, Hacker T, Nolan J (1990a) NAD(P)H nitroblue tetrazolium reductase levels in apparently normoxic tissues: a histochemical study correlating enzyme activity with binding of radiolabelled misonidazole. Br J Cancer 61:524–529

Cobb LM, Nolan J, Butler SA (1990b) Tissue distribution of ^{14}C- and ^{3}H-labelled misonidazole in the tumor-bearing mouse. Int J Radiat Oncol Biol Phys 18:347–351

Di Rocco RJ, Kuczynski BL, Pirro JP, et al. (1993) Imaging ischemic tissue at risk of infarction during stroke. J Cereb Blood Flow Metab 13:755–762

Franko AJ (1985) Hypoxic fraction and binding of misonidazole in EMT6/ed multicellular tumor spheroids. Radiat Res 103:89–97

Franko AJ (1986) Misonidazole and other hypoxia markers: metabolism and applications. Int J Radiat Oncol Biol Phys 12:1195–1202

Franko AJ, Chapman JD, Koch CJ (1982) Binding of misonidazole to EMT6 and V79 spheroids. Int J Radiat Oncol Biol Phys 8:737–739

Garrecht BM, Chapman JD (1983) The labelling of EMT-6 tumors in BALB/c mice with ^{14}C-misonidazole. Br J Radiol 56:745–753

Groshar D, McEwan AJB, Parliament MB, et al. (1993) Imaging tumor hypoxia and tumor perfusion. J Nucl Med 34:885–888

Hirst DG, Hazlehurst JL, Brown JM (1985) Changes in misonidazole binding with hypoxic fraction in mouse tumors. Int J Radiat Oncol Biol Phys 11:1349–1355

Hodgkiss RJ, Jones GW, Long A, Middleton RW, Parrick J, Stratford MR, Wardman P, Wilson GD (1991a) Fluorescent markers for hypoxic cells: a study of nitroaromatic compounds, with fluorescent heterocyclic side chains, that undergo bioreductive binding. J Med Chem 34:2268–2274

Hodgkiss RJ, Jones G, Long A, Parrick J, Smith KA, Stratford MRL, Wilson GD (1991b) Flow cytometric evaluation of hypoxic cells in solid experimental tumors using fluorescence immunodetection. Br J Cancer 63:119–125

Hodgkiss RJ, Middleton RW, Parrick J, Rami HK, Wardman P, Wilson GD (1992) Bioreductive fluorescent markers for hypoxic cells: a study of 2-nitroimidazoles with 1-substituents containing fluorescent, bridgehead-nitrogen, bicyclic systems. J Med Chem 35:1920–1926

Jette DC, Wiebe LI, Flanagan RJ, Lee J, Chapman JD (1986) Iodoazomycin riboside (1-(5′-iodo-5′-deoxyribofuranosyl)-2-nitroimidazole), a hypoxic cell marker. I. Synthesis and in vitro characterization. Radiat Res 105:169–179

Jin GY, Li SJ, Moulder JE, Raleigh JA (1990) Dynamic measurements of hexafluoromisonidazole (CCI-103F) retention in mouse tumors by ^{1}H/^{19}F magnetic resonance spectroscopy. Int J Radiat Biol 58:1025–1034

Joseph P, Jaiswal AR, Stobbe CC, Chapman JD (1994) The role of specific reductases in the intracellular activation and binding of 2-nitroimidazoles. Int J Radiat Oncol Biol Phys 29:351–355

Koch CJ, Stobbe CC, Baer KA (1984) Metabolism induced binding of ^{14}C-misonidazole to hypoxic cells: kinetic dependence on oxygen concentration and misonidazole concentration. Int J Radiat Oncol Biol Phys 10:1327–1331

Koh WJ, Rasey JS, Evans ML, et al. (1992) Imaging of hypoxia in human tumors with [F-18] fluoromisonidazole. Int J Radiat Oncol Biol Phys 22:199–212

Kusuoka H, Hasimoto K, Fukuchi K, Nishimura T (1994) Kinetics of a putative hypoxic tissue marker, technetium-99m-nitroimidazole (BMS181321), in normoxic, hypoxic, ischemic and stunned myocardium. J Nucl Med 35:1371–1376

Kwock L, Gill M, McMurry HL, Beckman W, Raleigh JA, Joseph AP (1992) Evaluation of a fluorinated 2-nitroimidazole binding to hypoxic cells in tumor-bearing rats by 19F magnetic resonance spectroscopy and immunohistochemistry. Radiat Res 129:71–78

Linder KE, Chan YW, Cyr JE, Nowotuik DP, Eckelman WC, Nunn AD (1993) Synthesis, characterization and in vitro evaluation of nitroimidazole-BATO complexes: new technetium compounds designed for imaging hypoxic tissue. Bioconjugate Chem 4:326–333

Linder KE, Chan Y, Cyr JE, Malley MF, Nowotuik DP, Nunn AD (1994) TCO (PnA.0-1-(2-nitroimidazole)) [BMS181321], a new technetium-containing nitroimidazole complex for imaging hypoxia: synthesis, characterization and xanthine-oxidase-catalyzed reduction. J Med Chem 37:9–17

Ling L, Sutherland R (1986) Low glucose protects against the hypoxic toxicity of misonidazole. Int J Radiat Oncol Biol Phys 12:1237–1240

Lord EM, Harwell L, Koch CJ (1993) Detection of hypoxic cells by monoclonal antibody recognizing 2-nitroimidazole adducts. Cancer Res 53:5721–5726

MacManus MP, Maxwell AP, Abram WP, Bridges, JM (1989) The effect of hypobaric hypoxia on misonidazole binding in normal and tumor-bearing mice. Br J Cancer 59:349–352

Mannan RH, Somayaji VV, Lee J, Mercer JR, Chapman JD, Wiebe LI (1991) Radioiodinated 1-(5-iodo-5-deoxy-beta-D-arabinofuranosyl)-2-nitroimidazole (iodoazomycin arabinoside: IAZA): a novel marker of tissue hypoxia. J Nucl Med 32:1764–1770

Mannan RH, Mercer JR, Wiebe LI, Kumar P, Somayaji VV, Chapman JD (1992) Radioiodinated azomycin pyranoside

(IAZP): a novel non-invasive marker for the assessment of tumor hypoxia. J Nucl Biol Med 36:60–67

Martin GV, Biskupiak JE, Caldwell JH, Rasey JS, Krohn KA (1993) Characterization of iodovinylmisonidazole as a marker for myocardial hypoxia. J Nucl Med 34:918–924

Mason RP (1982) Free-radical intermediates in the metabolism of toxic chemicals. In: Pryor WA (ed) Free radicals in biology, vol 5. Academic, New York, pp 161–222

Mason RP, Holtzman JL (1975) The role of catalytic superoxide formation in the O_2 inhibition of nitroreductase. Biochem Biophys Res Commun 67:1267–1274

Maxwell AP, MacManus MP, Gardiner TA (1989) Misonidazole binding in murine liver tissue: a marker for cellular hypoxia in vivo. Gastroenterology 93:1300–1303

Maxwell RJ, Workman P, Griffiths JR (1989) Demonstration of tumor-selective retention of fluorinated nitroimidazole probes by 19F magnetic resonance spectroscopy in vivo. Int J Radiat Oncol Biol Phys 16:925–929

Mercer JR, Mannan RH, Somayaji VV, Lee J, Chapman JD, Wiebe LI (1990) Sugar-coupled 2-nitroimidazoles: novel in vivo markers for hypoxic tumor tissue. In: Maddalena DJ, Snowdon GM, Boniface GR (eds) Advances in radiopharmacology. 6th International Symposium on Radiopharmacology, Wollongong University, Wollongong, p 104

Miller GG, Ngan-Lee I, Chapman JD (1982) Intracellular localization of radioactively labelled misonidazole in EMT6 tumor cells in vitro. Int J Radiat Oncol Biol Phys 8:741–744

Miller GG, Best MW, Franko AJ, Koch CJ, Raleigh JA (1989) Quantitation of hypoxia in multicellular spheroids by video image analysis. Int J Radiat Oncol Biol Phys 16:949–952

Moore RB, Chapman JD, Mokrzanowkski AD, Arnfield MR, McPhee MS, McEwan AJB (1992) Non-invasive monitoring of photodynamic therapy with 99mTc-HMPAO. Br J Cancer 65:491–497

Mueller-Klieser W, Schlenger KH, Walenta S, Gross M, Karbach U, Hoeckel M, Vaupel P (1991) Pathophysiological approaches to identifying tumor hypoxia in patients. Radiother Oncol Suppl 20:21–28

Olive PL, Durand RE (1981) Uptake and mutagenicity of AF-2 in Chinese hamster V-79 spheroids under aerobic and hypoxic conditions. Environ Mutagen 3:659–670

Olive PL, Rasey JS, Durand RE (1986) Comparison between the binding of ^{3}H-misonidazole and AF-2 in Chinese hamster V79 spheroids. Radiat Res 105:105–114

Parliament MB, Wiebe LI, Franko AJ (1992a) Nitroimidazole adducts as markers for tissue hypoxia: mechanistic studies in aerobic normal tissues and tumor cells. Br J Cancer 66:1103–1108

Parliament MB, Chapman JD, Urtasun RC, et al. (1992b) Non-invasive assessment of human tumor hypoxia with ^{123}I-iodoazomycin arabinoside: preliminary report of a clinical study. Br J Cancer 65:90–95

Prekeges JL, Rasey JS, Grunbaum Z, Krohn KH (1991) Reduction of fluoromisonidazole, a new imaging agent for hypoxia. Biochem Pharmacol 42:2387–2395

Raleigh JA, Koch CJ (1990) Importance of thiols in the reductive binding of 2-nitroimidazoles to macromolecules. Biochem Pharmacol 40:2457–2464

Raleigh JA, Miller GG, Franko AJ, Koch CJ, Fuciarelli AF, Kelley DA (1987) Fluorescence immunohistochemical detection of hypoxic cells in spheroids and tumors. Br J Cancer 56:395–400

Raleigh JA, Franko AJ, Treiber EO, Lunt JA, Allen PS (1989) Covalent binding of a fluorinated 2-nitroimidazole to

EMT-6 tumors in BALB/c mice: detection by F-19 nuclear magnetic resonance at 2.35T. Int J Radiat Oncol Biol Phys 10:1337–1340

Raleigh JA, Franko AJ, Kelly DA, Trimble LA, Allen PS (1991) Development of an in vitro 19F magnetic resonance method for measuring oxygen deficiency in tumors. Magn Reson Med 22:451–466

Raleigh JA, Zeman EM, Rathman M, La Dine JK, Cline JM, Thrall DE (1992) Development of an ELISA for the detection of 2-nitroimidazole hypoxia markers bound to tumor tissue. Int J Radiat Oncol Biol Phys 22:403–405

Raleigh JA, La Dine JK, Cline JM, Thrall DE (1994) An enzyme-linked immunosorbent assay for hypoxia marker binding in tumors. Br J Cancer 69:66–71

Rampling R, Cruickshank G, Lewis AD, Fitzsimmons SA, Workman P (1994) Direct measurement of pO_2 distribution and bioreductive enzymes in human malignant brain tumors. Int J Radiat Oncol Biol Phys 29:427–431

Rasey JS, Krohn KA, Freauff S (1982) Bromomisonidazole: synthesis and characterization of a new radiosensitizer. Radiat Res 91:542–554

Rasey JS, Krohn KA, Grunbaum Z, Conroy PJ, Bauer K, Sutherland RM (1985) Further characterization of 4-bromomisonidazole as a potential detector of hypoxic cells. Radiat Res 102:76–85

Rasey JS, Nelson NJ, Clin L, Evans ML, Grunbaum Z (1990) Characteristics of the binding of labeled fluorom-isonidazole in cells in vitro. Radiat Res 122:301–308

Rauth AM (1984) Pharmacology and toxicology of sensitizers: mechanism studies. Int J Radiat Oncol Biol Phys 10:1293–1300

Rojas A, Joiner MC, Hodgkiss RJ, Carl U, Kjellen E, Wilson GD (1992) Enhancement of tumor radiosensitivity and reduced hypoxia-dependent binding of a 2-nitroimidazole with normobaric oxygen and carbogen: a therapeutic comparison with skin and kidneys. Int J Radiat Oncol Biol Phys 23:361–366

Rowell NP, McCready VR, Tait D, et al. (1989) Technetium-99m HMPAO and SPECT in the assessment of blood flow in human lung tumors. Br J Cancer 59:135–141

Rumsey WL, Cyr JE, Raju N, Narra RK (1993) A novel [99m] technetium-labelled nitroheterocycle capable of identification of hypoxia in heart. Biochem Biophys Res Commun 193:1239–1246

Smith BR, Born JL (1984) Metabolism and excretion of [^{3}H]-misonidazole by hypoxic rat liver. Int J Radiat Oncol Biol Phys 10:1365–1370

Stone HB, Brown JM, Phillips TL, Sutherland RM (1993) Oxygen in human tumors: correlations between methods of measurement and response to therapy. Summary of a workshop held November 19–20, 1992, at the National Cancer Institute, Bethesda, Maryland [Review]. Radiat Res 136:422–434

Suga K, Honma Y, Uchisako H, et al. (1991) Assessment of 99mTc-HMPAO tumor scintigraphy using VX-2 tumors implanted in a lower limb muscle of rabbits. Nucl Med Commun 12:611–619

Urtasun RC (1992) Tumor hypoxia, its clinical detection and relevance. In: Dewey WC, Edington M, Fry RJM, Hall EJ, Whitmore GF (eds) Radiation research: a twentieth century perspective. Academic, San Diego, p 727

Urtasun RC, Koch CJ, Franko AJ, Raleigh JA, Chapman JD (1986) A novel technique for measuring human tissue pO_2 at the cellular level. Br J Cancer 54:453–457

Valk PE, Mathis CA, Prados MD, Gilbert JC, Budinger TF (1992) Hypoxia in human gliomas: demonstration by PET with fluorine-18-fluoromisonidazole. J Nucl Med 33:2133–2137

Van Os-Corby DJ, Chapman JD (1986) In vitro binding of ^{14}C-misonidazole to hepatocytes and hepatoma cells. Int J Radiat Oncol Biol Phys 12:1251–1254

Van Os-Corby DJ, Koch CJ, Chapman JD (1987) Is misonidazole binding to mouse tissues a measure of cellular pO_2? Biochem Pharmacol 36:3487–3494

Varghese AJ, Whitmore GF (1980) Binding to cellular macromolecules as a possible mechanism for the cytotoxicity of misonidazole. Cancer Res 40:2165–2169

Wiebe LI, Jette DC, Chapman JD, Flanagan RJ, Meeker BE (1986) Iodoazomycin riboside [1-(5′-iodo-5′-deoxyribofuranosyl)-2-nitroimidazole], a hypoxic cell marker in vivo evaluation in experimental tumors. In: Winkler C (ed) Nuclear medicine in clinical oncology. Springer, Berlin Heidelberg New York, pp 402–407

Zeman EM, Calkins DP, Cline JM, Thrall DE, Raleigh JA (1993) The relationship between proliferative and oxygenation status in spontaneous canine tumors. Int J Radiat Oncol Biol Phys 27:891–898

10 Significance of Hemoglobin Concentration for Treatment Outcome

C. Grau[1] and J. Overgaard[2]

10.1 Introduction

The presence of hypoxic cells in solid tumors and its importance for radioresistance has been widely acknowledged (Thomlinson and Gray 1955; Brown 1979; Dische 1989; Overgaard 1993). Much of the indirect support for the influence of hypoxia on radiation response comes from observations of a correlation between tumor control and hemoglobin level (Dische 1991; Overgaard 1989; Overgaard and Horsman 1996). This review will give a brief overview of the influence of hemoglobin on tumor oxygenation and radiotherapy results, as well as a discussion of potential ways of modifying the hemoglobin effect.

The dynamics of the reaction of hemoglobin with oxygen make it a particularly suitable oxygen carrier. Hemoglobin is a protein made up of four subunits, each of which contains a heme moiety attached to a polypeptide chain. Heme is a complex consisting of a porphyrin and one atom of ferrous iron. Each of the four iron atoms can reversibly bind a single oxygen molecule. The binding is rapid, requiring less than 10 ms. The quaternary structure of hemoglobin determines its affinity for oxygen; by shifting the rela-

tionship of its four-component polypeptide chains, the molecule fosters either uptake or delivery of oxygen. When hemoglobin takes up a small amount of oxygen, structural change of the molecule facilitates additional uptake. This gives rise to the characteristic sigmoid shape of the oxygen–hemoglobin dissociation curve shown in Fig. 10.1. The position of the oxyhemoglobin dissociation curve is conveniently described by the p_{50}, i.e., the pO_2 value at which hemoglobin is 50% oxygenated. The p_{50} is physiologically altered by red blood cell 2,3-diphosphoglycerate (2,3-DPG), carbon dioxide, carbon monoxide (CO), pH and temperature.

The influence of hemoglobin level on oxygen utilization by tumors has been directly measured in vivo (Gullino et al. 1967). Under normal conditions, the experimental tumor was able to remove about 50% of the oxygen carried to the tumor by the afferent blood. Changes in hemoglobin level, brought about by acute or chronic anemia, caused a corresponding linear change in the total amount of utilized oxygen. This elegant study showed that the capacity of the tumor to utilize blood could not be saturated in vivo, neither by blood transfusions nor by oxygen breathing.

10.2 The Hemoglobin Effect in Radiotherapy

10.2.1 Clinical Studies with Anemic Patients

In 1909, Gottwald Schwarz in a simple but elegant experiment demonstrated that the radiation response of skin was markedly decreased if radiation was applied to an area with reduced blood flow due to compression (Schwarz 1909). These findings gave rise to the first impression of the magnitude of the oxygen enhancement ratio, and the concept of "Kompressionsanämie" was soon introduced. The radiation protection obtained by local anemia was initially explained by changes in metabolism in the

[1,2] C. Grau, MD and J. Overgaard, MD, Department of Experimental Clinical Oncology, Aarhus University Hospital, 44 Nørrebrogade, DK-8000 Aarhus C, Denmark

anemic tissue or reduced backscatter resulting from a decrease in hemoglobin-bound iron atoms within the irradiated area. It was not until the mid-1950s that the role of oxygen in radiation response was fully acknowledged (GRAY et al. 1953). The results were immediately taken into the clinic by the introduction of hyperbaric oxygen (CHURCHILL-DAVIDSON 1968). Since then, a considerable amount of clinical and experimental studies on the influence of hemoglobin level on hypoxia and radiosensitivity have been presented, and the topic has been reviewed several times (DISCHE 1991; OVERGAARD and HORSMAN 1996; OVERGAARD 1988, 1989). Most of the clinical studies published have shown better tumor control in patients with higher hemoglobin levels than in patients with hemoglobin in the lower part of – or below – normal range (Fig. 10.2). In a literature search of studies analyzing the prognostic relationship of hemoglobin concentration and local control, 39 studies with more than 14000 patients showed an effect, compared to 12 studies with a total of 2790 patients in which such a hemoglobin effect could not be found (Table 10.1). There seems to be equally good documentation for the effect of hemoglobin on radiation response in carcinoma of the uterine cervix, head and neck, bronchus and bladder.

Anemia due to bleeding is a common feature in carcinoma of the uterine cervix, and the first clinical observation of the hemoglobin effect in radiotherapy also comes from a study of patients with this type of

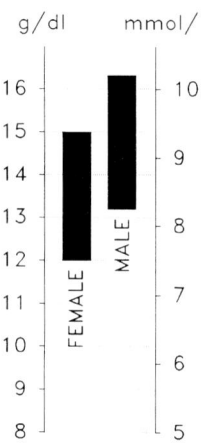

Fig. 10.2. The commonly accepted normal ranges for hemoglobin concentrations in males and females in g/dl (*left*) and mmol/l (*right*)

cancer (EVANS and BERGSJÖ 1965). Since blood transfusions were not readily available during World War II, a large proportion of the patients were treated with a hemoglobin value below 11 g/dl. For Stage II and III disease, a significant difference in crude survival based on hemoglobin level was observed, as shown in Fig. 10.3. This effect has subsequently been confirmed in a large majority of studies on cervical cancer. In a recent Danish retrospective study, a correlation between the initial hemoglobin concentration and treatment outcome was found (PEDERSEN et al. 1995), and this correlation continues throughout the entire normal hemoglobin range (Fig. 10.4).

In head and neck cancer, a similar correlation between hemoglobin concentration and locoregional tumor control has been established (OVERGAARD 1988), as shown in Fig. 10.5. In the Danish DAHANCA 2 trial, the hemoglobin effect was most pronounced in the subgroup of patients with pharyngeal and supraglottic laryngeal cancers, the same subgroup that was shown to benefit from the hypoxic cell radiosensitizer misonidazole (OVERGAARD et al. 1986, 1989b). It has been speculated that the apparent lack of hemoglobin effect in glottic laryngeal cancers could be attributed to the small size of these tumors (OVERGAARD 1988). However, other studies have shown a hemoglobin effect in laryngeal carcinoma as well. In a retrospective study of 306 patients with laryngeal cancer, VAN ACHT et al. (1992) examined the influence of hemoglobin values at different

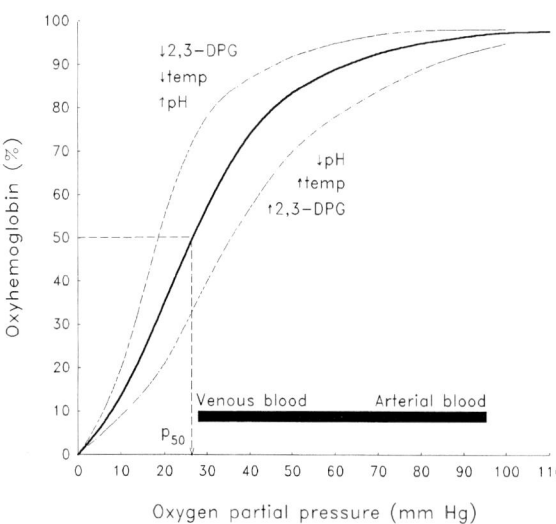

Fig. 10.1. The oxyhemoglobin dissociation curve and the principal factors that influence its position. The p_{50} and the normal range of pO_2 in arterial and venous blood are indicated

Table 10.1. Clinical studies – does hemoglobin level influence radiotherapy outcome?

Study	Yes (No. of patients)	No (No. of patients)	Endpoint
Carcinoma of uterine cervix			
Evans and Bergsjö 1965	1385		LC, S
Vigario et al. 1973	398		S
Thomson and Spratt 1977	91		S
Bush 1986	1055		LC
Kapp et al. 1983	910		LC, S
Mendenhall et al. 1984	264		LC, S
Revesz and Balmukhanov 1987	111		R
Freedman et al. 1987	150		S
Gonzales Gonzales et al. 1988	105		LC, S
Overgaard et al. 1989a	331		LC
Girinski et al. 1989	386		LC
Solberger and Sorbe 1990	151		LC, S
Rader et al. 1990	307		S
Tan et al. 1991	125		S
Hong et al. 1992	428		R
Chatani et al. 1994	200		S
Pedersen et al. 1995	424		LC, S
Werner-Wasik et al. 1995	125		LC
Johnson et al. 1983		295	LC, S
Awwad et al. 1986		44	LC
Obralic et al. 1990		121	S
Tsang et al. 1995		52	S
Head and neck carcinoma			
Taskinen 1969	999		S
Hansen 1975	1144		S
Blitzer et al. 1984	324		LC
Freedman et al. 187	258		S
Overgaard et al. 1989b	626		LC
Bentzen et al. 1991	181		LC
Overgaard et al. 1991	442		LC
Johansen et al. 1992	167		LC
van Acht et al. 1992	306		S
Fein et al. 1995	109		LC, S
Regueiro et al. 1995	90		LC, S
Schwaibold et al. 1988		58	LC
Fazekas et al. 1989		306	S
Sham and Choy 1990		759	S
Lindelov et al. 1990		145	LC
Johansen et al. 1990		213	LC
Hong et al. 1991		76	LC
Bronchogenic carcinoma			
Dische et al. 1986, 1988	754		S
Sasai et al. 1989	42		R, S
Oehler et al. 1990	264		R
Macchiarini et al. 1991	49		S
Hong et al. 1991		65	LC
Bladder carcinoma			
Quilty and Duncan 1986a,b; Quilty et al. 1986	889		LC, S
Gospodarowicz et al. 1989	121		LC, S
Greven et al. 1990	116		S
Wijkström et al. 1992	115		S
Hannisdal et al. 1993	202		S
Cole et al. 1995; Pollack et al. 1994	338		LC
Prostate carcinoma			
Dunphy et al. 1989		656	S
Total number of studies	39	12	
Total number of patients	14 482	2790	

LC, local control; S, survival; R, clinical response.

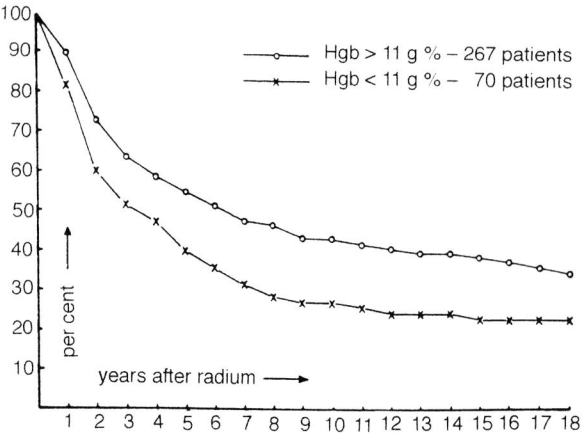

Fig. 10.3. Survival curves for clinical stage II carcinoma of the uterine cervix in relation to hemoglobin at the time of radium treatment. (Reprinted with permission from EVANS and BERGSJÖ 1965)

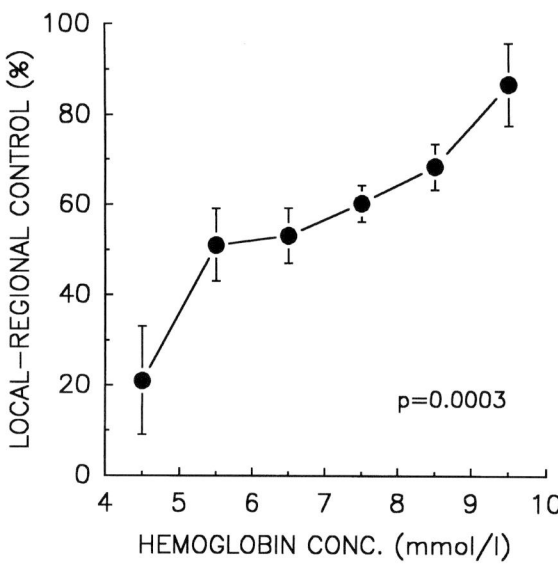

Fig. 10.4. Locoregional tumor control (i.e., tumor control within the irradiated area) as a function of pretreatment hemoglobin value in 424 patients treated with radiotherapy for squamous cell carcinoma of the uterine cervix. (From PEDERSEN et al. 1995)

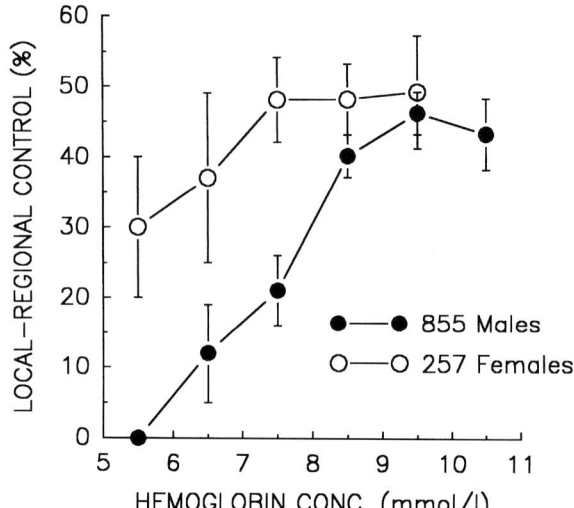

Fig. 10.5. Locoregional tumor control (i.e., tumor control within the irradiated area) as a function of sex and pretreatment hemoglobin value in 1112 patients treated with radiotherapy for squamous cell carcinoma of the larynx and pharynx. (Reprinted with permission from OVERGAARD 1988)

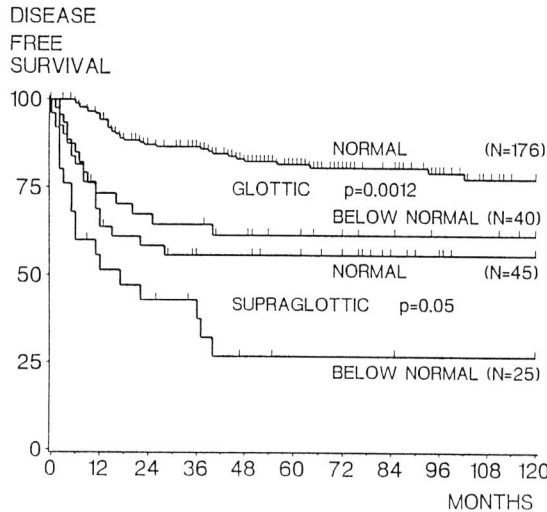

Fig. 10.6. Disease-free survival for glottic and supraglottic squamous cell carcinoma divided into normal and low hemoglobin values at day 35 after start of radiotherapy. (Reprinted with permission from VAN ACHT et al. 1992)

times during radiotherapy. They noted that in glottic carcinoma disease-free survival was significantly worse in patients with a hemoglobin value below normal either at the start or at the end of therapy. In patients with supraglottic carcinoma, this effect was only significant for hemoglobin values at the end of treatment (Fig. 10.6). In a multivariate analysis, only the hemoglobin value at the end of treatment – and not the pretreatment value – was found to be prog-

nostic next to T and N classification. These data are of considerable interest since most previous clinical reports have focussed on initial hemoglobin values. The data are not fully in agreement with the experimental knowledge discussed below, and the issue deserves further investigation. If a progressive drop in hemoglobin during therapy causes radiation resistance, this could imply that measures to increase hemoglobin should be tailored to counteract this drop continuously, and erythropoietin could be a suitable agent.

The probability of normal tissue damage, such as myelitis, also appears to increase with increasing hemoglobin concentration (DISCHE et al. 1986), a finding reminiscent of the damage to some normal tissues (such as laryngeal cartilage) exposed to hyperbaric oxygen (DISCHE 1991).

10.2.2
Acute Versus Chronic Anemia

The radiobiology of the hemoglobin effect has been studied in several experimental systems and has been reviewed by HIRST (1991). Mottram and Edinow already noted in 1932 that skin and tumor reactions were diminished in rats rendered acutely anemic by heart puncture (MOTTRAM and EDINOW 1932). Acute anemia may also be brought about by puncturing the orbital sinus, and chronic anemia may be induced by kidney irradiation, progressive tumor growth and low iron diet. Restoration of normal hemoglobin levels may be achieved by blood transfusion or erythropoietin therapy. It is evident from results of the experimental studies that a distinction between acute and chronic anemia must be made. Mice made acutely hypoxic by bleeding generally have been found to have very radioresistant tumors, whereas tumors in mice that had been chronically anemic (more than 6–48h) showed radiosensitivity close to that of tumors in nonanemic

mice, and transfusion of chronically anemic mice seems to lead to increased radiosensitivity (HIRST 1986; HIRST and WOOD 1987b). The difference between acute and chronic anemia in terms of influence on radiosensitivity led to the introduction of the two adaptation theories presented in Fig. 10.7 (HIRST 1986). The *reduced cord radius theory* is based on histological examination of tumors in animals exposed to low oxygen tension for several days, during which a reduced thickness of the perivascular tumor cord has been observed (HIRST et al. 1991). The transient increase in severe diffusion-limited hypoxia led to cell death and cord shrinkage. It is speculated that when these tumors are retransfused, the radius of the tumor cord is closer to or within the diffusion distance of oxygen, resulting in an increased percentage of cells receiving an oxygen supply sufficient for improvement not only in radiosensitivity, but also in increased proliferation. In the *increased oxygen availability theory*, the systemic response to anemia, an increase in red blood cell 2,3-DPG and subsequent increased p_{50} lead to adaptation of normal tumor oxygenation due to improved oxygen unloading. Retransfusion results in reduced hypoxia, but this effect is soon lost due to compensatory reduction in 2,3-DPG levels. To date, the experimental data do not particularly favor either of the two theories. For the clinical situation, it is important to realize that radiation-induced proliferation blockage may hinder the adaptive response integrated in the reduced cord theory. A practical consequence of this theory would be that any correc-

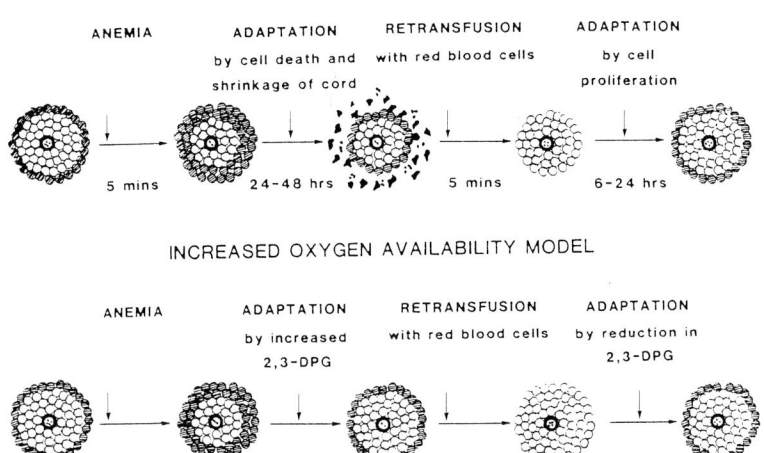

Fig. 10.7. Diagrammatic representation of a tumor blood vessel surrounded by its dependent volume of tumor cells. The effects on this model of anemia and subsequent transfusion with red blood cells are shown for two independent mecha-

nisms of tumor adaptation, one or both of which may be important in determining the radiobiological hypoxic fraction. Radiobiological hypoxic cells are depicted by *shading*. (Reprinted with permission from HIRST 1986)

tion of anemia (transfusions or erythropoietin) should not commence until a sufficient blockage of tumor cell proliferation has been established at the start of radiotherapy.

10.2.3
Blood Transfusions

The observation that anemic patients had poorer treatment outcome naturally led to the question of whether blood transfusion to patients with low hemoglobin would be beneficial in terms of improving tumor oxygenation and response to radiotherapy. This topic has to some extent been studied in women with advanced squamous cell carcinoma of the uterine cervix. In a small, but careful study (EVANS and BERGSJÖ 1965) it was demonstrated that transfusion led to increased oxygen tension within the tumor, as measured directly by oxygen electrode. The same study was also the first to show that transfusion to a hemoglobin level ≥11 g/dl improved survival significantly. This has been confirmed by Bush and co-workers in a small prospective randomized trial, also in carcinoma of the uterine cervix (BUSH 1986), and the results are illustrated in Fig. 10.8. Control patients were not transfused unless their hemoglobin level dropped below 10 g/dl. The other group had transfusions as required to maintain their hemoglobin level above 13.5 g/dl. There was a significantly improved local control rate in the low-hemoglobin patients if transfusion was given prior to radiotherapy. This represents the only published randomized study with retransfusion of anemic patients. Together with the retrospective studies showing no increase in the risk of distant metastasis for anemic patients (OVERGAARD et al. 1989a), the transfusion

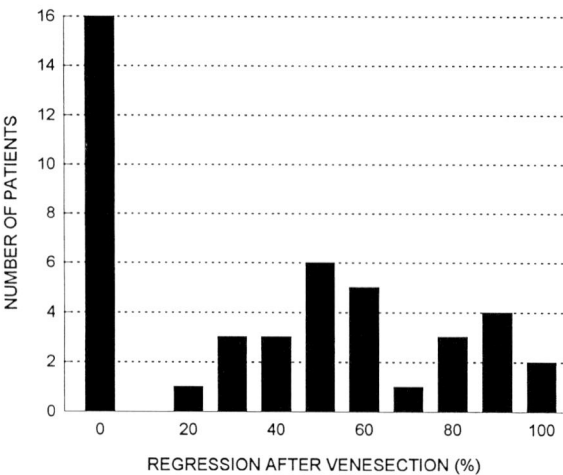

Fig. 10.9. Observed tumor regression after 1 week of induced anemia (<7.28 g/dl), obtained by isovolemic venesection in 53 patients with advanced head and neck squamous cell carcinoma. (Adapted from SEALY et al. 1989)

results suggest that anemia is related to increased radiation resistance rather than aggressive disease.

The theoretically attractive concept of autologous retransfusion after induced anemia has been studied in patients with squamous cell carcinoma of the head and neck (SEALY et al. 1989). Before radiotherapy, the patients were venesected twice to different hemoglobin levels, either 7.3 g/dl (37 patients) or 5.8 g/dl (16 patients). One week after the first drainage, patients were autotransfused to a hemoglobin range of 12.5–14.0 g/dl. Most interestingly, the week of anemia was found to cause both subjective pain relief and measurable tumor regression (Fig. 10.9). The patients commenced radiotherapy in hyperbaric oxygen chambers the day after retransfusion. Based on a comparison with historical controls, the overall treatment results were, despite the initial volume reduction, not improved. The authors note that, at least in some tumors, rapid regrowth was observed after retransfusion.

10.2.4
Erythropoietin

Anemia in cancer patients is associated with low serum levels of erythropoietin, a hormone secreted by the kidneys in response to tissue hypoxia (MILLER et al. 1990). Because of the recognized risk of chronic or acute transfusion reactions, as well as infections and immunosuppression associated with blood transfusions, there has been increasing focus on the use of

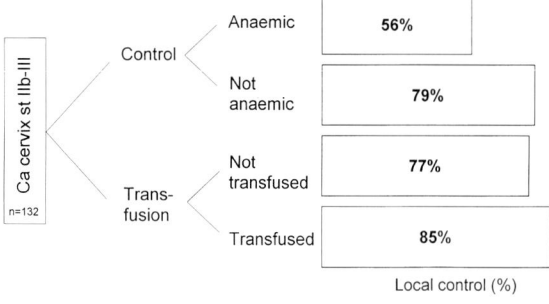

Fig. 10.8. The local failure rate in 132 patients (*pts*) with stage IIB and III squamous cell carcinoma (*ca*) of the uterine cervix receiving blood transfusions (*filled bars*) or no transfusion (*open bars*) before radiotherapy. (Adapted from BUSH 1986)

recombinant human erythropoietin in the treatment of anemia. So far, erythropoietin has been approved for treatment of anemia associated with chronic renal failure or AIDS. However, erythropoietin has also been investigated in cancer patients. In a phaseI/II study of 40 patients with hemoglobin levels below 13.5 g/dl undergoing radiotherapy for malignant tumors above the diaphragm, LAVEY and DEMPSEY (1993) showed a significant increase in hemoglobin in the erythropoietin-treated group as compared to controls (Fig. 10.10). Erythropoietin was administered subcutaneously three times per week in a dose of 150–300 u/kg. In 80% of the erythropoietin-treated patients, the hemoglobin level increased from a mean of 11.9 g/dl to >14 g/dl. The treatment was well tolerated; only one patient experienced a skin rash. In a preliminary report of a phase II randomized trial, VIJAYAKUMAR et al. (1993) showed similar results after 200-u/kg injections five times weekly in 14 patients. The same dosage was used in a study of 20 patients with carcinoma of the uterine cervix (DUSENBERY et al. 1994). The mean hemoglobin in the 15 patients treated with erythropoietin increased from 10.3 to 13.2 g/dl. Larger phase III randomized trials are needed to establish whether the rather slow increase in hemoglobin is sufficient to result in a therapeutic benefit. The sparse animal data so far do not indicate that this will be the case. There was no effect on radiosensitivity in anemic mice treated with erythropoietin (JOINER et al. 1993). In chronically anemic rats with DS-sarcomas, erythropoietin com-

pletely corrected anemia, but only partially reversed tumor hypoxia as measured by pO_2 (KELLEHER et al. 1995). One problem with erythropoietin in experimental systems may be the rapid tumor growth and short treatment temporal window compared to the clinical situation. Continuous erythropoietin treatment in a 6- to 8-week radiotherapy course may prove beneficial despite the lack of success in experimental systems.

Although hemoglobin is the major oxygen carrier, it should be briefly mentioned that increasing the amount of dissolved oxygen in the blood may also be important for tissue oxygenation. In addition to the early experiences with hyperbaric oxygen, the effect of breathing normobaric oxygen or carbogen on tumor oxygenation and radiation sensitivity has been widely documented. Recent studies have shown that the amount of physically dissolved oxygen may be further increased by combining oxygen breathing with chemical substances, mostly perfluorocarbon emulsions, that have a high oxygen carrying capacity. Combinations of carbogen breathing and nicotinamide, an agent known to prevent perfusion-limited hypoxia, have also been successful in the laboratory. Several of these treatments have reached clinical trials, and may prove to be a realistic way of sensitizing hypoxic cells in solid tumors to irradiation. These aspects are discussed elsewhere in this book.

10.3
Modification of Hemoglobin-Oxygen Affinity

The red blood cell 2,3-DPG content is one of the most important allosteric factors controlling the position of the oxygen dissociation curve (BUNN and FORGET 1986). The radiobiological effects of increased p_{50} due to elevated red blood cell 2,3-DPG was investigated in experimental tumor-bearing mice exposed to 12% oxygen for 36h before the tumors were irradiated under aerobic conditions (SIEMANN and MACLER 1986). Increased 2,3-DPG caused a considerable decrease of the hypoxic fraction, and thus increased radiation response. Similarly, when air-breathing mice were exchange-transfused with 2,3-DPG-enriched blood obtained from mice kept at 10% oxygen, a significant increase in radiation response was observed compared to mice exchange-transfused with blood from normal air-breathing mice (HIRST and WOOD 1987a). In the same setup, exchange transfusion with blood from pure-oxygen-

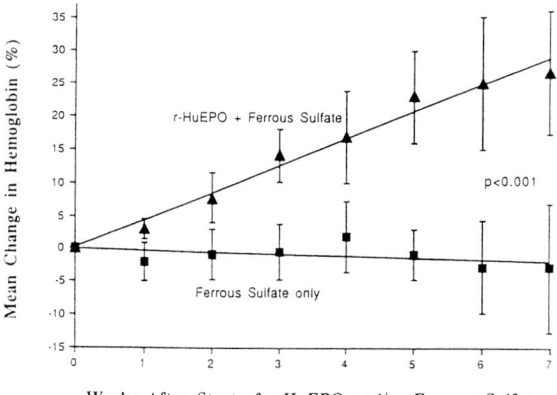

Fig. 10.10. Mean percent change in hemoglobin concentration during radiation therapy from the start of treatment for patients receiving erythropoietin (*r-HuEPO*) and ferrous sulfate (*filled triangles*) or ferrous sulfate only (*filled squares*). The *error bars* indicate 95% confidence intervals for each week's set of hemoglobin measurements. The difference between least-square lines is highly significant (*p* < 0.001). (Reprinted with permission from LAVEY and DEMPSEY 1993)

breathing mice (with low 2,3-DPG levels) reduced the radiation response (Hirst and Wood 1987a). Although these elegant experiments demonstrate a potentially exploitable biological mechanism, it must be realized that many other factors influence hemoglobin oxygen affinity, and it is still questionable whether a therapeutic manipulation of 2,3-DPG is possible in a clinical environment. As discussed by Hirst and Wood (1991), it is probably unrealistic to consider transfusing patients with large volumes of blood with reduced oxygen affinity even though techniques involving in vitro incubation with inosine, pyruvate and phosphate aimed at stimulating 2,3-DPG production or including the potent allosteric modifier inositol hexaphosphate have been successful in animals (Siemann and Macler 1986). It is probably also unrealistic to consider maintaining radiotherapy patients in a reduced-oxygen environment during an entire course of radiotherapy except at the time of irradiation. An alternative approach, which may be more attractive to the clinician, has been to develop drugs that reduce hemoglobin–oxygen affinity directly (Hirst and Wood 1991; Siemann et al. 1996). In a search for possible agents against sickle cell anemia, it was found that clofibric acid shifted the oxygen dissociation curve to the right (Abraham et al. 1983; Siemann et al. 1996). A wide range of structurally related antihyperlipidemic drugs, primarily chlorophenoxy acetic acid derivatives, have been tested for their right-shifting ability (Wootton 1984; Perutz and Poyart 1983) and potential use as radiation sensitizers (Hirst and Wood 1989; Hirst et al. 1987; Calais and Hirst 1991). Most of the agents tested were found to increase the p_{50} for substantial time, but the results in terms of radiosensitization have been divergent. So far, none of the agents have been tested as radiosensitizers in a clinical setting.

Modification of hemoglobin–oxygen affinity to *increase* hypoxia has been of some interest for use in combination with specific anticancer agents with hypoxic toxicity. The most widely studied agents are BW12C and its derivatives (Adams et al. 1986, 1989). These agents significantly modify oxygen–hemoglobin affinity in a time- and dose-dependent fashion. The effect on radiation response in both normal tissues and various experimental tumors has also been well established (Adams et al. 1986, 1989; Honess et al. 1989; Horsman and Overgaard 1992). However, recent studies have suggested that the observed radiation modification may be related more to BW12C-induced blood flow reductions than

to the change in p_{50} (Horsman and Overgaard 1992; Honess et al. 1991).

10.4
Tobacco Smoking

The physiological effects of tobacco smoking on oxygen transport to tumors constitute a special problem in radiotherapy. Smokers inhale small amounts of carbon monoxide which reacts with hemoglobin to form carboxyhemoglobin (HbCO), and heavy smokers may have up to 16%–18% HbCO (Overgaard et al. 1992; Bunn and Forget 1986; Nordenberg et al. 1990). This reaction decreases the amount of hemoglobin available for oxygen transport, and furthermore causes a left shift of the oxygen dissociation curve (Roughton and Darling 1944). The effects of HbCO on murine tumors have been evaluated in different systems in which the mice breathed clinically relevant concentrations of CO (Grau et al. 1992, 1994a,b; Siemann et al. 1978). A time- and dose-dependent formation of HbCO, a significant reduction of p_{50}, and also a significant reduction in tumor blood perfusion have been consistently found. The sum of these factors was a 30%–40% decrease in tumor oxygen supply within the clinically relevant HbCO range (Grau et al. 1994a). In good agreement with this, direct oxygen electrode measurement of tumor pO_2 has shown a clear relationship between CO concentration and low pO_2 measurements (Grau et al. 1994a; Horsman et al. 1993). CO-induced tumor hypoxia also resulted in significantly poorer radiation response (Siemann et al. 1978; Grau et al. 1992, 1994a,b). The effect of tobacco smoking on oxygen unloading capacity and treatment outcome is currently being analyzed in a prospective study of patients with squamous cell carcinoma of the head and neck treated with curative radiotherapy alone (Overgaard and Grau 1995, unpublished). Patients were questioned about their smoking habits. Prior to and weekly during treatment, a venous blood gas analysis was performed comprising estimation of total hemoglobin, HbCO and p_{50}. In some patients, arterial pO_2 was also estimated prior to treatment. A total of 229 consecutive patients (49 females and 180 males) were included in the study between November 1988 and December 1991. Radiotherapy was given with conventional fractionation (62–68 Gy, 2 Gy/fraction, 5 fractions/week). The tumor sites were glottic (100 patients), supraglottic (40 patients), pharynx (51 patients), subglottic (3 patients) and oral cavity (35 patients). The UICC classifications were: stage I: 80; stage II: 63; stage III: 50 and stage IV: 36. Smoking habits were classified as: never (10 patients), long-time quitters (34 patients), recent quitters (25 patients) moderate (less than a pack per day) smokers (57 patients) and heavy smokers (103 patients). Blood gas analysis showed no difference between the different groups of "nonsmokers" and the two groups of "smokers," respectively, and the final evaluation was therefore simplified to only two categories. No differences in total Hb concentrations were observed between smokers and nonsmokers, but smokers had significantly higher HbCO values (mean: 2.8% vs. 5.3%, $p < 0.001$), resulting in significantly lower values of effective hemoglobin, arterial oxygen delivery and tumor oxygen unloading capacity (OVERGAARD et al. 1992). All patients completed treatment, and smoking did not significantly affect treatment compliance. Actuarial univariate analysis showed that favorable locoregional tumor control, and corrected and crude survival were significantly associated with: laryngeal tumor, low stage, low T classification, no nodes and no heavy smoking during therapy. The 5-year locoregional tumor control was 59% in heavy smokers compared to 41% in nonsmokers (Fig. 10.11). A Cox multivariate analysis confirmed that T and N classification, tumor site and smoking were the most important independent prognostic factors. When included in the analysis, the various hemoglobin parameters measured were of high significance, the most important being the arterial oxygen supply to the tumor. It is concluded that smoking during radiotherapy of head and neck cancer has a signifi-

cant detrimental effect on both locoregional tumor control and survival. The effect can to a large extent be explained by a reduced tumor oxygen supply caused by the increased HbCO concentration in smokers. In addition to death due to insufficient treatment of the cancer in question, smoking patients also had a significantly increased risk of early death from other causes, including new primary cancers. Another study in head and neck cancer has shown a similar effect of tobacco smoking on survival after radiotherapy (BROWMAN et al. 1993). The available data strongly suggest that smoking during – and after – radiotherapy should be avoided in order to improve the therapeutic efficacy of radiotherapy. If a complete stop is not possible, patients should refrain from smoking in the mornings before radiotherapy, since clinical data suggest that even a 12-h refrain from smoking allows HbCO and p_{50} to revert to normal values (KAMBAM et al. 1986).

10.5
Summary

During the past 60 years, the vast majority of clinical studies in carcinoma of the uterine cervix, head and neck, bladder and bronchus have shown a significant influence of hemoglobin level on the outcome of radiotherapy. This effect has been confirmed in a few blood transfusion studies, but further randomized studies are necessary. Experimental radiobiological knowledge suggests that hypoxia resulting from anemia caused by bleeding may increase radioresistance, whereas chronic anemia causes little change in the response to radiotherapy. Transfusion to anemic tumor-bearing mice only transiently increases radiosensitivity. The biological findings have been explained by adaptation in tumor cord size and/or hemoglobin oxygen binding. Attempts to chemically or physiologically change hemoglobin-oxygen affinity have been successful in the laboratory, but none of the manipulations have reached clinical significance so far. Tobacco smoking constitutes a special problem for oxygen transport to tumors due to carbon monoxide in the tobacco smoke. Carbon monoxide binds strongly to hemoglobin, decreases the amount of effective hemoglobin, increases hemoglobin affinity for oxygen and reduces tumor blood flow. The sum of these effects is a significant increase in tumor hypoxia and radioresistance. Clinical observations also suggest that heavily smoking patients with squamous cell carcinoma of the head and neck have poorer treat-

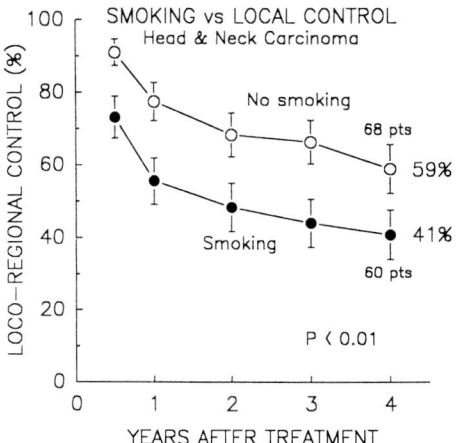

Fig. 10.11. Influence of smoking during treatment on the outcome of radiotherapy in 128 patients (*pts*) with advanced head and neck carcinoma. (Reprinted with permission from OVERGAARD and HORSMAN 1996)

ment outcome after primary radiotherapy, and increased risk of new cancers. It is therefore important to encourage these patients to quit smoking, at least during radiotherapy.

References

Abraham DJ, Perutz MF, Philips SEV, Philips SV (1983) Physiological and x-ray studies of potential antisickling agents. Proc Natl Acad Sci USA 80:324–328

Adams GE, Barnes DW, du-Boulay C, et al. (1986) Induction of hypoxia in normal and malignant tissues by changing the oxygen affinity of hemoglobin – implications for therapy. Int J Radiat Oncol Biol Phys 12:1299–1302

Adams GE, Stratford IJ, Nethersell AB, White RD (1989) Induction of severe tumor hypoxia by modifiers of the oxygen affinity of hemoglobin. Int J Radiat Oncol Biol Phys 16:1179–1182

Awwad HK, el Naggar M, Mocktar N, Barsoum M (1986) Intercapillary distance measurement as an indicator of hypoxia in carcinoma of the cervix uteri. Int J Radiat Oncol Biol Phys 12:1329–1333

Bentzen SM, Johansen LV, Overgaard J, Thames HD (1991) Clinical radiobiology of squamous cell carcinoma of the oropharynx. Int J Radiat Oncol Biol Phys 20:1197–1206

Blitzer PH, Wang CC, Suit HD (1984) Blood pressure and hemoglobin concentration: multivariate analysis of local control after irradiation for head and neck cancer. Int J Radiat Oncol Biol Phys 10:98

Browman GP, Wong G, Hodson I, et al. (1993) Influence of cigarette smoking on the efficacy of radiation therapy in head and neck cancer. N Engl J Med 328:159–163

Brown JM (1979) Evidence for acutely hypoxic cells in mouse tumours, and a possible mechanism of reoxygenation. Br J Radiol 52:650–656

Bunn HF, Forget BG (1986) Hemoglobin: molecular, genetic and clinical aspects. Saunders, Philadelphia

Bush RS (1986) The significance of anemia in clinical radiation therapy. Int J Radiat Oncol Biol Phys 12:2047–2050

Bush RS, Jenkin RD, Allt WE, Beale FA, Bean H, Dembo AJ, Pringle JF (1978) Definitive evidence for hypoxic cells influencing cure in cancer therapy. Br J Cancer Suppl 37:302–306

Calais G, Hirst DG (1991) In situ tumour radiosensitization induced by clofibrate administration: single dose and fractionated studies. Radiother Oncol 22:99–103

Chatani M, Matayoshi Y, Masaki N, Teshima T, Inoue T (1994) Long term follow-up results of high-dose rate remote afterloading intracavitary radiation therapy for carcinoma of the uterine cervix. Strahlenther Onkol 170:269–276

Churchill-Davidson I (1968) The oxygen effect in radiotherapy – historical review. Front Radiat Ther Oncol 1:1–15

Cole CJ, Pollack A, Zagars GK, Dinney CP, Swanson DA, von Eschenbach AC (1995) Local control of muscle-invasive bladder cancer: preoperative radiotherapy and cystectomy versus cystectomy alone. Int J Radiat Oncol Biol Phys 32:331–340

Dische S (1989) In The biological basis of radiotherapy. In: Steel GG, Adams GE, Horwich A (eds) The clinical consequences of the oxygen effect, 2nd edn. Elsevier, Amsterdam, p 135

Dische S (1991) Radiotherapy and anaemia – the clinical experience. Radiother Oncol 20:35–40

Dische S, Saunders MI, Warburton MF (1986) Hemoglobin, radiation, morbidity and survival. Int J Radiat Oncol Biol Phys 12:1335–1337

Dische S, Warburton MF, Saunders MI (1988) Radiation myelitis and survival in the radiotherapy of lung cancer. Int J Radiat Oncol Biol Phys 15:75–81

Dunphy EP, Petersen IA, Cox RS, Bagshaw MA (1989) The influence of initial hemoglobin and blood pressure levels on results of radiation therapy for carcinoma of the prostate. Int J Radiat Oncol Biol Phys 16:1173–1178

Dusenbery KE, McGuire WA, Holt PJ, Carson LF, Fowler JM, Twiggs LB, Potish RA (1994) Erythropoietin increases hemoglobin during radiation therapy for cervical cancer. Int J Radiat Oncol Biol Phys 29:1079–1084

Evans JC, Bergsjö P (1965) The influence of anemia on the results of radiotherapy in carcinoma of the cervix. Radiology 84:709–717

Fazekas JT, Scott C, Marcial V, Davis LW, Wasserman T, Cooper JS (1989) The role of hemoglobin concentration in the outcome of misonidazole-sensitized radiotherapy of head and neck cancers: based on RTOG trial #79-15. Int J Radiat Oncol Biol Phys 17:1177–1181

Fein DA, Lee WR, Hanlon AL, Ridge JA, Langer CJ, Curran WJJ, Coia LR (1995) Pretreatment hemoglobin level influences local control and survival of T1–T2 squamous cell carcinomas of the glottic larynx. J Clin Oncol 13:2077–2083

Freedman LS, Honess DJ, Bleehen NM, Adams GE, Dische S (1987) Does initial haemoglobin level modify the efficacy of radiosensitizers? An analysis of the MRC misonidazole studies in head and neck cancer and cervix cancer. Int J Radiat Biol 52:963–967

Girinski T, Pejovic-Lenfant MH, Bourhis J, et al. (1989) Prognostic value of hemoglobin concentrations and blood transfusions in advanced carcinoma of the cervix treated by radiation therapy: results of a retrospective study of 386 patients. Int J Radiat Oncol Biol Phys 16:37–42

Gonzales Gonzales D, van Dijk JD, van Klinken JW (1988) Further evidence on the role of the hemoglobin as prognostic factor in carcinoma of the uterine cervix stage IIB and III (abstract). 7th Annual Meeting of ESTRO, 5–8 September, Der Haag, The Netherlands, no. 82

Gospodarowicz MK, Hawkins NV, Rawlings GA, et al. (1989) Radical radiotherapy for muscle invasive transitional cell carcinoma of the bladder: failure analysis. J Urol 142:1448–1453

Grau C, Horsman MR, Overgaard J (1992) Influence of carboxyhemoglobin level on tumor growth, blood flow, and radiation response in an experimental model. Int J Radiat Oncol Biol Phys 22:421–424

Grau C, Khalil AA, Nordsmark M, Horsman MR, Overgaard J (1994a) The relationship between carbon monoxide breathing, tumour oxygenation and local tumour control in the C3H mammary carcinoma in vivo. Br J Cancer 69:50–57

Grau C, Nordsmark M, Khalil AA, Horsman MR, Overgaard J (1994b) Effect of carbon monoxide breathing on hypoxia and radiation response in the SCCVII tumor in vivo. Int J Radiat Oncol Biol Phys 29:449–454

Gray LH, Conger AD, Ebert M, Hornsey S, Scott OCA (1953) The concentration of oxygen dissolved in tissues at the time of irradiation as a factor in radiotherapy. Br J Radiol 26:638–648

Greven KM, Solin LJ, Hanks GE (1990) Prognostic factors in patients with bladder carcinoma treated with definitive irradiation. Cancer 65:908–912

Gullino PM, Grantham FH, Courtney AH (1967) Utilization of oxygen by transplanted tumors in vivo. Cancer Res 27:1020–1030

Hannisdal E, Fossa SD, Host H (1993) Blood tests and prognosis in bladder carcinomas treated with definitive radiotherapy. Radiother Oncol 27:117–122

Hansen HS (1975) Neoplasma malignum laryngis, Polyteknisk, Copenhagen

Hirst DG (1986) Anemia: a problem or an opportunity in radiotherapy? Int J Radiat Oncol Biol Phys 12:2009–2017

Hirst DG (1991) What is the importance of anaemia in radiotherapy? The value of animal studies. Radiother Oncol 20 [Suppl 1]:29–33

Hirst DG, Wood PJ (1987a) The influence of haemoglobin affinity for oxygen on tumour radiosensitivity. Br J Cancer 55:487–491

Hirst DG, Wood PJ (1987b) The adaptive response of mouse tumours to anaemia and retransfusion. Int J Radiat Biol Relat Stud Phys Chem Med 51:597–609

Hirst DG, Wood PJ (1989) Chlorophenoxy acetic acid derivatives as hemoglobin modifiers and tumor radiosensitizers. Int J Radiat Oncol Biol Phys 16:1183–1186

Hirst DG, Wood PJ (1991) Could manipulation of the binding affinity of haemoglobin for oxygen be used clinically to sensitize tumours to radiation? Radiother Oncol 20:53–57

Hirst DG, Wood PJ, Schwartz HC (1987) The modification of hemoglobin affinity for oxygen and tumor radiosensitivity by antilipidemic drugs. Radiat Res 112:164–172

Hist DG, Hirst VK, Joiner B, Prise V, Shaffi KM (1991) Changes in tumour morphology with alterations in oxygen availability: further evidence for oxygen as a limiting substrate. Br J Cancer 64:54–58

Honess DJ, White RD, Nethersell ABW, Bleehen NM (1989) Effects of manipulation of oxyhaemoglobin status by BW12C on tumor thermosensitivity and on blood flow in tumor and normal tissues in mice. Int J Radiat Oncol Biol Phys 16:1187–1190

Honess DJ, Hu DE, Bleehen NM (1991) BW12C: effects on tumour hypoxia, tumour radiosensitivity and relative tumour and normal tissue perfusion in C3H mice. Br J Cancer 64:715–722

Hong A, Dische S, Saunders MI, Lockwood P, Crocombe K (1991) Lung function and radiation response. Br J Radiol 64:1134–1139

Hong JH, Chen MS, Lin FJ, Tang SG (1992) Prognostic assessment of tumor regression after external irradiation for cervical cancer. Int J Radiat Oncol Biol Phys 22:913–917

Horsman MR, Overgaard J (1992) BW12C-induced changes in haemoglobin-oxygen affinity in mice and its influence on the radiation response of a C3H mouse mammary carcinoma. Radiother Oncol 25:43–48

Horsman MR, Khalil AA, Nordsmark M, Grau C, Overgaard J (1993) Relationship between radiobiological hypoxia and direct estimates of tumour oxygenation in a mouse tumour model. Radiother Oncol 28:69–71

Johansen LV, Overgaard J, Overgaard M, Birkler N, Fisker A (1990) Squamous cell carcinoma of the oropharynx: an analysis of 213 consecutive patients scheduled for primary radiotherapy. Laryngoscope 100:985–990

Johansen LV, Mestre M, Overgaard J (1992) Carcinoma of the nasopharynx: analysis of treatment results in 167 consecutively admitted patients. Head Neck 14:200–207

Johnson DW, Cox RS, Billingham G, Ung N, Martinez A (1983) Survival, prognostic factors, and relapse patterns in uterine cervical carcinoma. Am J Clin Oncol 6:407–415

Joiner B, Hirst VK, McKeown SR, McAleer JJ, Hirst DG (1993) The effect of recombinant human erythropoietin treatment on tumour radiosensitivity and cancer-associated anaemia in the mouse. Br J Cancer 68:720–726

Kambam JR, Chen LH, Hyman SA (1986) Effect of short-term smoking halt on carboxyhemoglobin levels and p_{50} values. Anesth Analg 65:1186–1188

Kapp DS, Fischer D, Gutierrez E, Kohorn EI, Schwartz PE (1983) Pretreatment prognostic factors in carcinoma of the uterine cervix. Int J Radiat Oncol Biol Phys 9:445–455

Kelleher DK, Matthiensen U, Thews O, Vaupel P (1995) Tumor oxygenation in anemic rats: effects of erythropoietin treatment versus red blood cell transfusion. Acta Oncol 34:379–384

Lavey RS, Dempsey WH (1993) Erythropoietin increases hemoglobin in cancer patients during radiation therapy. Int J Radiat Oncol Biol Phys 27:1147–1152

Lindelov B, Lauritzen AF, Hansen HS (1990) Stage I glottic carcinoma: an analysis of tumour recurrence after primary radiotherapy. Clin Oncol (R Coll Radiol) 2:94–96

Macchiarini P, Silvano G, Janni A, Mussi A, Chella A, Angeletti CA (1991) Results of treatment and lessons learned from pathologically staged T4 non-small cell lung cancer. J Surg Oncol 47:209–214

Mendenhall WM, Thar TL, Bova FJ, Marcus RB, Morgan LS, Million RR (1984) Prognostic and treatment factors affecting pelvic control of stage IB and IIA-B carcinoma of the intact uterine cervix treated with radiation therapy alone. Cancer 53:2649–2654

Miller C, Jones RJ, Piantadosi S, Abeloff MD, Spivak JL (1990) Decreased erythropoietin response in patients with the anemia of cancer. N Engl J Med 322:1689–1692

Mottram JC, Edinow A (1932) On the effect of anaemia on the reactions of the skin and of tumours to radium exposure. Br J Surg 19:481–487

Nordenberg D, Yip R, Binkin NJ (1990) The effect of cigarette smoking on hemoglobin levels and anemia screening. JAMA 264:1556–1559

Obralic N, Bilenjki D, Bilbija Z (1990) Prognostic importance of anemia related parameters in patients with carcinoma of the cervix uteri. Acta Oncol 29:199–201

Oehler W, Fischer J, Merkle K (1990) Beeinflusst der initiale Haemoglobinwert die Primärtumorreaktion? Eine Untersuchung an 264 bestrahlten Bronchialkarzinomen. Radiobiol Radiother (Berl) 31:325–331

Overgaard J (1988) The influence of haemoglobin concentration on the response to radiotherapy. Scand J Clin Lab Invest 48:49–53

Overgaard J (1989) Sensitization of hypoxic tumour cells – clinical experience. Int J Radiat Biol 56:801–811

Overgaard J (1993) Advances in clinical applications of radiobiology: phase III studies of radiosensitizers and novel fractionation schedules. In: Johnson JT, Didolkar MS (eds) Head and neck cancer, vol 3. Elsevier, Amsterdam

Overgaard J, Horsman MR (1996) Modification of hypoxia-induced radioresistance in tumors by the use of oxygen and sensitizers. Semin Radiat Oncol 6:10–21

Overgaard J, Hansen HS, Jorgensen K, Hjelm-Hansen M (1986) Primary radiotherapy of larynx and pharynx carcinoma – an analysis of some factors influencing local control and survival. Int J Radiat Oncol Biol Phys 12:515–521

Overgaard J, Bentzen SM, Kolstad P, et al. (1989a) Misonidazole combined with radiotherapy in the treatment of carcinoma of the uterine cervix. Int J Radiat Oncol Biol Phys 16:1069–1072

Overgaard J, Hansen HS, Andersen AP, et al. (1989b) Misonidazole combined with split-course radiotherapy in the treatment of invasive carcinoma of larynx and pharynx: report from the DAHANCA 2 study. Int J Radiat Oncol Biol Phys 16:1065–1068

Overgaard J, Hansen HS, Lindeløv B, Overgaard M, Jørgensen K, Rasmusson B, Berthelsen A (1991) Nimorazole as a hypoxic radiosensitizer in the treatment of supraglottic larynx and pharynx carcinoma. First report from the Danish Head and Neck Cancer Study (DAHANCA) protocol 5–85. Radiother Oncol [Suppl] 20:143–149

Overgaard J, Nielsen JE, Grau C (1992) Effect of carboxyhemoglobin on tumor oxygen unloading capacity in patients with squamous cell carcinoma of the head and neck. Int J Radiat Oncol Biol Phys 22:407–410

Pedersen D, Sogaard H, Overgaard J, Bentzen SM (1995) Prognostic value of pretreatment factors in patients with locally advanced carcinoma of the uterine cervix treated by radiotherapy alone. Acta Oncol 34:787–795

Perutz MF, Poyart C (1983) Bezafibrate lowers affinity of haemoglobin (preliminary communication). Lancet October 15:881–882

Pollack A, Zagars GK, Dinney CP, Swanson DA, von Eschenbach AC (1994) Preoperative radiotherapy for muscle-invasive bladder carcinoma. Long term follow-up and prognostic factors for 338 patients. Cancer 74:2819–2827

Quilty PM, Duncan W (1986a) Primary radical radiotherapy for T3 transitional cell cancer of the bladder: an analysis of survival and control. Int J Radiat Oncol Biol Phys 12:853–860

Quilty PM, Duncan W (1986b) The influence of hemoglobin level on the regression and long term local control of transitional cell carcinoma of the bladder following photon irradiation. Int J Radiat Oncol Biol Phys 12:1735–1742

Quilty PM, Kerr GR, Duncan W (1986) Prognostic indices for bladder cancer: an analysis of patients with transitional cell carcinoma of the bladder primarily treated by radical megavoltage X-ray therapy. Radiother Oncol 7:311–321

Rader JS, Haraf DJ, Halpern DJ, et al. (1990) Radiation therapy in the treatment of cervical cancer: the University of Chicago/Michael Reese Hospital experience. J Surg Oncol 44:157–165

Regueiro CA, Millan I, de la Torre A, Valcarcel FJ, Magallon R, Fernandez E, Aragon G (1995) Influence of boost technique (external beam radiotherapy or brachytherapy) on the outcome of patients with carcinoma of the base of the tongue. Acta Oncol 34:225–233

Revesz L, Balmukhanov SB (1987) Anaemia as a prognostic factor for the therapeutic effect of radiosensitizers. Int J Radiat Biol Relat Stud Phys Chem Med 51:591–595

Roughton FJW, Darling RC (1944) The effect of carbon monoxide on the oxyhemoglobin dissociation curve. Am J Physiol 141:17–31

Sasai K, Ono K, Hiraoka M, et al. (1989) The effect of arterial oxygen content on the results of radiation therapy for epidermoid bronchogenic carcinoma. Int J Radiat Oncol Biol Phys 16:1477–1481

Schwaibold F, Scariato A, Nunno M, et al. (1988) The effect of fraction size on control of early glottic cancer. Int J Radiat Oncol Biol Phys 14:451–454

Schwarz G (1909) Über Desensibilisierung gegen Röntgen- und Radiumstrahlen. Munchener Med Wochenschr 24:1–2

Sealy R, Jacobs P, Wood L, Levin W, Barry L, Boniaszczuk J, Blekkenhorst G (1989) The treatment of tumors by the induction of anemia and irradiation in hyperbaric oxygen. Cancer 64:646–652

Sham JS, Choy D (1990) Prognostic factors of nasopharyngeal carcinoma: a review of 759 patients. Br J Radiol 63:51–58

Siemann DW, Macler LM (1986) Tumor radiosensitization through reductions in hemoglobin affinity. Int J Radiat Oncol Biol Phys 12:1295–1297

Siemann DW, Hill RP, Bush RS (1978) Smoking: the influence of carboxyhemoglobin (HbCO) on tumor oxygenation and response to radiation. Int J Radiat Oncol Biol Phys 40:657–662

Siemann DW, Horsman MR, Chaplin DJ (1996) Modification of Oxygen Supply. In: Molls M, Vaupel P (eds) Medical radiology: blood perfusion and microenvironment of human tumors. Springer, Berlin Heidelberg New York

Solberger O, Sorbe B (1990) Fever, haemoglobin and smoking as prognostic factors during the treatment of cervical carcinoma by radiotherapy. Eur J Gynaecol Oncol 11:97–102

Tan R, Chung CH, Liu MT, Lai YL, Change KH (1991) Results of postoperative radiotherapy for clinical stage Ib uterine cervical carcinoma with evidence of microscopic involvement of surgical margin, parametrium and/or lymph node metastasis. J Formos Med Assoc 90:836–839

Taskinen PJ (1969) Radiotherapy and TNM classification of cancer of the larynx. Acta Radiol Suppl (Stockh) 287

Thomlinson RH, Gray LH (1955) The histological structure of some human lung cancers and the possible implications for radiotherapy. Br J Cancer 9:539–549

Thomson JM, Spratt JS (1977) Factors influencing survival in over 500 patients with stage II carcinoma of the cervix. Radiology 123:181–183

Tsang RW, Fyles AW, Kirkbride P, et al. (1995) Proliferation measurements with flow cytometry Tpot in cancer of the uterine cervix: correlation between two laboratories and preliminary clinical results (see comments). Int J Radiat Oncol Biol Phys 32:1319–1329

van Acht MJ, Hermans J, Boks DE, Leer JW (1992) The prognostic value of hemoglobin and a decrease in hemoglobin during radiotherapy in laryngeal carcinoma. Radiother Oncol 23:229–235

Vigario G, Kurohara SS, George FW (1973) Association of hemoglobin levels before and during radiotherapy with prognosis in uterine cervix cancer. Radiology 106:649–652

Vijayakumar S, Roach M, Wara W, et al. (1993) Effect of subcutaneous recombinant human erythropoietin in cancer patients receiving radiotherapy: preliminary results of a randomized, open-labeled, phase II trial. Int J Radiat Oncol Biol Phys 26:721–729

Werner-Wasik M, Schmid CH, Bornstein L, Ball HG, Smith DM, Madoc-Jones H (1995) Prognostic factors for local and distant recurrence in stage I and II cervical carcinoma. Int J Radiat Oncol Biol Phys 32:1309–1317

Wijkström H, Nilsson B, Tribukait B (1992) DNA analysis in predicting survival of irradiated patients with transitional cell carcinoma of bladder. Br J Urol 69:49–55

Wootton R (1984) Analysis of the effect of bezafibrate on the oxygen dissociation curve of human hemoglobin. FEBS Lett 171:187–191

11 Tumour pH

M. Stubbs[1]

11.1
Introduction

Tumours metabolise carbohydrates in order to grow. A consequence of this metabolism is the formation of hydrogen ions, which are actively transported out of the cell. The poorly organised vasculature of many tumours leads to sluggish flow and slow removal of these ions with consequent acidosis. It also leads to tissue hypoxia and is thus often responsible for the poor response of tumours to standard radiation and/ or chemotherapy. Since oncogene expression may alter the metabolism of tumours, it is important to understand the relationship between tumour metabolism and pH, and how it differs from that of normal tissue. Such an understanding may offer hope of therapy prior to the time we are able to manipulate gene expression in the living cell.

Determination of whether the pH gradient of cancer cells is lower or higher than normal is important in regard to mechanisms of pH regulation and to treatment concepts based on pH gradient. How does an understanding of pH and tumour metabolism affect strategies for therapeutic approaches?

11.2
Background

Warburg's classic work (WARBURG 1930) showed that tumour cells have a great capacity for aerobic glycolysis, converting glucose preferentially to lactic acid, even in the presence of oxygen. He also showed that extreme forms of rapidly growing ascites cells could convert glucose to lactate at a rate five times higher than normal human skeletal muscle performing strenuous anaerobic exercise. Because of this excessive lactic acid production it was assumed, for many decades after Warburg's observations, that the intracellular pH (pH_i) of tumours would be acid. Until relatively recently, microelectrodes were used most commonly for measuring pH in solid tissues (for reviews see WIKE-HOOLEY et al. 1984; VAUPEL et al. 1989a). pH as measured by the insertion of microelectrodes into tumours demonstrated mean acidic values (range from pH 5.6 to 7.6), whereas the pH of normal tissues measured by the same method did not go much below 6.8 (see Fig. 11.1). These values appeared to validate the assumption that tumours, in the continuously growing steady state, were "acidic". However, these measurements, which include an unknown component from damaged cells and blood released from ruptured microvessels, are now acknowledged to largely reflect pH_e, the pH of the extracellular fluid. Some normal tissues (e.g., exercising skeletal muscle) also form large amounts of lactic acid, but unlike the tumour with its compromised vasculature, this is rapidly removed.

11.3
Prevailing View of Tumour pH

The unfolding story of tumour pH and its consequences has recently become clearer (VAUPEL et al.

[1] Cancer Research Campaign Biomedical Magnetic Resonance Research Group, Division of Biochemistry, St. George's Hospital Medical School, Cranmer Terrace, London SW17 ORE, UK

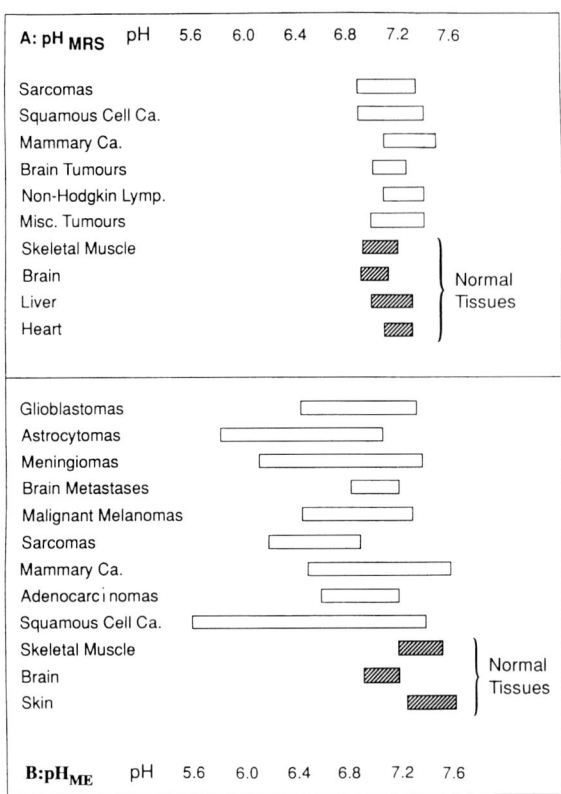

Fig. 11.1. Ranges of pH values measured in a variety of human tumours and normal tissues by ^{31}P magnetic resonance spectroscopy (pH_{MRS}) and microelectrodes (pH_{ME}). (From VAUPEL et al. 1989 as modified by GRIFFITHS 1991)

1989a,c; GRIFFITHS 1991; EVELHOCH 1992; STUBBS et al. 1994; GILLIES et al. 1994; MARTIN and JAIN 1994). Techniques for measuring pH without the invasiveness of inserting microelectrodes, most notably ^{31}P magnetic resonance spectroscopy (MRS), have now become available. The first totally noninvasive measurements of tumor pH by MRS were made in the early 1980s, firstly in an animal tumour (GRIFFITHS et al. 1981) and then in a human tumour (GRIFFITHS et al. 1983). Since then, many ^{31}P MRS studies in both humans and animals (see Fig. 11.1 and Table 11.1) have confirmed that the tumour pH measured by MRS (pH_{MRS}) is on average neutral (pH 7.0) to alkaline and similar to or even slightly higher (particularly in brain tumors) than that of normal tissue counterparts.

In normal tissue, pH_{MRS} was assumed to represent pH_i (GADIAN 1982), but because tumours tend to have higher inorganic phosphate (P_i) levels and larger extracellular volumes than normal tissues, this could not be automatically assumed for tumours. However, under many conditions it has now been

confirmed in experimental animal tumours that pH_{MRS} represents pH_i (STUBBS et al. 1992). This means that the pH gradient across the plasma membrane of tumours ($pH_i > pH_e$) is the reverse of normal tissues ($pH_i < pH_e$).

11.4
The Relationship Between Lactate⁻ and pH

Because H^+ and lactate⁻ move together on the monocarboxylate carrier (HALESTRAP and DENTON 1974), the distribution of H^+ and lactate⁻ across the plasma membrane tends to assume the relationship $[H^+]_i[lactate^-]_i/[H^+]_e[lactate^-]_e = 1$. In normal tissues, pH_i is lower than pH_e ($[H^+]_i/[H^+]_e > 1$), therefore $[lactate^-]_i/[lactate^-]_e$ must be <1. In solid tumours, however, the situation is reversed and pH_i is higher than pH_e ($[H^+]_i/[H^+]_e < 1$), so that $[lactate^-]_i/[lactate^-]_e$ becomes >1. This means that the rate at which lactic acid is extruded from the tumor cell is proportional to the difference between pH_i and pH_e; if pH_i stays the same while pH_e becomes more acid (as in tumors), lactic acid extrusion will be decreased and lactate⁻ ions will accumulate intracellularly (ALBERS et al. 1981; KALLINOWSKI et al. 1989; STUBBS et al. 1992). This calculation is based on the assumption that the distribution of lactate⁻ across the plasma membrane is similar for tumours as for several tissues including ascites tumour cells and that the lactate transporter exceeds the activity of all the enzymes responsible for the metabolism of lactate (SPENCER and LEHNINGER 1976; VEECH 1991). Correlations between high lactate levels and the incidence of metastasis in human cervical cancer have been noted (SCHWICKERT et al. 1995).

11.5
Regulation of Cellular pH

Cytoplasmic pH is strictly regulated in both normal (MADSHUS 1988; MOOLENAAR 1986) and tumour tissue (for reviews see TANNOCK and ROTIN 1989; VAUPEL et al. 1989b). If protons were passively distributed across the plasma membrane, pH_i would be about 0.6 to 1 pH unit lower than pH_e (assuming a membrane potential between −29 and −59 mV). The fact that pH_i at about 7.2 in steady state normal tissues is only about 0.2 pH units lower than pH_e (at 7.4) shows that there are mechanisms actively re-

Table 11.1. pH_i (intracellular pH) and pH_e (extracellular pH) of normal and tumour tissues in humans and animals

Tissue	pH_i	pH_e	Reference
Normal human tissue			
Resting skeletal muscle	7.1	7.35	VAUPEL et al. 1989a
	MRS	ME	
Brain	7.0	7.0	NEGENDANK 1992[a]
	MRS	ME	VAUPEL et al. 1989[b]
Normal animal tissue			
Liver (rat)	7.26	7.4	STUBBS et al. 1994[a]
	MRS	ME	WIKE-HOOLEY et al. 1984[b]
Skeletal muscle (rat)	7.1	7.42	GADIAN 1982[a]
	MRS	ME	WIKE-HOOLEY et al. 1984[b]
Human tumour			
Sarcoma	7.14	6.7	KOUTCHER et al. 1990[a]
	MRS	ME	VAUPEL et al. 1989[b]
Meningioma	7.14	6.6	NEGENDANK 1992[a]
	MRS	ME	VAUPEL et al. 1989[b]
Animal tumour			
Morris hepatoma 7777 (rat)	7.12	6.8	STUBBS et al. 1994
	MRS	MRS-APP	
RIF-1 (mouse)	7.25	6.7	GILLIES et al. 1994
	MRS	MRs-APP	
CaNT (mouse)	7.1	6.7	McCOY et al. 1995
	MRS	MRS-APP	

MRS, magnetic resonance spectroscopy; MRS-APP, MRS using 3-aminopropylphosphonate as an extracellular marker; ME, pH-sensitive microelectrodes.
[a] Author(s) used MRS.
[b] Authors used ME.

moving acid equivalents across the plasma membrane. Buffering capacity, determined in normal tissues to be about 25 mM per pH unit, also plays a role, and has been shown to be decreased in tumours (STUBBS et al. 1992).

Tissue pH_i is regulated by means of a transport system in the plasma membrane that mediates the electroneutral exchange of external Na^+ for internal H^+. This exchanger (or antiporter) responds to a fall in pH_i by rapidly extruding protons out of the cell. The Na^+/H^+ antiport is driven by the steep transmembrane Na^+ gradient which in turn is generated by the Na^+/K^+-ATPase. In addition to the Na^+/H^+ antiport there are other antiports such as Na^+-dependent Cl^-/HCO_3^- and the cation-independent Cl^-/HCO_3^- (Roos and BORON 1981; MADSHUS 1988; MOOLENAAR 1986) which also play a role in the regulation of pH_i. It has been shown in several cancer cell lines that Na^+/H^+ antiport was activated in response to a drop in pH_i (TANNOCK and ROTIN 1989) and that both the Na^+/H^+- and the Cl^-/HCO_3^- exchangers were present (BOYER and TANNOCK 1992). Vacuolar H^+-ATPase, often found in specialised epithelial cells, is also present in some tumour cells (GILLIES et al. 1992).

11.6
Methods for Measuring Tissue pH

Tissue pH, whether normal or malignant, is a composite of intra- and extracellular contributions. In normal tissues the intracellular space is about 75%–80% and the extracellular space 20%–25% of total tissue water. Since the extracellular space is relatively small in comparison to the intracellular space, many of the techniques used for measuring pH are largely a measure of pH_i, because any contribution from the extracellular compartment would be small or negligible.

Tumours, however, consist of normal host cells from the stroma and various other cell types such as macrophages, as well as actively growing tumour cells. The tumour cells often occupy less than one half the volume of a tumour (BRAUNSCHWEIGER and SCHIFFER 1986; STUBBS et al. 1992). One to ten percent of the volume is contributed by the blood vessels weaving through the tumour mass, and the remaining space is occupied by the interstitium, a collagen-rich matrix that surrounds the cancer cells (JAIN 1994). There may also be areas of necrosis that are fluid-filled. All these fluid-filled spaces, which are outside the plasma membrane of the cell (i.e., not intracellular), contribute to the extracellular fluid.

Potentiometric methods, which mainly represent H^+ activity, are based on the insertion of pH-sensitive electrodes with tip diameters ranging from $0.5\,\mu m$ to 2 mm into solid tumours, and have been in use for a long time. Until recently, electrodes were the only technique available for making pH measurements of *human* tumours and largely reflect pH_e.

^{31}P MRS, which is now installed in many centres, allows repeated monitoring of the patient (or animal), before and after treatment. In addition, it is noninvasive, nonperturbing and painless, and largely reflects pH_i.

Other techniques available for use, mostly in an experimental setting, are as follows: (a) Fluorescent pH indicators such as umbelliferone can be used for fluoroscopic imaging of intact cryostat sections of tumours (HOSSMANN et al. 1992); (b) fluorescence ratio imaging microscopy (MARTIN and JAIN 1994) that measures interstitial pH (or pH_e) used in conjunction with a rabbit ear tumour model; (c) sampling interstitial fluid (GULLINO 1970) from micropore chambers implanted within tumours grown in animals; (d) use of an MRS-visible extracellular marker 3-aminopropyl-phosphonate (GILLIES et al. 1994) so that pH_e measurements can be made in vivo in solid tumours of animals (see Table 11.1); (e) use of ^{19}F-labelled pH_i markers such as F-Quene (BEECH and ILES 1991) suitable for use in vivo in animals, 2-amino-3,3′-difluoroisobutyric acid (BENTAL and DEUTSCH 1994) suitable for use in cells, and 3-[N-(4-fluoro-2-trifluoromethylphenyl)-sulfamoyl]-propionic acid, said to distribute mainly in the extracellular space (FRENZEL et al. 1995) and therefore to be of potential use as a pH_e marker both in vivo and in vitro; (f) measuring the distribution of a weak acid such as ^{14}C-labelled 5,5-dimethyl-2,4-oxazolidione (DMO) across the cell membrane (for example, see ALBERS et al. 1981) or ^{11}C DMO in combination with positron emission tomography (HAWKINS and PHELPS 1988), or the distribution of fluorescent dyes such as SNARF-1 (MARTINEZ-ZAGUILAN et al. 1991).

The methods mentioned under (f) are suitable for measuring pH under experimental conditions in cultured cells. In such preparations it is virtually impossible to mimic the conditions attained in vivo, and as such, the experiments are never condition-independent. However, pH_i can be calculated (using the Henderson-Hasselbach equation) from the equilibrium distribution of the labelled weak acid or dye on either side of the cell membrane on the assumption that only the uncharged form is membrane-permeant.

11.7
pH_i Measurement by MRS

pH_{MRS}, which is largely synonymous with pH_i, is measured from the difference in chemical shift (i.e., frequency) of the MRS signal between the P_i signal, which is pH-sensitive, and a pH-insensitive endogenous reference signal, either phosphocreatine (which is present in several normal tissues but often absent in tumors; NG et al. 1989) or the signal from the α-phosphate of adenosine triphosphate (αATP), the main energy currency of the cell which is present in all living tissues, normal or malignant. Tissue P_i exists mainly as HPO_4^{2-} and $H_2PO_4^-$ at all physiological pH values. Because these two species are in fast exchange, the chemical shift position of the signal that is observed in the MR spectrum is an average of the relative amounts of the two species. The pH can then be determined from a standard titration curve, constructed at similar temperature and ionic strength to that in vivo (PRICHARD et al. 1983; MORRIS 1986). The protons from the water resonance used for shimming in MRS have also been considered as a suitable pH reference (MADDEN et al. 1990).

Theoretically, two P_i peaks can be resolved in the MR spectrum, since the MRS measurement of tissue pH is a composite value of pH_i and pH_e. In normal tissues the signal from P_i present in the extracellular fluid is negligible. On the other hand, tumours, with their larger extracellular volume (and increased P_i signal), might be expected to give split P_i signals. However, although split P_i peaks have occasionally been observed (BHUJWALLA et al. 1990), generally the signal to noise ratio of the tumour spectra is not good enough to resolve them (for details see STUBBS et al. 1992).

When the first pH_{MRS} measurements of human tumours were made in the 1980s, the only localisation technique available for ensuring that the signal was coming mostly from the tumour was use of a surface coil, and consequently only superficial tumours were suitable for interrogation. However, there was always the possibility that some signal might have been coming from tissues (especially skeletal muscle) nearby (SMITH et al. 1989). In more recent years, various localisation techniques to perform spatially encoded spectroscopy (ORDIDGE et al. 1986; BROWN et al. 1982) have been devised. Such techniques allow a sophisticated degree of localisation of tumour tissue, even in deep-seated lesions (LEACH 1992). However, with present MRS instruments, particularly in a clinical setting, the spatial resolution of pH *within* a tumour is still poor.

11.8
pH Effects on Therapeutic Modalities

The effects of acidosis on cells in tissue culture and in animal tumour systems have been much studied (for a review see TANNOCK and ROTIN 1989). Many cellular processes depend on pH, including the synthesis of macromolecules, cell proliferation, DNA synthesis, the activity of various enzymes, such as enzymes of glycolysis, and the transport of metabolites and drugs. The action of therapeutic agents may therefore depend on both pH_i and pH_e. Ionizing radiation and hyperthermia have both been shown to be more effective in cultured cell lines at low pH although treatment of human tumours in vivo by these two modalities suggests that the converse may occur (VAN DEN BERG et al. 1989). Of course, since the effect of ionising radiation is dependent on tissue oxygen levels (THOMLINSON and GRAY 1955), this is not surprising as many tumours contain areas of hypoxia that would not be responsive to radiation in the same way as cultured cells bathed in well-oxygenated medium. Low pH inhibits the development of thermotolerance and increases thermosensitivity (SONG et al. 1993).

The activity of certain anticancer drugs may depend on pH_i and pH_e. pH_e influences active transport and passive diffusion of drugs into cells and thereby influences intracellular drug concentrations. Now that it is clear that pH_i is more alkaline than pH_e, we would predict that drugs that are weak acids or bases with low pk, will tend to partition preferentially across the cell membrane into the intracellular compartment, whereas drugs with high pk will tend to accumulate in the more acid extracellular compartment (GERWECK et al. 1991). Conjugates (LAVIE et al. 1991) and nontoxic prodrugs (TIETZE et al. 1989) that release free or cytotoxic drugs at acid pH would release them into the extracellular fluid.

pH will alter the chemistry and therefore the reactivity of certain drugs; low pH increases the cytotoxicity of several alkylating agents and platinum-containing drugs (PARKINS et al. 1994; MAIDORN et al. 1993; ATEMA et al. 1993). It also enhances the cytotoxicity of bis-chloroethylating drugs (JÄHDE et al. 1989).

Manipulation of pH to enhance the effectiveness of anticancer drugs has been attempted in several studies. Glucose, with or without insulin to stimulate its cellular uptake, has been used as a tumour pH modifier. A consistent acidification has been found in tumour pH_i (OKUNIEFF et al. 1989; HWANG et al. 1991; GERWECK et al. 1991) and pH_e (VOLK et al. 1993; JÄHDE et al. 1992). However, i.p. glucose also leads to decreased blood flow (VAUPEL and OKUNIEFF 1988), and may therefore impair delivery of the drug that one is trying to potentiate. Some investigators have tried to get around this by using moderate hyperglycaemia in combination with other drugs (KUIN et al. 1994). Along similar lines, investigators have used blood flow modifiers such as flavone acetic acid to increase tumour acidity and thus to increase the cytotoxicity of chemotherapeutic agents (PARKINS et al. 1993, 1994).

Another approach to cancer therapy through pH is to block the exchangers (e.g., Na^+/H^+), which would impair pH regulation. This stems from the observations that mutant cells that lack the Na^+/H^+ exchanger fail to generate tumours (ROTIN et al. 1989). Amiloride and its analogues have been used by Tannock and coworkers (NEWELL et al. 1992; MAIDORN et al. 1993; LUO and TANNOCK 1994) to inhibit the regulation of tumour pH_i as a possible mechanism for tumour-selective therapy. These approaches, which lead to increased cellular acidity and increased cell killing in vitro, also have some effects in solid animal tumours in vivo.

11.9
Effects of Treatment on Cellular pH

pH_i and pH_e of tumours tend to decrease in many solid tumours after hyperthermia due to several pathogenetic mechanisms, including increased ATP hydrolysis, changes in the chemical equilibria of the cellular buffering systems and an increase in tissue pCO_2 (VAUPEL et al. 1990; VAUPEL 1990). However, THISTLETHWAITE et al. (1985) found that pH_e tended to *rise* with the number of treatment sessions in human tumours, probably due to increasing areas of necrosis which are known to be associated with more alkaline pH values. BCNU treatment of brain tumours is also associated with alkaline pH_i as measured by MRS (ARNOLD et al. 1987). Similarly, radiation therapy causes an alkalinisation of the tumour cell (NG et al. 1989), and tumour pH has also been found to be a useful prognostic indicator in thermoradiotherapy (ENGIN et al. 1994).

11.10
Source of Protons That Cause Tumour pH$_e$ To Be Acid

Since tumours tend to produce excess lactate from glycolysis, it has been assumed that this causes the acidity of the extracellular fluid. However, it has now been shown that a glycolysis-deficient cell line is still capable of producing protons and developing low pH$_e$ (NEWELL et al. 1993). DENNIS et al. (1991) suggest in cardiac ischemia that ischemic acidosis is predominantly due to the retention of protons from glycolytic ATP turnover, CO_2 accumulation and eventually net ATP breakdown. However, in hypoxic tumours, which are continuously producing lactic acid and regenerating ATP from glycolytic metabolism, the major source of protons is most likely to be the production of lactic acid with a small contribution from other proton-producing pathways.

11.11
Comparisons Between Techniques for Measuring pH

All the techniques for measuring pH have limitations of some sort, besides usually only being able to measure *either* pH$_i$ *or* pH$_e$. The DMO and dye distribution methods which require pH$_e$ to be determined by an independent method can only be used in isolated cells, not in solid tumours. The two main techniques available for human tumour investigation in vivo interrogate widely different sample volumes, different in size by orders of magnitude. Electrode measurements sample a few microenviroments within a tumour, whereas MRS produces volume-average measurements, from between a few hundred milligrams of tissue in a rodent tumour up to a few grams of tissue in a human tumour. Since tumours are known to be heterogeneous, neither measuring a few microenviroments nor measuring an average over a large part of the tumour is ideal. However, there are some advantages to both techniques. Volume averaging, particularly when, in the whole tumour, comparisons are to be made with measurements of other ions or metabolites by more classical techniques, has an advantage. The values for ion or metabolite concentrations are then directly comparable to the MRS measurement and any heterogeneity is averaged out. On the other hand, tumours *are* heterogeneous and if microenvironments are to be investigated, for instance in relation to studying tissue oxygenation

with oxygen electrodes, then microelectrodes would be the method of choice.

These differences have to be borne in mind when conclusions are to be drawn about the pH values obtained and, more importantly, which tumour compartment (intracellular or extracellular) each of these methods interrogates. The biggest advantage of MRS is that it is totally noninvasive and can therefore be easily used in patients. A "noninvasive" technique for measuring pH$_e$ (albeit one that can only be used in specially prepared animal tumours), fluorescence ratio imaging microscopy (MARTIN and JAIN 1994), gives values that are similar to both microelectrode measurement of pH$_e$ (suitable for human use) and measurement of pH$_e$ using 3-aminopropylphosphonate (not suitable for human use). This suggests that all three techniques sample the same compartment. Although fluorescence ratio imaging microscopy has the strength of good spatial resolution, it has the limitation that it can only be used in conjunction with a particular tumour model.

11.12
Conclusions

The fact that the pH gradient across the plasma membrane of tumors (pH$_i$ > pH$_e$) is the reverse of that in normal tissue (pH$_i$ < pH$_e$) has implications for both tumor metabolism and therapeutic modalities. Because of an inadequate blood supply, tumors are often hypoxic with impaired Krebs cycle activity, low ATP and increased P$_i$, and they rely mainly on glycolysis for energy (VAUPEL et al. 1994). The rapid production and subsequent export of lactate and H$^+$ from the tumor cell could account for reversal of the proton gradient, activation of the Na$^+$/H$^+$ exchange and increased [Na$^+$]$_i$ (STUBBS et al. 1994). This in turn would decrease the Na$^+$/Ca^{2+} exchange, causing the accumulation of Ca^{2+} which would precipitate as calcium phosphate, a very common feature of tumor pathology. The change in gradient of one ion (H$^+$) involves alterations in the linked equilibria of many ions and also of energy metabolites. These tumor characteristics are important both diagnostically and therapeutically. Indeed, GATENBY (1995) has proposed a model which postulates that the metabolic changes induced by transformation cause microenvironmental consequences that allow the tumor to invade and destroy the host.

References

Albers C, Van Den Kerkhoff W, Vaupel P, Mueller-Klieser W (1981) Effect of CO_2 on intracellular pH of ascites tumor cells. Respir Physiol 45:273–285

Arnold DL, Shoubridge EA, Feindel W, Villemure JG (1987) Metabolic changes in a cerebral glioma within hours of treatment with intra-arterial BCNU demonstrated by phosphorus magnetic resonance spectroscopy. Can J Neurol Sci 14:570–575

Atema A, Buurman KJ, Noteboom E, Smets LA (1993) Potentiation of DNA-adduct formation and cytotoxicity of platinum-containing drugs by low pH. Int J Cancer 54:166–172

Beech JS, Iles RA (1991) Hepatic intracellular pH in vivo using F-Quene 1 and ^{19}F NMR Spectroscopy. Magn Reson Med 19:386–392

Bental M, Deutsch C (1994) 19F-NMR study of primary human T lymphocyte activation: effects of mitogen on intracellular pH. Am J Physiol 266:541–551

Bhujwalla ZM, Blackband SJ, Wehrle JP, Grossman S, Eller S, Glickson JD (1990) Metabolic heterogeneity in RIF-1 tumors detected in vivo by ^{31}P NMR spectroscopy. NMR Biomed 3:233–238

Boyer MJ, Tannock IF (1992) Regulation of intracellular pH in tumor cell lines: influence of microenvironmental conditions. Cancer Res 52:4441–4447

Braunschweiger PG, Schiffer LM (1986) Effect of dexamethasone on vascular function in RIF-1 tumor. Cancer Res 46:3299–3303

Brown TR, Kincaid BM, Ugurbil K (1982) NMR chemical shift imaging in three dimensions. Proc Natl Acad Sci USA 79:3523–3526

Dennis SC, Gevers W, Opie LH (1991) Protons in ischemia: where do they come from; where do they go to? J Mol Cell Cardiol 23:1077–1086

Engin K, Leeper DB, Thistlethwaite AJ, Tupchong L, McFarlane JD (1994) Tumor extracellular pH as a prognostic factor in thermoradiotherapy. Int J Radiat Oncol Biol Phys 29:125–132

Evelhoch JL (1992) The pH of human tumors: facts and problems. In: Dewey EC, Edington M, Fry RJM, Hall EJ, Whitmore GF (eds) Radiation research, a twentieth century perspective, vol 2. Academic, San Diego, pp 778–783

Frenzel T, Koszler S, Bauer H, Niedballa U, Weinmann HJ (1995) Noninvasive in vivo pH measurements using a fluorinated pH probe and fluorine-19 magnetic resonance spectroscopy. Invest Radiol 29:220–222

Gadian DG (1982) Nuclear magnetic resonance spectroscopy and its application to living systems. Oxford University Press, Oxford

Gatenby RA (1995) The potential role of transformation-induced metabolic changes in tumor-host interaction. Cancer Res 55:4151–4156

Gerweck LE, Rhee JG, Koutcher JA, Song CW, Urano M (1991) Regulation of pH in murine tumor and muscle. Radiat Res 126:206–209

Gillies RJ, Martinez-Zaguilan R, Peterson EP, Perona R (1992) Role of intracellular pH in mammalian cell proliferation. Cell Physiol Biochem 2:159–179

Gillies RJ, Liu Z, Bhujwalla Z (1994) 31P-MRS measurements of extracellular pH of tumors using 3-aminopropylphosphonate. Am J Physiol 267:195–203

Griffiths JR (1991) Are cancer cells acidic? Br J Cancer 64:425–427

Griffiths JR, Stevens AN, Iles RA, Gordon RE, Shaw D (1981) 31P-NMR investigation of solid tumors in the living rat. Biosci Rep 1:319–325

Griffiths JR, Cady E, Edwards RHT, McCready VR, Wilkie DR, Wiltshaw E (1983) ^{31}P-NMR studies of a human tumor in situ. Lancet i:1435–1436

Gullino PM (1970) Techniques for the study of tumor physiopathology. In: Busch H (ed) Methods in cancer research, vol 8. Academic, New York, pp 45–91

Halestrap AP, Denton RM (1974) Specific inhibition of pyruvate transport in rat liver mitochondria and human erythrocytes by alpha-cyano-4-hydroxycinnamate. Biochem J 138:313–316

Hawkins RA, Phelps ME (1988) PET in clinical oncology. Cancer Metastasis Rev 7:119–142

Hossmann KA, Linn F, Okada Y (1992) Bioluminescence and fluoroscopic imaging of tissue pH and metabolites in experimental brain tumors of cat. NMR Biomed 5:259–264

Hwang YC, Kim SG, Evelhoch JL, Seyedsadr M, Ackerman JJ (1991) Modulation of murine radiation-induced fibrosarcoma-1 tumor metabolism and blood flow in situ via glucose and mannitol administration monitored by ^{31}P and ^{2}H nuclear magnetic resonance spectroscopy. Cancer Res 51:3108–3118

Jähde E, Glusenkamp KH, Klunder I, Hulser DF, Tietze LF, Rajewsky MF (1989) Hydrogen ion-mediated enhancement of cytotoxicity of bis-chloroethylating drugs in rat mammary carcinoma cells in vitro. Cancer Res 49:2965–2972

Jähde E, Volk T, Atema A, Smets LA, Glusenkamp KH, Rajewsky MF (1992) pH in human tumor xenografts and transplanted rat tumors: effect of insulin, inorganic phosphate, and m-Iodobenzylguanidine. Cancer Res 52:6209–6215

Jain RK (1994) Barriers to drug delivery in solid tumors. Sci Am 271:42–49

Kallinowski F, Tyler G, Mueller-Klieser W, Vaupel P (1989) Growth-related changes of oxygen consumption rates of tumor cells grown in vitro and in vivo. J Cell Physiol 138:183–191

Koutcher JA, Ballon D, Graham M, Healey JH, Casper ES, Heelan R, Gerweck LE (1990) ^{31}P NMR spectra of extremity sarcomas: diversity of metabolic profiles and changes in response to chemotherapy. Magn Reson Med 16:19–34

Kuin A, Smets L, Volk T, et al. (1994) Reduction of intratumoral pH by the mitochondrial inhibitor m-Iodobenzylguanidine and moderate hyperglycemia. Cancer Res 54:3785–3792

Lavie E, Hirschberg DL, Schreiber G, Thor K, Hill L, Hellstrom I, Hellstron KE (1991) Monoclonal antibody L6-daunomycin conjugates constructed to release free drug at the lower pH of tumor tissue. Cancer Immunol Immunother 33:223–230

Leach MO (1992) Practicalities of localisation in animal and human tumors. NMR Biomed 5:244–252

Luo J, Tannock IF (1994) Inhibition of the regulation of intracellular pH: potential of 5-(N,N-hexamethylene) amiloride in tumor-selective therapy. Br J Cancer 70:617–624

Madden A, Glaholm J, Leach MO (1990) An assessment of the sensitivity of in vivo ^{31}P nuclear magnetic resonance spectroscopy as a means of detecting pH heterogeneity in tumors: a simulation study. Br J Radiol 63:120–124

Madshus IH (1988) Regulation of intracellular pH in eukaryotic cells. Biochem J 250:1–8

Maidorn RP, Cragoe EJJ, Tannock IF (1993) Therapeutic potential of analogues of amiloride: inhibition of the regulation of intracellular pH as a possible mechanism of tumor selective therapy. Br J Cancer 67:297–303

Martin GR, Jain RK (1994) Noninvasive measurement of interstitial pH profiles in normal and neoplastic tissue using fluorescence ratio imaging microscopy. Cancer Res 54: 5670–5674

Martinez-Zaguilan R, Martinez GA, Lattanzio F, Gillies RJ (1991) Simultaneous measurement of intracellular pH and Ca^{2+} using the fluorescence of SNARF-1 and fura-2. Am J Physiol 29:297–307

McCoy CL, Parkins CS, Chaplin DJ, Griffiths JR, Rodrigues LM, Stubbs M (1995) The effect of blood flow modification on intra- and extracelluar pH measured by ^{31}P MRS in murine tumours. Br J Cancer 72:905–911

Moolenaar WH (1986) Regulation of cytoplasmic pH by Na^+/H^+ exchange. Trends Biochem Sci 11:141–143

Morris PG (1986) NMR imaging in medicine and biology. Oxford University Press, Oxford

Negendank W (1992) Studies of human tumors by MRS: a review. NMR Biomed 5:303–324

Newell KW, Wood P, Stratford I, Tannock I (1992) Effects of agents which inhibit the regulation of intracellular pH on murine solid tumors. Br J Cancer 66:311–317

Newell K, Franchi A, Pouyssegur J, Tannock I (1993) Studies with glycolysis-deficient cells suggest that production of lactic acid is not the only cause of tumor acidity. Proc Natl Acad Sci USA 90:1127–1131

Ng TC, Majors AW, Vijayakumar S, et al. (1989) Human neoplasm pH and response to radiation therapy: P-31 MR spectroscopy studies in situ. Radiology 170:875–878

Okunieff P, Vaupel P, Sedlacek R, Neuringer LJ (1989) Evaluation of tumor energy metabolism and microvascular blood flow after glucose or mannitol administration using ^{31}P Nuclear Magnetic Resonance spectroscopy and Laser Doppler flowmetry. Int J Radiat Oncol Biol Phys 16:1493–1500

Ordidge RJ, Connelly A, Lohman JAB (1986) Image-selected in vivo spectroscopy (ISIS). A new technique for spatially selective NMR spectroscopy. J Magn Reson 66:283–294

Parkins CS, Denekamp D, Chaplin DJ (1993) Enhancement of mitomycin-C cytotoxicity by combination with flavone acetic acid in a murine tumor. Anticancer Res 13:1437

Parkins CS, Chadwick J, Chaplin DJ (1994) Enhancement of chlorambucil cytotoxicity by combination with flavone acetic acid in a murine tumor. Anticancer Res 14:1603

Prichard JW, Alger JR, Behar KL, Petroff OA, Shulman RG (1983) Cerebral metabolic studies in vivo by ^{31}P NMR. Proc Natl Acad Sci USA 80:2748–2751

Roos A, Boron WF (1981) Intracellular pH. Physiol Rev 61:296–434

Rotin D, Steele-Norwood D, Grinstein S, Tannock I (1989) Requirement of the Na^+/H^+ exchanger for tumor growth. Cancer Res 49:205–211

Schwickert G, Walenta S, Sundfor K, Rofstad EK, Mueller-Klieser W (1995) Correlation of high lactate levels in human cervical cancer with incidence of metastasis. Cancer Res 55:4757–4759

Smith SR, Griffiths RD, Martin PA, Edwards RHT (1989) Measurement of tumor pH with in vivo MR spectroscopy. Br J Radiol 173:572–573

Song CW, Lyons JC, Griffin RJ, Makepeace CM, Cragoe EJ (1993) Increase in thermosensitivity of tumor cells by lowering intracellular pH. Cancer Res 53:1599–1601

Spencer TL, Lehninger AL (1976) L-Lactate transport in Ehrlich ascites-tumor cells. Biochem J 154:405–414

Stubbs M, Bhujwalla ZM, Tozer GM, et al. (1992) An assessment of ^{31}P MRS as a method of measuring pH in rat tumors. NMR Biomed 5:351–359

Stubbs M, Rodrigues LM, Howe FA, Wang J, Jeong KS, Veech RL, Griffiths JR (1994) The metabolic consequences of a reversed pH gradient in rat tumors. Cancer Res 54:4011–4016

Tannock IF, Rotin D (1989) Acid pH in tumors and its potential for therapeutic exploitation. Cancer Res 49:4373–4384

Thistlethwaite AJ, Leeper DB, Moylan DJ, Nerlinger RE (1985) pH distribution in human tumors. Int J Radiat Oncol Biol Phys 11:1647–1652

Thomlinson RH, Gray LH (1955) The histological structure of some human lung cancers and the possible implications for radiotherapy. Br J Cancer 9:539–549

Tietze LF, Neumann M, Mollers T, Fischer R, Glusenkamp KH, Rajewsky MF, Jähde E (1989) Proton-mediated liberation of aldophosphamide from a nontoxic prodrug: a strategy for tumor-selective activation of cytocidal drugs. Cancer Res 49:4179–4184

van den Berg AP, Wike-Hooley JL, Broekmeyer-Reurink MP, van der Zee J, Reinhold HS (1989) The relationship between the unmodified initial tissue pH of human tumors and the response to combined radiotherapy and local hyperthermia treatment. Eur J Cancer Clin Oncol 25:73–78

Vaupel P (1990) Pathophysiological mechanisms of hyperthermia in cancer therapy. In: Gautherie M (ed) Biological basis of oncologic thermotherapy. Springer, Berlin Heidelberg New York, pp 73–134 (Clinical thermology. Subseries thermotherapy)

Vaupel P, Okunieff P (1988) Role of hemoconcentration in dose-dependent flow decline observed in murine tumors after intra-peritoneal administration of glucose or mannitol Cancer Res 48:7102–7106

Vaupel P, Kallinowski F, Okunieff P (1989a) Blood flow, oxygen and nutrient supply, and metabolic microenvironment of human tumors: a Review. Cancer Res 49:6449–6465

Vaupel P, Okunieff P, Neuringer LJ (1986b) Blood flow, tissue oxygenation, pH distribution, and energy metabolism of murine mammary adenocarcinomas during growth. Adv Exp Med Biol 248:835–845

Vaupel P, Okunieff P, Neuringer LJ (1989c) Blood flow, tissue oxygenation, pH distribution, and energy metabolism of murine mammary adenocarcinomas during growth. In: Rakusan K, Biro GP, Goldstick TK, Turek Z (eds) Oxygen transport to tissue. Plenum, New York, pp 835–845

Vaupel P, Okunieff P, Neuringer LJ (1990) In vivo ^{31}P-NMR spectroscopy of murine tumours before and after localised hyperthermia. Int J Hyperthermia 6:15–31

Vaupel P, Schaefer C, Okunieff P (1994) Intracellular acidosis in murine fibrosarcomas coincides with ATP depletion, hypoxia, and high levels of lactate and total P_i. NMR Biomed 7:128–136

Veech RL (1991) The metabolism of lactate. NMR Biomed 4:53–58

Volk T, Jähde E, Fortmeyer HP, Glusenkamp K, Rajewsky MF (1993) pH in human tumor xenografts: effect of intravenous administration of glucose. Br J Cancer 68:492–500

Warburg O (1930) The metabolism of tumors. Constable, London

Wike-Hooley JL, Haveman J, Reinhold HS (1984) The relevance of tumor pH to the treatment of malignant disease. Radiother Oncol 2:343–366

12 Energy Status of Malignant Tumors in Patients and Experimental Animals

W. Mueller-Klieser[1], P. Vaupel[2], and C. Streffer[3]

CONTENTS

12.1 Introduction

Solid tumors are often relatively resistant to conventional nonsurgical treatment modalities. A variety of biological mechanisms is involved in the lack of responsiveness of these neoplasms, such as an intrinsic, genetically determined resistance as well as epigenetic factors which include interaction between cancer cells and their microenvironment. Besides cell-cell and cell-matrix interactions, specific physiological properties of the metabolic milieu in solid tumors can largely modulate tumor response to treatment. Such properties of the metabolic environment encompass microcirculatory parameters, tissue oxygen and nutrient supply, tumor pH, and bioenergetic status (VAUPEL and JAIN 1991; VAUPEL et al. 1989; VAUPEL 1992, 1993; SUTHERLAND 1988). The impact of these physiological properties can be mediated through *direct actions*, such as O_2 dependence of standard radiotherapy and some anticancer drugs and inadequate delivery of agents due to special features of intratumor pharmacokinetics, or through *indirect mechanisms*, e.g., the position of

tumor cells in different phases of the cell cycle, modulation of proliferation kinetics, or changes of pharmacodynamics through modulation of the metabolic microenvironment.

Despite the apparent importance of physiological properties for tumor growth, early tumor response to treatment, and, probably, prediction of long-term therapeutic outcome, reliable data on solid tumors are relatively scarce, although the number of clinical investigations dealing with this subject is rapidly increasing. Some excellent reviews and monographs are now available that address this field of cancer research.

This article is intended (a) to give a state-of-the-art summary of the current information concerning the energy status of tumors in patients and experimental animals and (b) to show that knowledge of the physiological properties of tumors might be beneficial for designing specifically tailored treatment protocols adapted to each individual clinical case.

12.2 Human Tumor Bioenergetics Monitored by ^{31}P NMR

^{31}P nuclear magnetic resonance (NMR) is able to provide important biochemical information on living tissues. Because the magnetic resonance spectroscopy (MRS) technique is non-invasive and painless, it allows the patient to be repeatedly monitored throughout the course of tumor treatment (NG et al. 1987; STEEN 1989a; EVELHOCH et al. 1990; NEGENDANK 1992). In vivo ^{31}P NMR spectroscopy has been employed to monitor the energy metabolism of human tumors since 1983 (GRIFFITHS et al. 1983). From the studies available, information has been gained that is potentially beneficial to the clinical treatment of cancer. Among other factors, clinical studies indicate that serial monitoring of tumor response can assist in optimizing the timing of treatments (NG et al. 1987).

[1] W. MUELLER-KLIESER, Dr. rer. nat. and [2] P. VAUPEL, Dr. med., Institute of Physiology and Pathophysiology, University of Mainz, Duesbergweg 6, D-55099 MAINZ, GERMANY
[3] C. STREFFER, Dr. rer. nat. Institute of Medical Radiobiology, University Clinics of Essen, Hufelandstrasse 55, D-45122 Essen, Germany

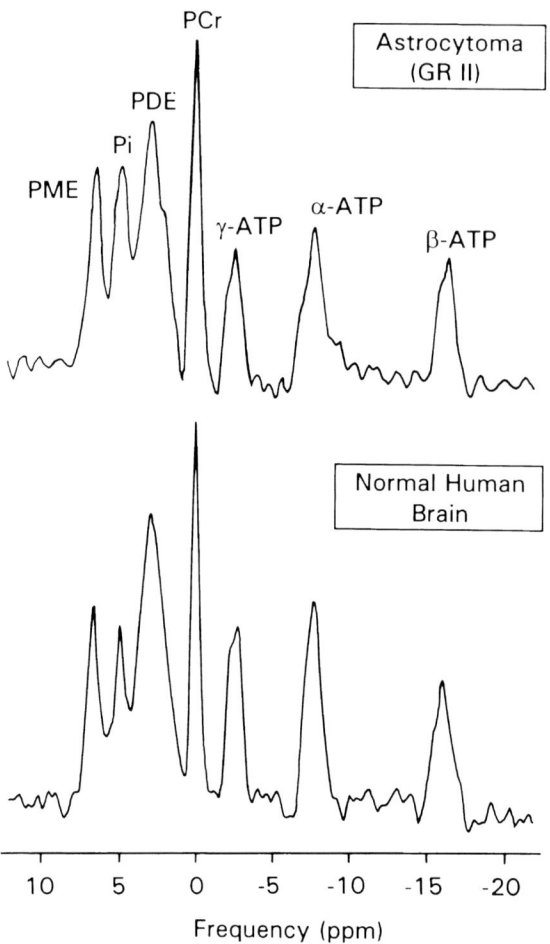

Fig. 12.1. Typical ^{31}P NMR spectra of a human grade II astrocytoma (*upper panel*) and of a normal human brain. (*lower panel*; modified according to VAUPEL 1994)

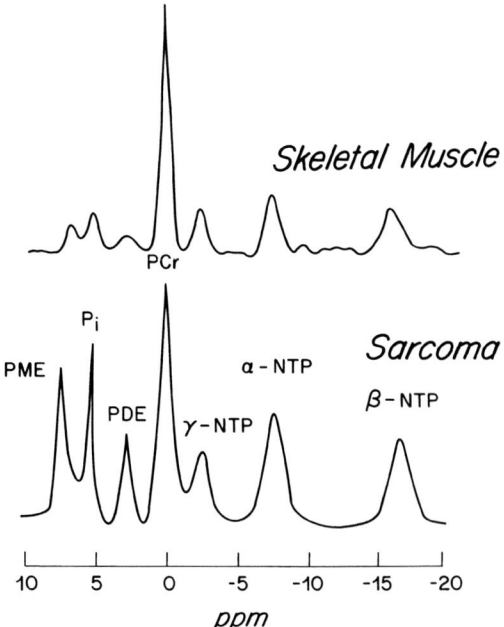

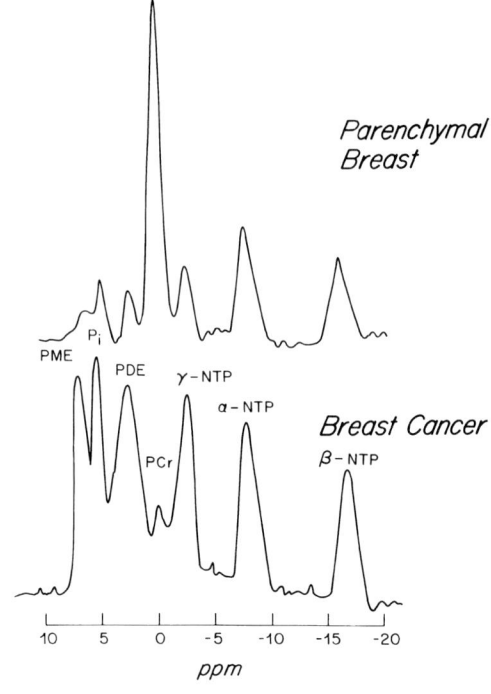

Fig. 12.3. Typical ^{31}P NMR spectra of parenchymal breast and a human breast cancer. (Spectra are redrawn from original recordings; GRIFFITHS et al. 1983; DEN HOLLANDER and LUYTON 1987; SEGEBARTH et al. 1987; NEGENDANK et al. 1988; SEMMLER et al. 1988a,b; OBERHAENSLI et al. 1986)

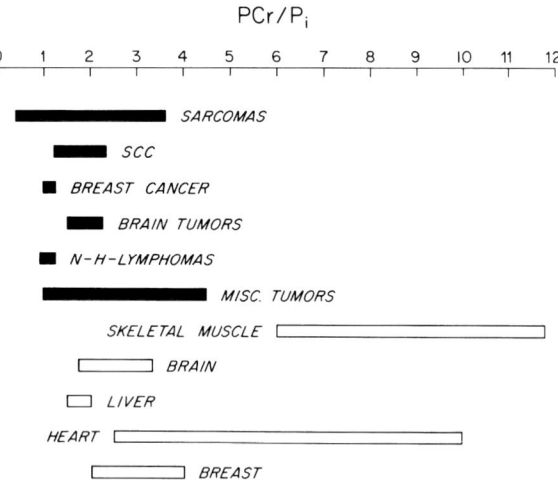

Fig. 12.4. Pooled PCr/Pi ratios calculated for human malignancies (*black bars*) and for normal tissues (*white bars*). *SCC*, squamous cell carcinomas. (Adapted from VAUPEL 1994)

◀──────────────────────────────

Fig. 12.2. Typical ^{31}P NMR spectra of resting skeletal muscle and a human sarcoma. (Spectra are redrawn from original recordings; GRIFFITHS et al. 1983; DEN HOLLANDER and LUYTON 1987; SEGEBARTH et al. 1987; NEGENDANK et al. 1988; SEMMLER et al. 1988a,b; OBERHAENSLI et al. 1986)

In Figs. 12.1–12.3, ^{31}P NMR spectra from normal tissues are compared with tumor spectra (spectra are redrawn from original recordings; GRIFFITHS et al. 1983; DEN HOLLANDER and LUYTON 1987; SEGEBARTH et al. 1987; NEGENDANK et al. 1988; SEMMLER et al. 1988a,b; OBERHAENSLI et al. 1986). These exemplary spectra illustrate that in many human malignancies, with the exception of brain tumors, very often high concentrations of phosphomonoesters (PME), phosphodiesters (PDE), and inorganic phosphate (P_i) and low creatine phosphate (PCr) levels are found. It is interesting to note that studies on human brain tumors often fail to show any significant differences in the spectra of malignancies vs. normal brain tissue (DEN HOLLANDER and LUYTON 1987).

The PME signal primarily represents phosphocholine and phosphethanolamine, which are membrane phospholipid precursors. In addition, phosphorylated sugars (glucose-6-phosphate, fructose-6-phosphate, and fructose-1,6-diphosphate) might be present and fall within the spectral region of PME. The PDE peak in murine tumors was identified to be caused mainly by glycerophosphocholine and glycerophosphoethanolamine, which are both membrane phospholipid decomposition products (EVANOCHKO et al. 1984; NG et al. 1986). Thus, PME and PDE signals may be indicative of membrane turnover rates in living tissue.

For comparison of the bioenergetic status of normal tissues with that of malignancies, relevant metabolite ratios obtained with MRS are compiled in

Figs. 12.4 and 12.5. PCr/Pi ratios for various human malignancies and normal tissues are graphically displayed in Fig. 12.4. The data show that PCr/P_i in normal brain and in brain tumors are similar (DEN HOLLANDER and LUYTON 1987; SEGEBARTH et al. 1987; OBERHAENSLI et al. 1986; EVANOCHKO et al. 1984; NG et al. 1986; ARNOLD et al. 1987; LEVINE et al. 1987; ROTH et al. 1987; VERMEULEN et al. 1987; LUYTON et al. 1988; HUBESCH et al. 1988; CADOUX-HUDSON et al. 1988), whereas this ratio is significantly higher in skeletal muscle or myocardium relative to sarcomas, and in parenchymal breast vs. breast cancer ($p < 0.001$).

β-NTP/P_i ratios for human tumors and normal tissues are presented in Fig. 12.5. Here, again, no clear differences are seen between normal brain and brain tumors (DEN HOLLANDER and LUYTON 1987; SEGEBARTH et al. 1987; OBERHAENSLI et al. 1986; EVANOCHKO et al. 1984; NG et al. 1986; ARNOLD et al. 1987; LEVINE et al. 1987; ROTH et al. 1987; VERMEULEN et al. 1987; LUYTON et al. 1988; HUBESCH et al. 1988; CADOUX-HUDSON et al. 1988). The only significant differences are obtained for sarcomas vs. skeletal muscle. From these data, the preliminary conclusion can be drawn that, on average, the bioenergetic status with respect to adenine nucleotides may be similar in normal brain and in brain tumors. The latter data thus confirm the results on the ATP distribution obtained with quantitative bioluminescence (see below).

12.3
Tumor Bioenergetics Assessed by HPLC

Global ATP concentrations measured in experimental tumors using high-pressure liquid chromatography (HPLC) are typically in the range of 0.4–2.0 mM.

In experimental studies on animal tumors, global ATP concentrations and adenylate energy charge changed only marginally, as long as tumor masses did not exceed 1% of the body weight and thus biologically relevant tumor sizes (KRÜGER et al. 1991; VAUPEL et al. 1994a,b; SCHAEFER et al. 1993; VAUPEL 1996). This stable bioenergetic status coincided with a "physiological" tissue oxygenation (i.e., pO_2 distribution comparable with that of normal organs; median p$O_2 \geq 10$ mmHg; VAUPEL et al. 1994b), mean tissue glucose concentrations of up to 2 mM and mean lactate levels <10 mM (VAUPEL et al. 1994b; STREFFER 1990; BUSSE and VAUPEL 1996). Different findings were obtained at the microregional level,

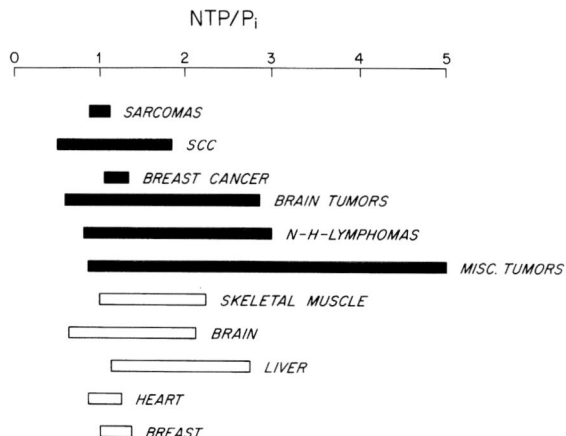

Fig. 12.5. Pooled β-NTP/P_i ratios for human tumors (*black bars*) and for normal tissues (*white bars*). *SCC*, squamous cell carcinomas. (Adapted from VAUPEL 1994)

where local ATP in tumors decreased considerably in areas with restricted blood perfusion (see below).

Increased ATP hydrolysis with enlarging tumor mass is a typical finding observed during tumor growth. As a result of increased ATP degradation, an accumulation of purine catabolites has to be expected, together with a formation of protons at several stages during degradation to the final product uric acid. The accumulation of the purine catabolites xanthine ($p < 0.005$), hypoxanthine ($p < 0.05$), and uric acid ($p < 0.005$) in rat DS sarcomas is shown in Fig. 12.6. During tumor growth, changes in ATP and adenosine concentrations were only marginal (VAUPEL 1996).

12.4
Microregional ATP Distribution

A comparison of the microcirculation in normal organs with that in solid tumors suggests that global information on the metabolic state of malignancies may often not be representative of the specific situation of metabolism in tumor microareas. Whereas global ATP concentrations are in the range of 0.4–

2.0 mM in most tumors, the concentration of this high-energy phosphate may be close to 0 mM in some distinct tumor regions that may still contain viable cancer cells. This can have implications for metabolism, growth, and therapeutic response of malignancies, and it illustrates that predictions concerning the biological behavior of tumors on the basis of global data can be afflicted with considerable error. A further aspect of tumor heterogeneity is the emergence of necrosis, since ATP concentration is expected to be low in such areas due to its rapid hydrolysis upon cell death. Unlike viable tumor regions with low ATP concentration, necrotic areas may not contribute to the behavior of tumors, e.g., regrowth, after treatment. An additional complication of the situation in the metabolic micromilieu of tumors is the infiltration of tumors by defense cells of the host. Such cells may "carry" ATP preferentially to those tumor areas in which necrosis is about to develop and in which ATP concentrations are expected to be low.

The discussion implies that techniques are required that make it possible to image metabolites such as ATP with high spatial resolution and enable registration of metabolites in relation to the architecture of the tumors. In addition, the determination of absolute tissue concentrations of metabolites is desirable to enable a direct comparison of individual tumors and their therapeutic response or their behavior during clinical follow-up. This should make it possible to evaluate the prognostic significance of such metabolic studies in patients.

A method has been established for metabolic imaging that can be applied to rapidly frozen tissues, such as rapidly excised experimental tumors immersed in liquid nitrogen or cryobiopsies taken from tumors in patients (MUELLER-KLIESER et al. 1988, 1991; MUELLER-KLIESER and WALENTA 1993). This technique enables the measurement of ATP, glucose, and lactate concentrations within cryosections in absolute terms with a spatial resolution at or near the cellular level using quantitative bioluminescence, single photon imaging, and computerized image analysis (MUELLER-KLIESER and WALENTA 1993; WALENTA et al. 1990; TAMULEVICIUS and STREFFER 1995). The use of serial sections makes it possible to determine the different metabolites in quasi-identical planes and to correlate metabolite distribution with histological structure. The bioluminescence technique can be combined with any imaging procedure based on tissue sectioning. For example, ATP imaging has been associated with imaging of local blood perfusion with the iodo-antipyrine autoradiographic technique (DELLIAN et al. 1993a,b,

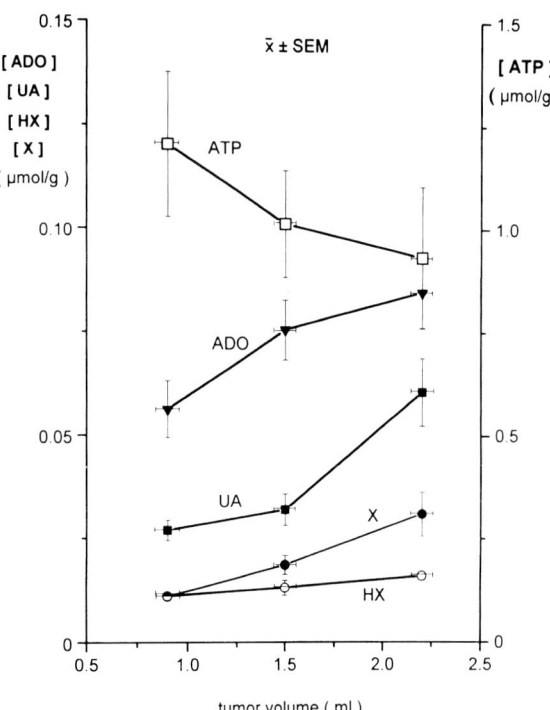

Fig. 12.6. Concentrations of ATP, adenosine (*ADO*), uric acid (*UA*), xanthine (*X*), and hypoxanthine (*HX*) in rat DS sarcomas as a function of tumor volume. Values are means ± SEM. (Adapted from VAUPEL 1994)

1994). A computer algorithm has been developed allowing the establishment of pixel-to-pixel correlations between the distribution of ATP and regional blood flow (KUHNLE et al. 1992; WALENTA et al. 1992) as well as among the images of the three metabolites measured (MUELLER-KLIESER et al. 1994). Finally, interactive computerized image analysis allowed the evaluation of metabolite concentrations in circumscribed tissue areas, such as viable tumor tissue, mostly necrotic and stromal areas, and tumor-adjacent normal tissue (MUELLER-KLIESER and WALENTA 1993). There is no other technique currently available that offers these specific advantages of bioluminescence imaging.

Quantitative bioluminescence and single photon imaging have been applied to several types of implantation tumors of the mouse, i.e., to two murine sarcoma lines (KHT, RIF-1) and two human ovarian carcinoma xenograft lines (MLS, OWI), in conjunction with ^{31}P MRS and cryospectrophotometry to characterize the energy and oxygenation status of the tumors, respectively (MUELLER-KLIESER et al. 1990). Global ATP concentration determined either with bioluminescence or with NMR decreased in these tumors with increasing tumor volume from 100 to 600 mm³. A positive correlation was obtained between the intensity of β-NTP in MRS and global ATP concentration measured with bioluminescence, whereas the correlation between P_i concentration measured with NMR and ATP concentration determined from bioluminescence was negative. Both correlations were true across the various tumor lines, yet the data indicated that the slopes of the individual correlations within each line may differ from one another to some extent. The findings indicate that metabolic imaging in parts of these tumors by bioluminescence can generate results that are representative of global data averaged over the whole tumor as obtained by the integrative MRS technique. The combination of ATP imaging with cryospectrophotometry showed that there is a negative correlation between ATP concentration and the proportion of potentially hypoxic tissue regions in these tumors. This is in good agreement with observations using ^{31}P NMR, bioluminescence, and oxygen-sensitive needle electrodes in a murine fibrosarcoma (FSaII; VAUPEL et al. 1994b). The study demonstrated a decrease in ATP with decreasing oxygen tension in the range below 10 mmHg.

Measurements of ATP, glucose, and lactate by imaging bioluminescence were performed in human melanoma xenografts (MUELLER-KLIESER et al. 1991; MUELLER-KLIESER and WALENTA 1993) including an evaluation of distinct tissue regions. Distributions of all three metabolites were characterized by marked heterogeneities. Concentrations of ATP and glucose were found to be significantly lower ($p < 0.05$) and lactate content was significantly higher ($p < 0.05$) in these tumors than in the adjacent normal tissue. The content of both ATP and glucose was significantly less ($p < 0.05$) in necrotic than in viable tumor regions, whereas no such difference was found for lactate. There was a tendency towards a decreasing ATP concentration with increasing tumor size, which was only obvious in the center and not in peripheral areas of these malignancies (MUELLER-KLIESER et al. 1991). This finding is in accordance with the general observation that most but not all implanted tumors show a size-dependent shift of their metabolic milieu during growth. For example, a strong dependency on tumor volume of ATP, PCr, glucose, lactate, and pH has been documented in FSaII fibrosarcomas by VAUPEL and colleagues (1994b). Five of six human melanoma xenografts exhibited a decrease in pH, blood flow, and bioenergetic status with increasing tumor weight, although the dependencies were rather weak, as mirrored by relatively shallow slopes of the respective regression lines (LYNG et al. 1993).

In contrast to tumors of nonneurological origin, measurements with quantitative bioluminescence in brain tumors of rats and cats revealed ATP concentrations similar to those in normal brain, i.e., 2.6 vs. 2.5 µmol/g (HOSSMANN et al. 1986; PASCHEN et al. 1987; PASCHEN 1985; KOGURE and ALONSO 1978). Whereas glucose concentration was only slightly lower in tumors than in brain (2.4 and 2.8 µmol/g, respectively), lactate concentration was substantially higher in malignancies than in healthy tissue (6.4 vs. 1.2 µmol/g). Heterogeneities in brain tumors were pronounced to an extent similar to that found in nonneurological cancers (HOSSMANN et al. 1992). Measurements of ATP in 19 human glioblastomas and 30 meningiomas by conventional biochemical methods resulted in values of about 0.5 µmol/g tissue in both tumor entities. However, the glucose levels were lower in glioblastomas (0.5 µmol/g) than in meningiomas (1.2 µmol/g), and the lactate levels were higher in glioblastomas (7.5 µmol/g) than in meningiomas (4.7 µmol/g) (STREFFER and coworkers, unpublished data). This is in agreement with other findings that lactate levels in particular are higher in more malignant tumors (WALENTA et al. 1997).

Bioluminescence imaging of ATP combined with autoradiographic determination of local blood perfusion using iodo-[^{14}C]antipyrine was applied to a subcutaneous amelanotic hamster melanoma (KUHNLE et al. 1992; WALENTA et al. 1992). Both parameter values could be imaged in necrotic and viable tumor regions and in surrounding normal tissue. The images obtained were obviously correlated with each other (Fig. 12.7a–c), which could be substantiated by establishing pixel-to-pixel correlations between ATP and regional perfusion (Fig. 12.7d). Such correlations imply that the intracellular level of ATP in viable tumor areas is relatively independent of blood flow rates beyond $0.3–0.4\,\mathrm{ml\,g}^{-1}\,\mathrm{min}^{-1}$, and drops to very low values with decreasing flow below this perfusion range. This is in accordance with observations on FSaII fibrosarcomas, suggesting that ATP is relatively constant at perfusion rates beyond $0.5\,\mathrm{ml\,g}^{-1}\,\mathrm{min}^{-1}$ and declines with deterioration of perfusion below this "threshold" (VAUPEL et al. 1994). Furthermore, the findings are in agreement with earlier studies on gliomas of the rat, where similar correlations were obtained between blood perfusion and ATP measured in relative terms (MIES et al. 1990). The correlation between blood perfusion and ATP in necrotic tumor regions of hamster melanomas was linear, with both parameter values localized in a relatively low range compared to viable tissue (WALENTA et al. 1992).

Bioluminescence measurements have been performed in a number of cryobiopsies from cervix cancers in patients. The biopsies were taken immediately after oxygen tension measurements with needle probes adjacent to the electrode track (VAUPEL and MUELLER-KLIESER 1992). Although extended necrosis was not detectable in histological sections, the distributions of ATP, glucose, and lactate were very heterogeneous. There was no clear-cut relationship between global ATP concentrations and oxygen tension values in seven patients' tumors investigated under these conditions (VAUPEL and MUELLER-KLIESER 1992). This lack of correlation may be mainly due to the heterogeneous distribution patterns of oxygen and the different metabolites. In fact, it has been shown for ATP and blood flow imaging in hamster melanomas as described above that correlations found for regional values deteriorated substantially or vanished, when global values of ATP and perfusion were considered (WALENTA 1994).

Measurements with imaging bioluminescence were performed in ten cervix tumors before standard radiation treatment in order to establish correlations among the images of the different metabolites and to compare the results with clinical data, such as incidence of metastasis and patient survival. There was a positive linear correlation between the average concentration of glucose in tumors and in blood (Fig. 12.8). This finding implies that the metabolic milieu in these cancers was mainly determined by the nutritive supply. Distributions of ATP, glucose, and lactate exhibited pronounced heterogeneities, yet distribution patterns were obviously correlated with each other in the majority of the tumors investigated. This is illustrated by Fig. 12.9 showing the distribution of the three metabolites and an adjacent histological section. The distribution patterns of the metabolite seem to mirror the tissue architecture. With only a few exceptions, positive pixel-to-pixel correlations represented by a positive Spearman's coefficient among images of the three metabolites

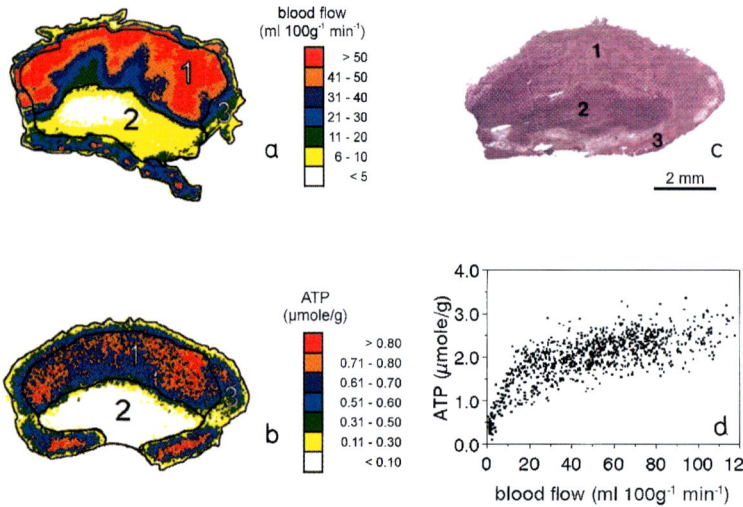

Fig. 12.7a–d. Color-coded images of **a.** local blood perfusion registered with the autoradiographic iodo-[^{14}C]antipyrine technique, **b.** of local ATP concentrations obtained from bioluminescence measurements, and **c.** of the histological structure in a hamster melanoma. *1*: viable tumor region; *2*: necrotic region; *3*: normal skin and subcutaneous tissue. A pixel-to-pixel correlation between the images of ATP and perfusion is shown in **d.** (Data adapted from WALENTA et al. 1992)

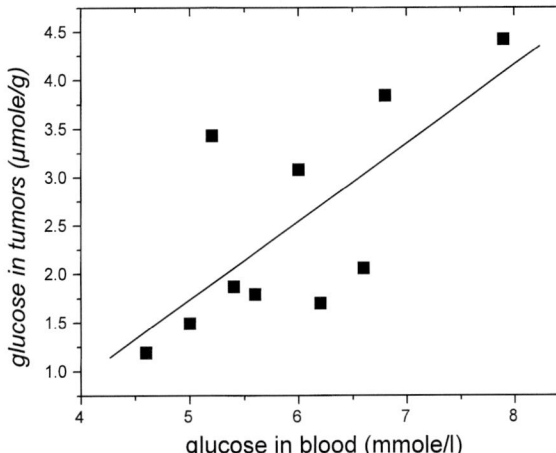

Fig. 12.8. Correlation between the glucose content in tumors and in the blood of patients with cervix carcinomas ($y = -2.31 + 0.808x$; $r = 0.715$; $p = 0.02$)

were found in the cervix tumors. This is exemplified by Fig. 12.10 including frequency distributions and mean (+SE) concentrations of the metabolites in different regions of the neoplastic tissue. Spearman's correlation coefficients for the relation between glucose and lactate or ATP and glucose were significantly less in patients without metastasis than in patients with documented tumor spread (0.386 vs. 0.159 and 0.366 vs. 0.175, respectively; $p < 0.01$). Accordingly, there was a tendency towards longer patient survival with decreasing Spearman's coefficient. These findings suggest that good correlation among distribution patterns of ATP, glucose, and lactate is a negative prognostic factor in these tumors. Furthermore, global lactate concentrations were significantly ($p < 0.05$) higher in malignancies with metastatic spread than those without. This finding has been confirmed in a recent study on primary lesions of head and neck where lactate concentrations in metastatic tumors were significantly higher than those in nonmetastatic malignancies (WALENTA et al. 1997).

The data obtained in cervix as well as head and neck tumors of patients illustrate that metabolic imaging with quantitative bioluminescence, single photon imaging, and computerized image analysis complements conventional clinical staging and pathohistological grading. The degree of metabolic heterogeneity and lactate concentrations seem to be prognostic factors in these malignancies. These preliminary findings need to be substantiated in a larger cohort of patients and in different tumor entities in the future.

12.5
Tumor Bioenergetics After Hyperthermia

Hyperthermic treatment leads to increased rates of metabolic processes in the heated tissues. Therefore a rapid turnover of metabolites takes place. Deregulation of metabolism in general and energy metabolism in particular is the consequence (STREFFER 1990). Investigations of a number of experimental tumors (transplantable rodent tumors as well as human xenografts in mice and rats) have shown that ATP levels decrease when heating of the tumor tissue is performed in the range of 42°–45°C (KRÜGER et al. 1991; SCHAEFER et al. 1993; DELLIAN et al. 1993a, 1994; TAMULEVICIUS and STREFFER 1995; STREFFER 1990). It has been demonstrated that this effect is due to decreased blood flow, reduced energy metabolism as a consequence of a restricted supply, and increased ATP turnover and degradation (STREFFER 1990; BUSSE and VAUPEL 1996). Several authors have studied ATP levels in tumors with ^{31}P NMR spectroscopy after a hyperthermic treatment (EVANOCHKO et al. 1984; LILLY et al. 1985; NARUSE et al. 1986; SIJENS et al. 1987; VAUPEL and KELLEHER 1995). In these studies the ATP levels decreased within minutes to hours after local heat treatment of experimental tumors. At the same time, a remarkable increase in the P_i level was observed. Temperature ranged between 42°C and 47°C. From the NMR spectra of the phosphorylated metabolites, including P_i, it is possible to determine the pH of the tissue. After heating of an adenocarcinoma transplanted on mice to 47°C for 30 min, the pH within the tumor decreased from 7.2 to 6.8 (EVANOCHKO et al. 1984). In other studies only a slight or no change of pH was observed (NARUSE et al. 1986; SIJENS et al. 1987). These changes and those of ATP levels are apparently closely connected to blood flow which also decreases after the heat treatment.

The measurements of ATP and blood perfusion in hamster melanomas mentioned earlier in this chapter were extended to tumors treated with hyperthermia (waterbath; 43.3 ± 0.1°C; DELLIAN et al. 1993b) alone or in combination with multifocal shock waves (700 waves at 1.67 Hz; 80 nF; 80 kV; DELLIAN et al. 1994). Hyperthermia induced a decrease in ATP and blood flow which was associated with a highly significant ($p < 0.001$) delay in tumor growth (DELLIAN et al. 1993b). The patterns of decrease in ATP and blood flow as a function of time after heat treatment followed different kinetics, which led to loss of the correlation between the two parameters as registered under control conditions.

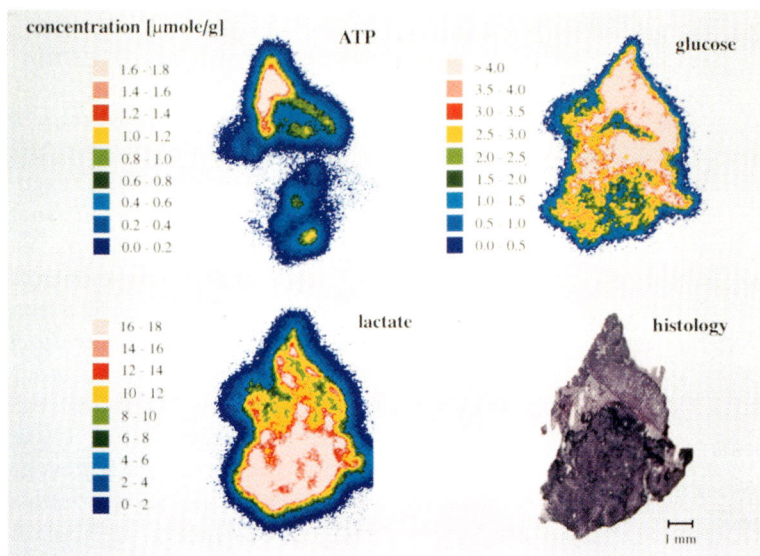

Fig. 12.9. Color-coded images of the local content of *ATP*, *glucose*, and *lactate* and a parallel section (*histology*) stained with hematoxylin/eosin in a human cervix tumor

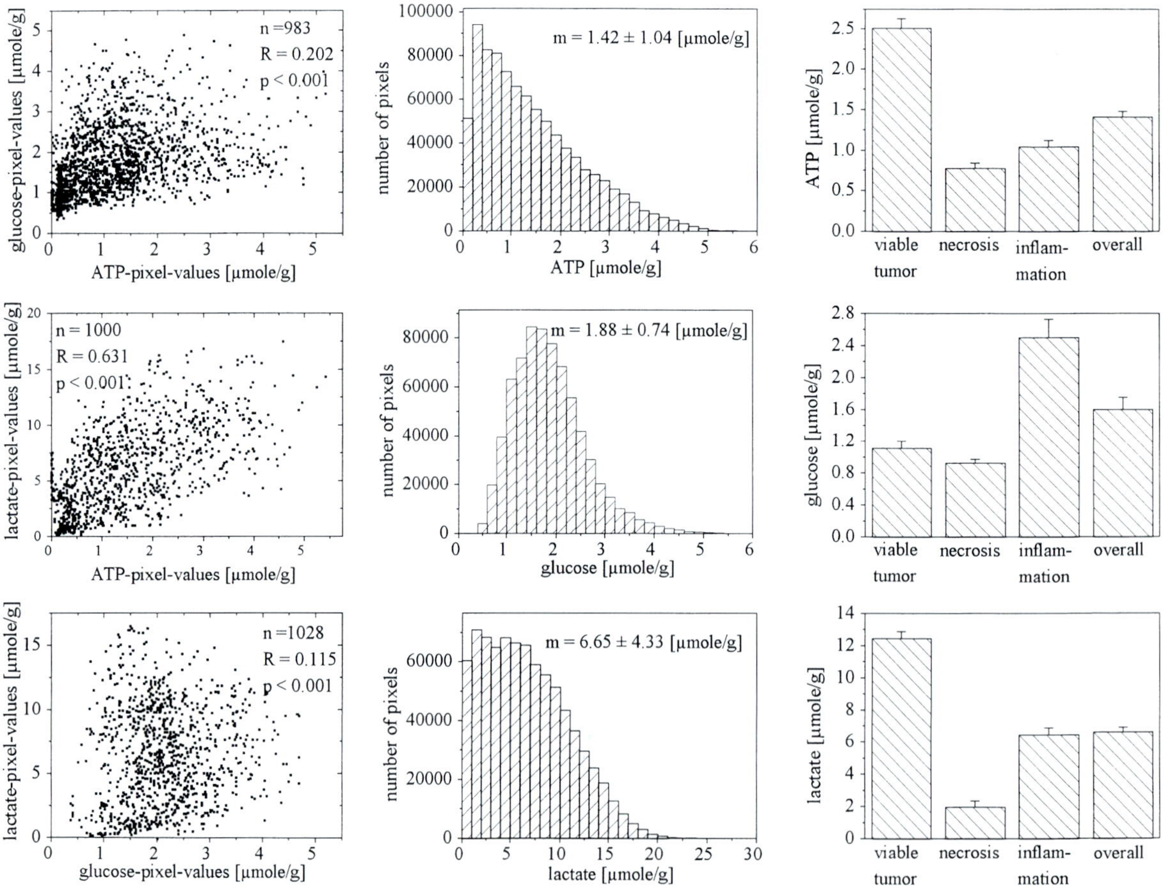

Fig. 12.10. Pixel-to-pixel correlations between ATP, glucose, and lactate distributions (*left column*) as well as frequency distributions (*central column*; *m* = mean ± SD) and mean (+SE) regional metabolite content (*right column*) of ATP (*top panels*), glucose (*middle panels*), and lactate (*bottom panels*) in a human cervix tumor

Shock waves enhanced the efficiency of hyperthemia considerably, as reflected by a complete tumor regression (DELLIAN et al. 1994).

The intensified ATP degradation upon localized hyperthermia also results in an accumulation of various purine catabolites. Thus, it has been reported that in a rat DS sarcoma a substantial increase in inosine, xanthine, hypoxanthine, and uric acid occurs after heat treatment at 44°C for 60 min. This increase in purine catabolism may contribute to an increased acidosis after heat treatment (BUSSE and VAUPEL 1996), since protons are produced by these processes. Besides the correlation of the ATP levels with blood flow, by way of ATP turnover as well as adenosine nucleotide degradation these metabolic processes are closely connected to glycolysis in tumors. Therefore glucose as well as lactate levels provide some information about the changes and their metabolic implication after hyperthermia. In many cases, with mild hyperthermia (up to 42°C) a decrease in the glucose levels is observed, as apparently the glucose turnover is increased. However, with severe hyperthermia (>43°C) quite often an increase in glucose levels is also found (STREFFER 1990), as the glycolytic rate may be reduced. However, these changes vary from tumor line to tumor line. They are mainly dependent on the supply of glucose as well as oxygen to the tumor, on blood flow, and on the turnover of glucose via the glycolytic pathway.

Much more consistent is the increase in lactate levels after hyperthermia which is generally found in most tumors (STREFFER 1990; VAUPEL and KELLEHER 1995). Lactate cannot be metabolized in mammalian tissues via the citrate cycle under anaerobic conditions. Instead, it must be transported from the tumor tissue to liver, where it can be used in oxidative energy metabolism or gluconeogenesis, thus avoiding lactate accumulation in tumor tissue. Drainage by blood flow is necessary for these processes. If blood flow decreases after heating, the degree of tumor hypoxia increases. The increase in hypoxia after severe hyperthermic treatment is further demonstrated by the increase in lactate/pyruvate ratios in tumors (STREFFER 1990). The increase in lactate concentration is found to be comparatively homogeneously distributed over the whole tumor tissue when measured by bioluminescence (Fig. 12.11), whereas the spatial distribution of ATP and glucose is very heterogeneous after hyperthermic treatment (TAMULEVICIUS and STREFFER 1997). The increase in lactate apparently contributes to the induced acidosis after hyperthermia and probably enhances cell killing through

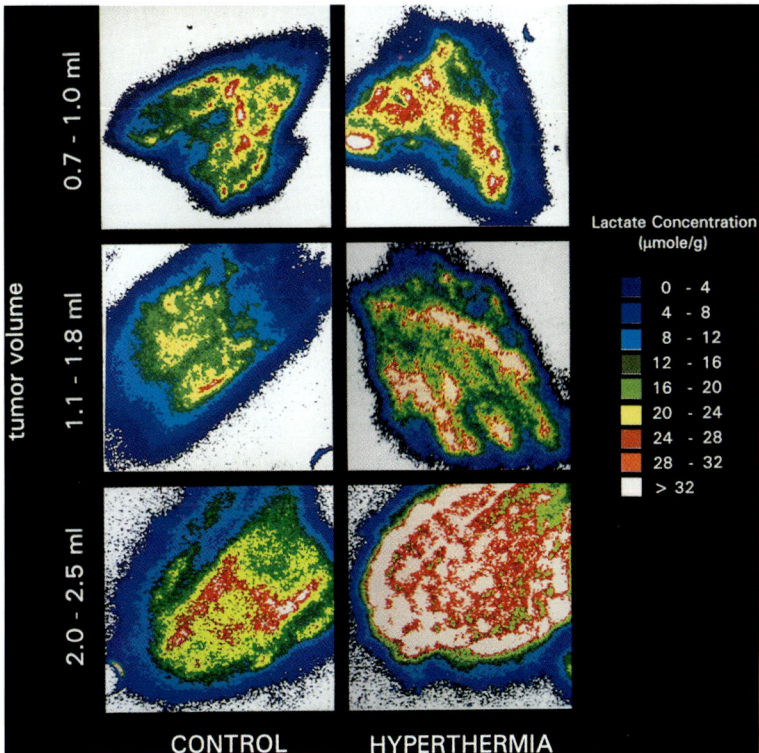

Fig. 12.11. Color-coded distribution of lactate concentrations in control (*left panels*) and heat-treated (44°C/60 min) DS sarcomas of three different volume ranges. (Adapted from VAUPEL and KELLEHER 1995)

this mechanism. As demonstrated by Fig. 12.11, this effect appears to be more pronounced in larger tumors than in smaller lesions.

In general, hyperthermia induces strong changes in energy metabolism by increased metabolic rates and turnover. It leads to a decrease in ATP levels by enhanced ATP turnover and degradation of adenosine nucleotides. In tumors, ATP production is mainly performed via glycolysis with lactate production under both hypoxic and normoxic conditions. In most cases, this leads to an increase in lactate levels in tumor tissue. These processes are closely connected to blood flow and oxygen as well as to glucose supply to the tumor tissue which may be reduced by hyperthermia. Besides a decrease in ATP, induced hypoxia is also documented by an increase in lactate/pyruvate ratios which mirror the redox equilibria in tumor tissues after heat treatments (STREFFER 1990).

12.6
Therapy Monitoring via Bioenergetic Status

It is a common observation that a therapeutic attack on solid tumors can induce changes in the bioenergetic status of the malignant tissue. This effect often precedes by far any changes in tumor volume which may be exploited as an early indicator of therapeutic efficiency. Bioenergetics following different treatment modalities – other than hyperthermia – have been monitored in numerous studies, some of which should be briefly mentioned here. Changes in the energetic status of tumors after *irradiation* have been investigated most intensively by different NMR techniques (NG et al. 1987; SIJENS et al. 1986; KOUTCHER et al. 1987; TOZER et al. 1989; KRISTJANSEN et al. 1992; MAHMOOD et al. 1995). SEMMLER and colleagues have monitored bioenergetics following *combined radio- and chemotherapy* (1988b). Studies with NMR or HPLC on energy-rich phosphates in tumors after treatment have been performed following *chemo- and/or endocrine* therapy (EVELHOCH et al. 1987; STEEN 1989b; NEEMAN and DEGANI 1989; STUBBS et al. 1990), application of *cytokines* such as interleukin 1α (CONSTANTINIDIS et al. 1989) or TNFα (SHINE et al. 1989; KLUGE et al. 1992), and *photodynamic therapy* (DODD et al. 1989). Most of the results obtained give rise to the hope that the treatment response of individual patients to various anticancer therapies may be tested and opti-

mized by the design of individual, highly efficient treatment protocols.

References

Arnold DL, Shoubridge EA, Feindel W, Villemure J-G (1987) Metabolic changes in cerebral gliomas within hours of treatment with intra-arterial BCNU demonstrated by phosphorus magnetic resonance spectroscopy. Can J Neurol Sci 14:570–575

Busse M, Vaupel P (1996) Accumulation of purine catabolites in solid tumors exposed to therapeutic hyperthermia. Experientia 52:469–473

Cadoux-Hudson T, Blackledge MJ, Rajagopalan B, Taylor DJ, Radda GK (1988) Measurement of phosphorus metabolites in patients with intracranial tumours. Proc 7th Annu Meeting Soc Magn Reson Med, San Francisco, vol 2, p 614

Constantinidis I, Braunschweiger PG, Wehrle JP, Kumar N, Johnson CS, Furmanski P, Glickson JD (1989) ^{31}P-nuclear magnetic resonance studies of the effect of recombinant human Interleukin 1α on the bioenergetics of RIF-1 tumors. Cancer Res 49:6379–6382

Dellian M, Walenta, S, Kuhnle GEH, Gamarra F, Mueller-Klieser W, Goetz AE (1993a) Relation between autoradiographically measured blood flow and ATP concentrations obtained from imaging bioluminescence in tumors following hyperthermia. Int J Cancer 53:785–791

Dellian M, Walenta S, Gamarra F, Kuhnle GEH, Mueller-Klieser W, Goetz AE (1993b) Ischemia and loss of ATP in tumors following treatment with focused high energy shock waves. Br J Cancer 68:26–31

Dellian M, Walenta S, Gamarra F, Kuhnle GEH, Mueller-Klieser W, Goetz AE (1994) High energy shock waves enhance hyperthermic response of tumors: effects on blood flow, energy metabolism and tumor growth. J Natl Cancer Inst 86:287–293

den Hollander JA, Luyton PR (1987) Image-guided localized ^{1}H and ^{31}P NMR spectroscopy of humans. Ann N Y Acad Sci 508:386–398

Dodd NJF, Moore JV, Poppitt DG, Wood B (1989) In vivo magnetic resonance imaging of the effects of photodynamic therapy. Br J Cancer 60:164–167

Evanochko WT, Ng TC, Glickson JD (1984) Application of in vivo NMR spectroscopy to cancer. Magn Reson Med 1:508–534

Evelhoch JL, Keller NA, Corbett TH (1987) Response-specific Adriamycin sensitivity markers provided by in vivo ^{31}P nuclear magnetic resonance spectroscopy in murine mammary adenocarcinomas. Cancer Res 47:3396–3401

Evelhoch JL, Negendank W, Valeriote FA, Baker LH (1990) Magnetic resonance in experimental and clinical oncology. Kluwer, Boston

Griffiths JR, Cady E, Edwards RHT, McCready VR, Wilkie DR, Wiltshaw E (1983) ^{31}P NMR studies of a human tumour in situ. Lancet 1:1435–1436

Hossmann K-A, Mies G, Paschen W, Szabo L, Dolan E, Wechsler W (1986) Regional metabolism of experimental brain tumors. Acta Neuropathol (Berl) 69:139–147

Hossmann K-A, Linn F, Okada Y (1992) Bioluminescence and fluoroscopic imaging of tissue pH and metabolites in experimental brain tumors of cat. NMR Biomed 5:259–264

Hubesch B, Sappey-Marinier D, Roth K, Sanuki E, Hodes JE, Matson GB, Weiner W (1988) Improved ISIS for studies of human brain and brain tumors. Proc 7th Annu Meeting Soc Magn Reson Med, San Francisco, vol 1, p 348

Kluge M, Elger B, Engel T, Schaefer C, Seega J, Vaupel P (1992) Acute effects of tumor necrosis factor α or lymphotoxin on global blood flow, laser Doppler flux, and bioenergetic status of subcutaneous rodent tumors. Cancer Res 52:2167–2173

Kogure K, Alonso OF (1978) A pictorial representation of endogeneous brain ATP by a bioluminescent method. Brain Res 154:273–284

Koutcher JA, Okunieff P, Neuringer L, Suit H, Brady T (1987) Size dependent changes in tumor phosphate metabolism after radiation therapy as detected by ^{31}P NMR spectroscopy. Int J Radiat Oncol Biol Phys 13:1851–1855

Kristjansen PEG, Pedersen AG, Quistorff B, Spang-Thomsen M (1992) Different early effect of irradiation in brain and small cell lung cancer examined by in vivo ^{31}P-magnetic resonance spectroscopy. Radiother Oncol 24:186–190

Kuhnle GEH, Dellian M, Walenta S, Mueller-Klieser W, Goetz AE (1992) Simultaneous high resolution measurement of ATP and blood flow in experimental tumors. J Natl Cancer Inst 84:1642–1647

Krüger W, Mayer W-K, Schaefer C, Stohrer M, Vaupel P (1991) Acute changes of systemic parameters in tumour-bearing rats, and of tumour glucose, lactate and ATP levels upon local hyperthermia and/or hyperglycaemia. J Cancer Res Clin Oncol 117:409–415

Levine SR, Welch KMA, Helpern JA, Bruce R, Ewing JR, Kensora T, Smith MB (1987) Cerebral cortical phosphate metabolism and pH in patients with multiple subcortical infarcts: a controlled study with 31-phosphorus NMR spectroscopy. Proc 6th Annu Meeting Soc Magn Reson Med, New York City, vol 2, p 1001

Lilly MB, Katholo CR, Ng TC (1985) Direct relationship between high-energy phosphate content and blood flow in thermally treated murine tumors. J Natl Cancer Inst 75:885–889

Luyton PR, den Hollander JA, van der Knaap MS, Valk J (1988) In vivo ^{31}P and ^{1}H NMR spectroscopy in patients with white matter disorders. Proc 7th Annu Meeting Soc Magn Reson Med, San Francisco, vol 1, p 257

Lyng H, Olsen DR, Southon TE, Rofstad EK (1993) ^{31}P-nuclear magnetic resonance spectroscopy in vivo of six human melanoma xenograft lines: tumour bioenergetic status and blood supply. Br J Cancer 68:1061–1070

Mahmood U, Alfieri AA, Ballon D, Traganos F, Koutcher JA (1995) In vitro and in vivo ^{31}P nuclear magnetic resonance measurements of metabolic changes post radiation. Cancer Res 55:1248–1254

Mies G, Paschen W, Ebhardt G, Hossmann K-A (1990) Relationship between bloodflow, glucose metabolism, protein synthesis, glucose and ATP content in experimentally induced glioma (RGT 2.2) of rat brain. J Neurooncol 9:17–28

Mueller-Klieser W, Walenta S (1993) Geographical mapping of metabolites in biological tissue with quantitative bioluminescence and single photon imaging (review). Histochem J 25:407–420

Mueller-Klieser W, Walenta S, Paschen W, Kallinowski F, Vaupel P (1988) Metabolic imaging in microregions of tumors and normal tissues with bioluminescence and photon counting. J Natl Cancer Inst 80:842–848

Mueller-Klieser W, Schaefer C, Walenta S, Rofstad EK, Fenton B, Sutherland RM (1990) Assessment of tumor energy and oxygenation status by bioluminescence, NMR, and cryospectrophotometry. Cancer Res 50:1681–1685

Mueller-Klieser W, Kroeger M, Walenta S, Rofstad EK (1991) Comparative imaging of structure and metabolites in tumors (invited review). Int J Radiat Biol 60:147–159

Mueller-Klieser W, Walenta S, Schwickert G (1994) Quantitative bioluminescence imaging – a method for the detection of metabolite distributions in frozen tissues. In: Proc Int Symp Biomed Optics Eur 93, Budapest. SPIE Proc 2083:34–40

Naruse SH, Higuchi T, Horikawa Y, Tanaka Cl, Nakamura K, Hirakawa K (1986) Radiofrequency hyperthermia with successive monitoring of its effects on tumors using NMR spectroscopy. Proc Natl Acad Sci USA 83:8343–8347

Neeman M, Degani H (1989) Early estrogen-induced metabolic changes and their inhibition by actinomycin D and cycloheximide in human breast cancer cells: ^{31}P and ^{13}C NMR studies. Proc Natl Acad Sci USA 86:5585–5589

Negendank W (1992) Studies of human tumors by MRS: a review. NMR Biomed 5:303–324

Negendank W, Crowley M, Keller N, Nussdorfer M, Evelhoch JL (1988) In vivo ^{31}P MRS of normal human breasts: age dependence and comparison with breast cancers. Proc 7th Annu Meeting Soc Magn Reson Med, San Francisco, vol 1, p 336

Ng TC, Majors AW, Meaney TF (1986) In vivo MR spectroscopy of human subjects with a 1.4-T whole body MR imager. Radiology 158:517–520

Ng TC, Vijayakumar S, Majors AW, Thomas FJ, Meaney TF, Baldwin NJ (1987) Response of a non-Hodgkin lymphoma to ^{60}Co therapy monitored by ^{31}P MRS in situ. Int J Radiat Oncol Biol Phys 13:1545–1551

Oberhaensli RD, Hilton-Jones D, Bore PJ, Hands LJ, Rampling RP, Radda GK (1986) Biochemical investigation of human tumours in vivo with phosphorus-31 magnetic resonance spectroscopy. Lancet 1:8–11

Paschen W (1985) Regional quantitative determination of lactate in brain sections. A bioluminescent approach. J Cereb Blood Flow Metab 5:609–612

Paschen W, Djuricic B, Mies G, Schmidt-Kastner R, Linn F (1987) Lactate and pH in the brain: association and dissociation in different pathophysiological states. J Neurochem 48:154–159

Roth K, Hubesch B, Naruse S, et al. (1987) Quantitation of metabolites in human brain using volume selected ^{31}P NMR. Proc 6th Annu Meeting Soc Magn Reson Med, New York City, vol 2, p 608

Schaefer C, Mayer W-K, Krüger W, Vaupel P (1993) Microregional distributions of glucose, lactate, ATP and tissue pH in experimental tumours upon local hyperthermia and/or hyperglycaemia. J Cancer Res Clin Oncol 119:599–608

Segebarth CM, Baleriaux DF, Arnold DL, Luyton PR, den Hollander JA (1987) MR image-guided P-31 MR spectroscopy in the evaluation of brain tumor treatment. Radiology 165:215–219

Semmler W, Gademann G, Bachert-Baumann P, Bier V, Zabel H-J, Lorenz WJ, van Kaick G (1988a) In vivo ^{31}Phosphor-Spektroskopie von Tumoren: prä-, intra- und posttherapeutisch. Fortschr Röntgenstr 149:369–377

Semmler W, Gademann G, Bachert-Baumann P, Zabel H-J, Lorenz WJ, van Kaick G (1988b) Monitoring human tumor response to therapy by means of P-31 MR spectroscopy. Radiology 166:533–539

Shine N, Palladino MA, Patton JS, Deisseroth A, Karczmar GS, Matson GB, Weiner MW (1989) Early metabolic response to tumor necrosis factor in mouse sarcoma: a phosphorus-31 nuclear magnetic resonance study. Cancer Res 49:2123–2127

Sijens PE, Bovée WMMJ, Seijkens D, Los G, Rutgers DH (1986) In vivo ^{31}P-nuclear magnetic resonance study of the response of a murine mammary tumor to different doses of γ-radiation. Cancer Res 46:1427–1432

Sijens PE, Bovee WMMJ, Seijkens D, Koole P, Los G, van Rijssel RH (1987) Murine mammary tumor response to hyperthermia and radiotherapy evaluated by in vivo ^{31}P-nuclear magnetic resonance spectroscopy. Cancer Res 47:6467–6473

Steen RG (1989a) Response of solid tumors to chemotherapy monitored by in vivo ^{31}P nuclear magnetic resonance spectroscopy: a review. Cancer Res 49:4075–4085

Steen RG (1989b) Response of solid tumors to chemotherapy monitored by in vivo ^{31}P nuclear magnetic resonance spectroscopy: a review. Cancer Res 49:4075–4085

Streffer C (1990) Biological basis of thermotherapy. In: Gautherie M (ed) Biological basis of oncologic thermotherapy. Springer, Berlin Heidelberg New York, pp 1–71

Stubbs M, Coomes RC, Griffiths JR, Maxwell RJ, Rodrigues LM, Gusterson BA (1990) ^{31}P-NMR spectroscopy and histological studies of the response of rat mammary tumours to endocrine therapy. Br J Cancer 61:258–262

Sutherland RM (1988) Cell and environment interactions in tumor microregions: the multicell spheroid model. Science 240:177–184

Tamulevicius P, Streffer C (1995) Metabolic imaging in tumours by means of bioluminescence. Br J Cancer 72:1102–1112

Tamulevicius P, Streffer C (1997) Bioluminescence imaging of metabolites in a human tumour xenograft after treatment with hyperthermia and/or the radiosensitizer pimonidazole. Int J Hyperthermia, in press

Tozer GM, Bhujwalla ZM, Griffiths JR, Maxwell RJ (1989) Phosphorus-31 magnetic resonance spectroscopy and blood perfusion of the RIF-1 tumor following x-irradiation. Int J Radiat Oncol Biol Phys 16:155–164

Vaupel P (1992) Physiological properties of malignant tumours. NMR Biomed 5:2220–2225

Vaupel P (1993) Oxygenation of solid tumors. In: Teicher BA (ed) Drug resistance in oncology. Dekker, New York, pp 53–85

Vaupel PW (1994) Blood flow, oxygenation, tissue pH distribution, and bioenergetic status of tumors. Schering Research Foundation, Berlin

Vaupel P (1996) Is there a critical tissue oxygen tension for bioenergetic status and cellular pH regulation in solid tumors? Experientia 52:464–468

Vaupel P, Jain RK (1991) Tumor blood supply and metabolic microenvironment. Characterization and implications for therapy. Fischer, Stuttgart

Vaupel P, Kelleher DK (1995) Metabolic status and reaction to heat of normal and tumor tissue. In: Seegenschmiedt MH, Fessenden P, Vernon CC (eds) Thermoradiotherapy and thermochemotherapy, vol 1. Springer, Berlin Heidelberg New York, pp 157–176

Vaupel P, Mueller-Klieser W (1992) Oxygenation and bioenenergetic status of human tumors. In: Chapman JD, Dewey WC, Whitmore GF (eds) Radiation research – a twentieth century perspective, vol 2. Academic, San Diego, pp 772–777

Vaupel P, Kallinowski F, Okunieff P (1989) Blood flow, oxygen and nutrient supply, and metabolic microenvironment of human tumors: a review. Cancer Res 49:6449–6465

Vaupel P, Kelleher DK, Engel T (1994a) Stable bioenergetic status despite substantial changes in blood flow and tissue oxygenation in a rat tumour. Br J Cancer 69:46–49

Vaupel P, Schaefer C, Okunieff P (1994b) Intracellular acidosis in murine fibrosarcomas coincides with ATP depletion, hypoxia, and high levels of lactate and total P_i. NMR Biomed 7:128–136

Vermeulen J, Luyton PR, den Hollander JA (1987) Determination of metabolite concentrations from localized ^{31}P NMR spectra of the human brain. Proc 6th Annu Meeting Soc Magn Reson Med, New York City, vol 1, p 136

Walenta S (1994) Entwicklung eines abbildenden Biolumineszenzverfahrens für die Beschreibung des lokalen Energiestoffwechsels in Gewebeschnitten. Thesis, University of Mainz, Germany

Walenta S, Dötsch J, Mueller-Klieser W (1990) ATP concentrations in multicellular tumor spheroids assessed by single photon imaging and quantitative bioluminescence. Eur J Cell Biol 52:389–393

Walenta S, Dellian M, Goetz AE, Kuhnle GEH, Mueller-Klieser W (1992) Pixel-to-pixel correlation between images of absolute ATP concentrations and blood flow in tumours. Br J Cancer 66:1099–1102

Walenta S, Salameh A, Lyng H, Evensen JF, Mitze M, Rofstad EK, Mueller-Klieser W (1997) Correlation of high lactate levels in head and neck tumors with incidence of metastasis. Am J Pathol 150:409–415

13 Therapeutic Significance of Microenvironmental Factors

D.J. Chaplin[1], M.R. Horsman[2], M.J. Trotter[3], and D.W. Siemann[4]

CONTENTS

13.1 Introduction

Advancements in cancer research, specifically in the field of tumour biology, are realized at several levels of inquiry: molecular, cellular, organ/tissue and whole animal. The rapid evolution of the broad discipline of molecular biology has shifted the experimental emphasis from the pathophysiological realm to that of important cellular and subcellular disease mechanisms. Nevertheless, structure/function abnormalities of neoplasia at the tissue level remain only partly understood and continued research is essential. Elucidation of the therapeutic implications of aberrant solid tumour structure, physiology and metabolism may provide important insights into enhancing the potential of both conventional and novel treatment modalities.

Solid malignant tumours are composed of both cancerous cells and normal host components. Tumour growth, resulting from uncontrolled neoplastic cell division, is absolutely dependent on a parallel proliferation of the nonmalignant cells which comprise the tumour vasculature. Thus, neovascularization, uncommon in normal adult tissues, is a universal characteristic of all solid tumours larger than 1–2mm in diameter (FOLKMAN 1986, 1990). In nonneoplastic tissue, the vascular supply of an organ is balanced with the need for blood flow imposed by the metabolic activity of its cells. In general terms, WEIBEL (1984) called this principle "symmorphosis", defining it as "a state of structural design commensurate to functional needs resulting from regulated morphogenesis". Neoplasms, by definition, exemplify abnormal morphogenesis and there is a disproportionate relationship between tumour tissue and its vascular supply. Tumours are said to "outgrow" their blood supply; neovascularization lags behind the increase in the number of neoplastic cells (TANNOCK 1970) and consequently the vasculature is unable to meet the increasing nutrient demands of the expanding tumour mass. In solid tumour tissue, the principle of symmorphosis is not operative.

Since the geometry of a capillary network is dictated by the shape and arrangement of the cells it supplies (BAEZ 1977; WEIBEL 1984), it is not surprising that tumour vasculature, like the cancerous tissue it serves, is morphologically and functionally abnormal. Tumour blood supply is characterized by spatial and temporal heterogeneity in both structure and function. A decline in average blood flow with tumour growth can lead to randomly distributed regions of altered microenvironment (VAUPEL et al. 1981, 1989; KALLINOWSKI et al. 1989). The development of hypoxia, acidosis and nutrient depletion can appreciably alter the tumour response to nonsurgi-

[1] D.J. CHAPLIN, PhD, Tumour Microcirculation Group, Gray Laboratory Cancer Research Trust, Mount Vernon Hospital, Northwood, Middlesex HA6 2JR, UK
[2] M.R. HORSMAN, PhD, Danish Cancer Society, Department of Experimental Clinical Oncology, Aarhus University Hospital, 44 Nørrebrogade, DK-8000 Aarhus C, Denmark
[3] M.J. TROTTER, MD, Department of Pathology, Vancouver General Hospital, 855 West 12th Avenue, Vancouver, British Columbia, Canada V5Z 1M9
[4] D.W. SIEMANN, PhD, Department of Radiation Oncology, Shands Cancer Center, University of Florida, P.O. Box 100385, Gainesville, FL 32610, USA

cal therapy (MOULDER and ROCKWELL 1987; WIKE-HOOLEY et al. 1984; TEICHER et al. 1981). For example, the deleterious effect of hypoxia on radiation cytotoxicity has been recognised for many years (CRABTREE and CRAMER 1933; MOTTRAM 1936; THOMLINSON and GRAY 1955) and recently tumour oxygenation has been shown to be an independent prognostic factor for radiation treatment of advanced cancer of the uterine cervix (HÖCKEL et al. 1993). The purpose of this chapter is to briefly review some of the consequences of the abnormal physiological microenvironment on tumour cells and their response to both new and established approaches to cancer treatment.

13.2
Microenvironment

The two main physiological characteristics of the microenvironment in tumours which impact on therapeutic outcome are oxygenation and pH.

13.2.1
Oxygenation

It has been established that solid tumours in both rodents and humans are characterised by regions of low oxygen tension. An example of the distribution of oxygen tensions in tumour and normal tissue in the mouse is shown in Fig. 13.1. In this figure the pO_2 histogram for normal subcutis in CBA mice is compared to that for the murine sarcoma F growing in the subcutis. It can clearly be seen that almost all the values for tumour tissue are below 15 mmHg (i.e. 2% O_2) whereas such values are rare in normal subcutis. Essential to understanding the influence of tumour oxygenation and how best to overcome it is to establish how hypoxia occurs in tumour tissue. For many years hypoxic regions were envisaged to occur distant from blood vessels as a result of oxygen utilization of the cells closer to the capillaries. As a result of such a process, cells would be hypoxic for many hours and also be subject to chronic depletion of nutrient supplies. However, there is increasing evidence that hypoxic cells can also result from dynamic changes in microregional perfusion (CHAPLIN et al. 1987; HILL et al. 1996). Such cells are subjected to relatively short periods of oxygen and nutrient deprivation and as a result would be healthy and actively proliferating. A schematic representation of the temporal changes in erythrocyte flux and

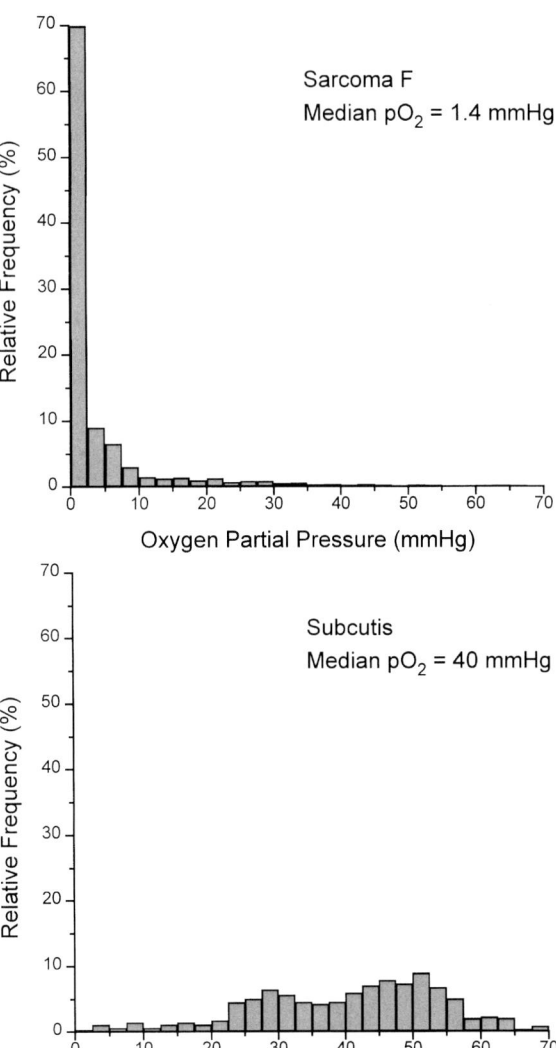

Fig. 13.1. Histograms showing pO_2 distributions in subcutis and SaF tumour implanted in the subcutis of CBA mice. (According to COLLINGRIDGE, HILL and CHAPLIN, unpublished data)

the resultant induction of hypoxia that can occur in a given vessel segment is shown in Fig. 13.2.

13.2.2
pH

In addition to possessing regions of reduced oxygenation, tumours have also been shown to have an acidic microenvironment compared to normal tissues (WIKE-HOOLEY et al. 1984). This reduced pH is often ascribed to the fact that tumour cells have an increased capacity for glycolysis with the resultant production of lactic acid. However, this is probably

Fig. 13.2a–d. Four of the many possible scenarios relating erythrocyte flux and hypoxia in one cross-sectional area of a tumour vessel at a given instant in time. For illustrative purposes, the cells with a nominal level of oxygenation ($pO_2 < 5$ mmHg) are *shaded blue*. **a, b, c** and **d** show the effect of varying erythrocyte flux. It is also possible to have a status similar to **d** with no erythrocyte flux, but retaining plasma perfusion

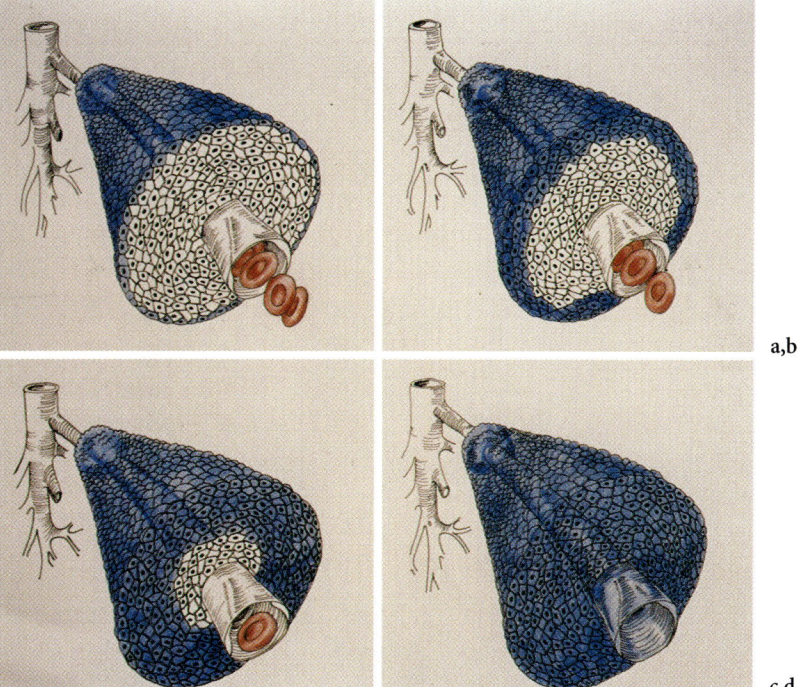

a,b

c,d

not the only process involved; lactate production via the breakdown of glutamine and CO_2 production as a result of cellular respiration could also contribute to the shift to acidic pH in malignant tissue (NEWELL et al. 1993). It is now well established that although tumour extracellular pH is acidic, the cells themselves maintain their intracellular pH close to neutrality (GRIFFITHS 1991; VAUPEL et al. 1989).

The cells maintain their pH via intracellular buffer systems and the use of efficient systems for the elimination of protons (proton pumps) and for the import of proton acceptors (e.g. HCO_3^-)

13.3
Therapeutic Consequences of the Tumour Microenvironment

13.3.1
Radiotherapy

It has been known for many years that as the oxygen tension is reduced below a partial pressure of 10 mmHg (1.3×10^4 ppm) cells become increasingly resistant to radiation damage. In fact, anoxic cells are approximately three times more resistant to radiation than those irradiated under aerobic conditions. The response of murine SaF cells irradiated under different oxygenation conditions is shown in Fig.

13.3. This effect, coupled with the finding that both human and rodent tumours possess regions of tissue oxygenation below 10 mmHg, indicates why tumour hypoxia remains a key focus of research in radiobiology and radiotherapy. The fact that tumour cells exist at a level of hypoxia consistent with that required for radiation resistance does not necessarily mean that such cells contribute to treatment outcome. For example, cells in such regions could be nonclonogenic and they could be reoxygenated during fractionated radiation treatment. However, evidence from animal tumour systems indicates that radiobiologically hypoxic cells are clonogenic and do contribute to therapeutic outcome during a fractionated treatment protocol (ROJAS et al. 1992). Evidence exists that such cells are also important in determining the outcome of clinical radiotherapy regimes (OVERGAARD and HORSMAN 1996).

Because of their potential importance for treatment outcome, a large effort has been afforded over the years to identifying strategies that will reduce or eliminate radiobiologically hypoxic cells within solid tumours. Breathing high oxygen content gases during radiation treatment under normobaric or hyperbaric conditions was shown to enhance the response of experimental tumours to radiation. However, as described in more detail in other chapters, the clinical results were disappointing with little or no benefit being observed in the majority of trials. A second

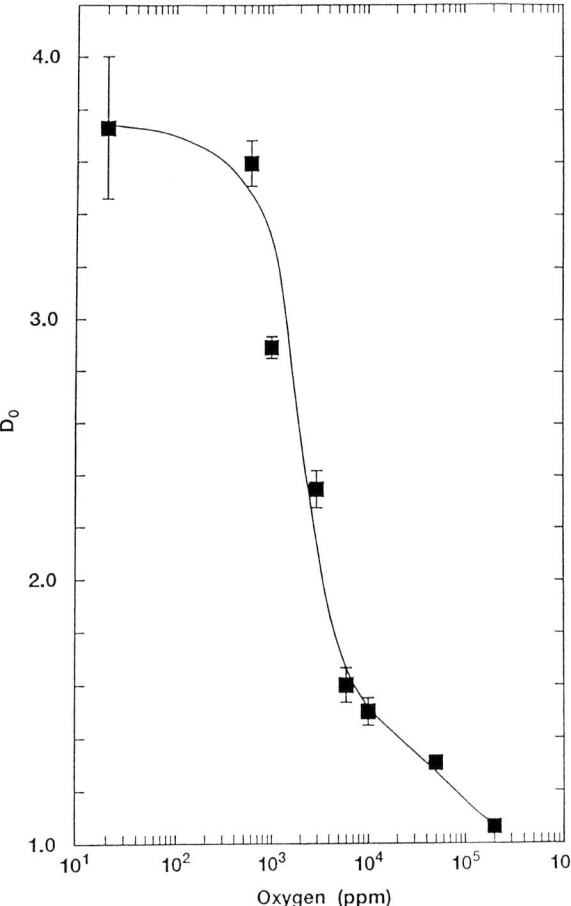

Fig. 13.3. The effect of oxygen concentration on the radiation sensitivity of murine sarcoma F cells in vitro. The graph shows the Do values obtained from full radiation dose-response curves obtained at each oxygen concentration. Do is defined at the dose of radiation in Gy required to reduce survival to 0.37. (Data provided by Hodgkiss and Webster from the Gray Laboratory)

does retain sensitizing potential even at the doses achieved in the clinic, has shown that improvement in local tumour control can be obtained with a chemical radiosensitizer used in clinical treatment regimes (OVERGAARD et al. 1991). As a result of the development of chemical radiosensitizers, in particular the nitromidazoles, it was discovered that such compounds, in addition to sensitizing hypoxic cells to radiation, could also selectively kill them (WORKMAN and STRATFORD 1993; BROWN and SIIM 1996). With the compounds that were originally identified as sensitizing agents, cytotoxic effects could only be achieved at high and not clinically relevant dose levels. Over the past decade, through the process of rational drug screening and drug development, several potent bioreductive cytotoxins, agents whose cytotoxic activity is dramatically enhanced when they are metabolised in a hypoxic environment, have been identified (Fig. 13.4). Two of these agents, E09 and Tirapazamine, are now being evaluated in clinical trials. The therapeutic advantages of killing hypoxic cells rather that sensitizing them have been described (BROWN and SIIM 1996). For such drugs to be effective and safe, several criteria need to be satisfied; these include (i) the particular enzyme profile required for reduction of the drug to the toxic species being present within the tumour to be treated and (ii) the oxygen dependency of production of the toxic species occurring over the same range as that for induction of radioresistance.

approach which received intense investigation during the 1970s and 1980s was the use of chemical radiosensitizers which could "fix" radiation damage in a manner similar to oxygen. Such compounds would not be utilized in the process of cellular respiration and therefore would be able to reach the hypoxic cells. Again, although such compounds showed great promise in experimental tumour systems, particularly with large single doses of radiation, the clinical results were in general disappointing (OVERGAARD 1994). This could be explained in part by the fact that the levels of radiosensitizer achieved within tumour tissue were not sufficient to elicit significant sensitization. However, a more recent study with the nitromidazole nimorazole, which

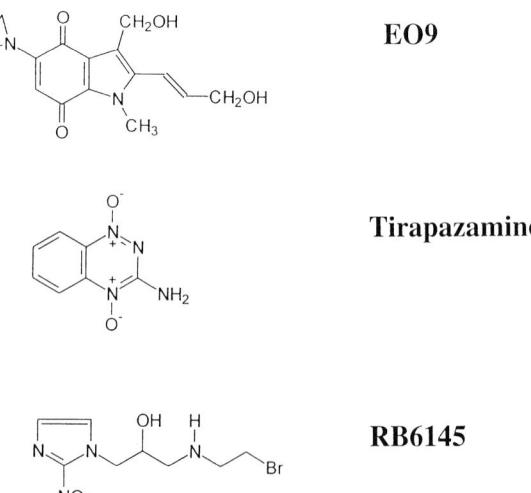

Fig. 13.4. Structures of lead compounds of three classes of bioreductive drugs

13.3.2
Chemotherapy

Although the importance of microenvironment to radiation response was characterized in the late 1950s, it was 20 years later before clear indication that microenvironmental features of solid tumours could modulate the response to chemotherapeutic agents. It was realised at this time that cells located distant from the functional blood supply could be resistant to drug therapy because of several factors. Firstly, limited penetration of plasma-borne anticancer agents would result in less of the drug reaching these cells. Secondly, as a result of a decline in nutrient and oxygen availability, cells further away from the vascular system would be dividing at a reduced rate and thus would be protected from the effects of chemotherapeutic agents whose activity is selective for rapidly dividing cell populations.

The large difference in proliferation rate between cells located in perivascular areas and those adjacent to regions of necrosis was elegantly demonstrated by TANNOCK (1968). These studies involved determining labelling indices in histological sections of tumours excised following treatment with [3H]thymidine. More recently it has become possible, using the staining gradient established within tumour tissue following intravenous injection of the DNA-binding fluorochrome Hoechst 33342, to separate and isolate cells relative to their distance from the functional vasculature. Using this technique has confirmed the reduced proliferation rate of tumour cells distant from the vasculature (MINCHINTON et al. 1990; OLIVE 1989). One of the approaches used in experimental systems to improve drug and nutrient delivery is to administer the chemotherapy drug along with an agent which selectively increases tumour blood flow. Most of the work to date has focused on the use of angiotensin II as the blood flow modifier and benefit has been observed using this approach in experimental tumours (SUZUKI et al. 1981). However, the variability in response of blood flow in different tumours and the same tumours at different sizes to angiotensin II complicates the routine use of such a protocol (JIRTLE 1988; TOZER et al. 1996).

Two additional features of the tumour microenvironment in vivo that have been shown to profoundly influence the response to chemotherapeutic drugs are oxygen tension and pH. Studies by ROIZEN-TOWLE and HALL (1978) demonstrated that exponentially growing cells in a culture which is rendered chronically hypoxic show more resistance than nor-mally proliferating cells to bleomycin. Studies with multicellular spheroids indicated that certain factors, in addition to drug penetration, drug uptake and cell cycle effects, were responsible for the resistance to Adriamycin in the centre of the spheroid (SUTHERLAND et al. 1979). The effect of chronic hypoxia on the response of cells in vitro to Adriamycin demonstrated that large increases in resistance were obtained if cells were maintained in hypoxic conditions for several hours prior to drug treatment (SMITH et al. 1980; BORN and EICHHOLTZ-WIRTH 1981). MARTIN and MCNALLY (1980) showed that cells in vitro, treated with the concentration of drug attained in the same cells grown as a solid tumour in vivo, were much more sensitive and associated this discrepancy at least in part to hypoxia. Subsequent studies with a range of drugs demonstrated that oxygenation status was an important factor influencing the cytoxicity of many established chemotherapeutic drugs (TEICHER et al. 1981; CHAPLIN et al. 1989; SIEMANN et al. 1991).

Attempts to improve the response of tumours in vivo to chemotherapeutic drugs by improving oxygenation via the breathing of high-oxygen-content gases and administration of perfluoro-chemical emulsions have indicated that therapeutic benefit can occur from such an approach (TEICHER and HOLDEN 1987). However, from the in vitro studies with Adriamycin, it is evident that drug resistance persists in cells up to 15h after being reoxygenated (WILSON et al. 1989). Therefore, prolonged improvement of tumour oxygenation prior to drug treatment may be required to achieve the maximum benefit.

A few years before hypoxia-induced cellular resistance to certain common chemotherapeutic drugs was shown, other compounds which were selectively more toxic in a hypoxic environment had been identified (HALL and ROIZEN-TOWLE 1975; MOORE et al. 1976). These findings eventually led to the development of bioreductive drugs, as described above, which are potent hypoxic cell cytotoxins (WORKMAN and STRATFORD 1993; BROWN and SIIM 1996). These compounds not only have potential utility in a therapeutic regime with radiation, but also in combination with other modalities including drugs to which hypoxia is known to induce treatment resistance (SIEMANN 1984).

In addition to hypoxia, the pH status of malignant tissue can dramatically influence drug activity, particularly that of compounds which are weak acids and bases. As mentioned previously, the extracellular pH in tumours is more acidic than the intracellu-

Weak Acid pKa 6.4 (AH)   Weak Base pKa 7.4 (B)

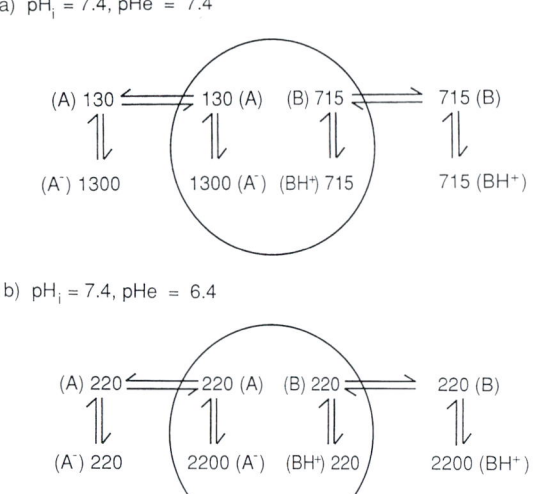

a) pH_i = 7.4, pHe = 7.4

(A) 130 ⇄ 130 (A)   (B) 715 ⇄ 715 (B)

(A⁻) 1300   1300 (A⁻) (BH⁺) 715   715 (BH⁺)

b) pH_i = 7.4, pHe = 6.4

(A) 220 ⇄ 220 (A)   (B) 220 ⇄ 220 (B)

(A⁻) 220   2200 (A⁻) (BH⁺) 220   2200 (BH⁺)

Fig. 13.5a,b. The calculated distribution of a given number of molecules (2860) of a weak acid (pK_a 6.4) and weak base (pK_a 7.4) between intracellular and extracellular compartments with **a** pH_i = 7.4 and pH_e = 7.4 and **b** pH_i = 7.4 and pH_e is 6.4. The latter may be more representative of conditions found within microregions of the tumour. The model assumes just passive diffusion of the nonionized form of the drug and that extracellular and intracellular compartments are of similar volume

lar pH; thus the degree of ionization of weak acids and bases will differ between the compartments.

Since only the unionized form of the drug will pass across the cell membrane and thus establish an equilibrium, the degree of ionization can dramatically influence drug distribution. Figure 13.5 shows a representative picture of drug distribution for a weak acid (pK_a 6.4) and weak base pK_a (7.4) between intra- and extracellular compartments with pH_i being 7.4 and pH_e being either 7.4 or 6.4.

In vitro studies have clearly demonstrated that selective decrease of the extracellular pH increases the uptake and cytotoxicity of weak acids such as chlorambucil, but decreases the uptake and activity of weak bases such as vinblastine (MIKKELSON and WALLACH 1982; GERWECK and SEETHARAMAN 1996; PARKINS et al. 1996). In vivo there exists another potential barrier, i.e. between the intravascular and tumour compartments. However, this is thought not to be such a critical barrier in tumours, as a result of abnormally high endothelial permeability and the presence of vasculature without intact endothelial lining.

There is evidence that the extracellular pH decreases as a function of distance from the vasculature (DELLIAN et al. 1996); therefore it would be expected that the uptake of weak acids would be increased in

Fig. 13.6. Fluorescence photomicrograph of a section from a SaF tumour following administration of 5-carboxyfluorescein diacetate (20 mg/kg IP) and the vascular marker Hoechst 33342 (10 mg/kg IV). The 5-carboxyfluorescein diacetate was injected 45 min and Hoechst 33342 5 min prior to tumour ex-

cision. Following excision, the tumours were embedded, frozen, sectioned and viewed on a fluorescence microscope using UV excitation to visualize the Hoechst distribution and visible light to visualize the carboxyfluorescein

these regions. The distribution of the weak acid 5-carboxyfluorescein diacetate relative to the vasculature within a tumour, following systemic administration of this agent to a mouse, can be seen in Fig. 13.6. There is clear evidence that the agent is selectively distributed in regions distant from the functional vasculature.

Clearly, the activity of weakly acidic and basic drugs could be improved by manipulation of the transmembrane pH gradient. For example, approaches such as breathing hypercapnic gases or intravenous administration of glucose, which can further decrease extracellular pH, could improve the anti-tumour effects of weak acids such as chlorambucil (GULLINO et al. 1965; VOLK et al. 1993; DELLIAN et al. 1996). Moreover, reducing the pH gradient could improve the activity of weak bases. The pH gradient is maintained by membrane-based ion exchange mechanisms (MOOLENAAR et al. 1984; MADSHUS 1988; TANNOCK and ROTIN 1989). One of the major membrane proton pumps is the Na^+/H^+ antiport. Amiloride and related compounds have been identified as potent inhibitors of this transporter (GRINSTEIN et al. 1989). Inhibition of proton efflux results in intracellular acidification of cells maintained in an acidic extracellular milieu. Similar effects can be obtained using nigericin, which acts as a K^+ ionophore and exchanges extracellular H^+ with intracellular K^+ ions (TANNOCK and ROTIN 1989). Reducing the cell's ability to maintain its pH in the face of extracellular acidification has, in experimental systems, been shown to reduce cell viability (TANNOCK and ROTIN 1989) and enhance the cytotoxicity of weakly basic drugs such as vinblastine (PARKINS et al. 1996).

However, altering the uptake of weak acids and bases is not the only impact that pH could have on the activity of chemotherapeutic agents. For example, pH-induced alterations in drug stability, active transport processes, drug reactivity, activity of enzymes involved in localized drug activation and the interaction of the drug with its molecular target could all significantly alter treatment efficacy. Decreasing pH from 7.4 to between 6.0 and 6.5 has been shown to enhance the cytotoxicity of many agents other than weak acids and bases. These include alkylating agents (JÄHDE et al. 1989; CHAPLIN et al. 1989), 5-fluorouracil (KUNG et al. 1963) and the bioreductive drugs Tirapazamine (SKARSGARD et al. 1993), EO9 (PHILLIPS et al. 1992) and RB6145 (SKARSGARD et al. 1995).

The cytotoxicity of some drugs is enhanced dramatically in vitro by the combination of hypoxia and an acidic microenvironment, e.g. melphalan (CHAPLIN et al. 1989; SIEMANN et al. 1991). Selective reduction of blood flow in tumours which will induce further decreases in oxygenation and acidity have been shown to enhance the anti-tumour effects of melphalan in vivo (CHAPLIN et al. 1989; SIEMANN 1990).

13.3.3
Biological Therapies

Although much experimental effort has been afforded to evaluating biological therapies in vitro and in vivo, relatively little attention has focused on how the complex microenvironmental and physiological features of tumours influence treatment efficiency. However, it is now evident that tumour-relevant oxygen tensions can dramatically alter the potency of certain cytokines. Studies by AUNE and POGUE (1989) demonstrated that changing the oxygen tension of the cellular environment from 12% to 4% reduced the anti-proliferative effects of IFN γ. The induction of lymphokine-activated killer (LAK) cell activity by interleukin 2 (IL-2) in human peripheral blood mononuclear cells is critically dependent on the oxygen tension in culture conditions (ISHIZAKA et al. 1992). This study showed that for a given degree of target cell killing, the IL-2 activated effector-to-target-cell ratio had to be increased by a factor of 3 if the effector cells were incubated with IL-2 at 5% instead of 21% oxygen and by a factor of 20 if the incubation with IL-2 was carried out at 2% oxygen. Several studies have demonstrated the critical importance of oxygen tension in determining the cytotoxic/anti-proliferative effects of TNF (SAMPSON and CHAPLIN 1994; LYNCH et al. 1995; PARK et al. 1992; NALDINI et al. 1994). In our own studies we have shown that exposure to 2% oxygen for 24 h prior to TNF treatment induces a 4- to 50-fold increase in resistance compared to cells incubated at 21% oxygen (SAMPSON and CHAPLIN 1994; LYNCH et al. 1995). Moreover, continued passage of cells at tumour-relevant oxygen tensions, prior to additions of TNF, further increased cellular resistance. These latter studies showed that pre-incubation oxygen tension is the important determining factor for the response to subsequent exposure to TNF. The underlying mechanism responsible for the induction of resistance has not yet been elucidated, although cell cycle effects are not responsible (LYNCH et al. 1995). Several mechanisms can be postulated. Firstly, there is convincing evidence that reactive oxygen species mediate, at least in part, the cytotoxic activities of TNF (Park et al. 1992; Zimmerman et al. 1989; Yamauchi et al. 1989). Thus, if hypoxia induces the production of proteins that protect the cell against such species, increased resistance would be evident. One such protein could be manganese superoxide dismutase (MnSOD), high levels of which are known to protect cells from TNF-induced cytotoxicity (Wong et al. 1989; Zyad et al. 1994). Secondly, endogenous TNF production has been reported to protect cells against exposure to exogenous TNF (Vanhaesebroeck et al. 1992; Himeno et al. 1990). Recent reports indicate that incubating macrophages in an atmosphere containing 2% oxygen can dramatically up-regulate the endogenous production of both TNF and its soluble receptors (Scannel et al. 1993). A third explanation is that receptor number, affinity, or rate of internalization of the receptor-ligand complex is altered by prior exposure to low oxygen environments.

13.3.4
Photodynamic Therapy

Photodynamic therapy (PDT) involves the systemic administration of phototoxic drugs (photosensitizers) which are activated locally in the tumour by treatment with light (Pass 1993; Fisher et al. 1995). The activation of the photosensitizer by light results in the production of reactive oxygen species, in particular singlet oxygen. These reactive oxygen species can kill the tumour cells directly or damage tumour endothelium, causing indirect tumour cell killing via induction of ischaemia. Thus, it is evident that the presence of oxygen is an essential pre-requisite for tumour cell killing via PDT.

It has been shown that oxygen levels above 5% (approximately 38 mmHg) are needed for full photodynamic cell inactivation (Henderson and Fingar 1987; Moan and Sommer 1985; Chapman et al. 1991). Decreasing oxygen levels below 5% progressively limits cellular photoinactivation with a half-value of about 1% (7.6 mmHg) O_2. The oxygen limitations for PDT are, therefore, similar to those for ionizing radiation. Limited sensitivity towards tumour cell inactivation by ionizing radiation due to insufficient oxygen supply would correlate with resistance towards photodynamic cell inactivation. However, while hypoxia renders cells up to three times less sensitive to ionizing radiation, it makes them completely resistant to PDT. Thus, the existence of hypoxic cells within tumours will be a critical

factor determining therapeutic outcome following PDT. The impact of hypoxia is further enhanced by the fact that oxygen is consumed during photoactivation; thus there is the potential, particularly at high light fluence rates, for inducing additional hypoxia and as a result further reducing treatment effectiveness (Foster et al. 1991; van Geel et al. 1996).

Attempts to improve oxygenation during PDT, using perfluorochemicals in combination with carbogen breathing, have provided little or no therapeutic benefit (Fingar et al. 1988). However, these results may reflect the fact that such a strategy would primarily reduce the hypoxia arising from diffusion limitations and will have little or no effect on hypoxia occurring as a result of microregional blood flow changes.

The combination of bioreductive drugs with PDT has been shown to result in enhanced anti-tumour effects in experimental tumour systems and in patients with skin metastases (Gonzales et al. 1986; Bremner et al. 1992; Baas et al. 1996). The use of bioreductive drugs with PDT not only should kill the treatment-resistant hypoxic cells, but also has the potential to exploit the large induction of hypoxia that occurs following treatment as a result of endothelial damage and subsequent blood flow reduction.

In contrast to hypoxia, pH is not itself a major determinant of tumour response to PDT (Varnes et al. 1996). However, recent studies have shown that the ionophore nigericin can enhance PDT induced cytotoxicity in vitro and that these effects are more pronounced at acidic pH (Varnes et al. 1992, 1996). The mechanisms of this effect have not been elucidated but, unlike the effects seen with hyperthermia, appear unrelated to changes in intracellular pH.

13.3.5
Hyperthermia

The use of heat as a therapeutic modality, either alone or in combination with radiation, has also been the subject of active investigation over the past three decades. It has been established that the major microenvironmental factor which predisposes tumour cells to the cytotoxic effects of hyperthermia is pH. Cells incubated in an acidic environment are much more sensitive to treatment (Gerweck 1977; Wike-Hooley et al. 1984). Moreover, it is generally accepted that it is the pH_i that is the critical parameter rather than pH_e. This is emphasized by the finding that the effects of hyperthermia under acidic extracellular conditions are dramatically enhanced by agents such as amiloride and nigericin which reduce the cell's ability to maintain a more neutral intracellular milieu (LYONS et al. 1992). In contrast to the importance of pH, oxygenation status has little effect on cellular response to heat treatment (GERWECK et al. 1974, 1979).

13.4
Summary

The occurrence within solid tumours of regions of hypoxia and acidic pH can influence many of the therapeutic approaches used in clinical practice. In particular, a level of oxygenation of less then 15 mmHg induces resistance to many conventional and emerging cancer treatments (Table 13.1). Strategies to improve tumour oxygenation could thus play an important role in many of these approaches.

It is to be hoped that recent advances in our knowledge of how hypoxia occurs will result in effective strategies to overcome both diffusion- and perfusion-limited hypoxia and thus improve treatment efficacy. In addition, the development of potent bioreductive drugs could also facilitate the eradication of treatment-resistant hypoxic cells.

Strategies to manipulate tumour pH have therapeutic potential, particularly with hyperthermia and certain chemotherapeutic agents. Further acidification of the extracellular milieu alone or in combination with agents which compromise the cell's ability to maintain its intracellular pH have been shown to produce therapeutic benefits in experimental systems. It is to be hoped that these advances can eventually be translated to produce improved therapeutic benefit in the clinic.

In conclusion, advances made over the past few decades have dramatically improved our understanding of tumour microenvironment and its potential importance for therapy. This emphasises the need for a continued focus on tumour pathophysiology alongside relevant studies of the cellular and molecular mechanisms involved in malignant disease.

Acknowledgement. This work was supported by grants from the British Cancer Research Campaign, the Danish Cancer Society and the American NIH (grant number CA 36858).

Table 13.1. Effect of hypoxia (≤ 15 mmHg) on treatment sensitivity compared to effects achieved at a higher pO_2 level

Treatment modality	Hypoxia during treatment	Prexposure to hypoxia
Radiation	Resistance (<10 mmHg)	–
Chemotherapeutics (agent dependent)	Resistance Sensitization (no detailed O_2 dependency, most studies use anoxia)	Resistance (≤ 15 mmHg)
Bioreductive drugs	Sensitization (exact O_2 dependency, varies with agent)	–
Photodynamic therapy	Resistance	–
Cytokine effects: TNF α cytotoxicity	Resistance (anoxia)	Resistance (≤ 15 mmHg)
Cytokine effects: IL-2-induced lymphocyte activation	Resistance (≤ 15 mmHg)	–

References

Aune TM, Pogue S (1989) Inhibition of tumour cell growth by interferon γ is mediated by two distinct mechanisms, dependent on tumour oxygenation, induction of tryptophan degradation and depletion of intracellular nicotinamide adenine dinucleotide. J Clin Invest 84:863–875

Baas P, van Geel IPJ, Oppelaar H, Meyer M, Beynen JH, van Zandwijk N, Stewart FA (1996) Enhancement of photodynamic therapy by mitomycin C: a preclinical and clinical study. Br J Cancer 73:945–951

Baez (1977) Microcirculation. Annu Rev Physiol 39:391–415

Born R, Eichholtz-Wirth H (1981) Effect of different physiological conditions on the action of adriamycin on Chinese hamster cells in vitro. Br J Cancer 44:241–246

Bremner JCM, Adams GE, Pearson JK, Sansom J, Stratford IJ (1992) Increasing the effect of photodynamic therapy on the RIF-1 murine sarcoma using the bioreductive drugs RSU 1069 and RB 6145. Br J Cancer 66:1070–1076

Brown JM, Siim BG (1996) Hypoxia specific cytotoxins in cancer therapy. Semin Radiat Oncol 6:22–36

Chaplin DJ, Olive PL, Durand RE (1987) Intermittent blood flow in a murine tumour: radiobiological effects. Cancer Res 47:597–601

Chaplin DJ, Acker B, Olive PL (1989) Potentiation of the tumor cytotoxicity of melphalan by vasodilating drugs. Int J Radiat Oncol Biol Phys 16:1131–1135

Chapman JD, Stobbe CC, Arnfield MR, Santus R, Lee J, McPhee MS (1991) Oxygen dependency of tumor cell killing in vitro by light-activated photofrin II. Radiat Res 126:73–79

Crabtree HG, Cramer W (1933) The action of radium on cancer cells. II. Some factors determining the susceptibility of cancer cells to radium. Proc R Soc Lond [Biol] 113:238–250

Dellian M, Helmlinger G, Yuan F, Jain RK (1996) Fluorescence ratio imaging of interstitial pH in solid tumours: effect of glucose on spatial and temporal gradients. Br J Cancer 74: 1206–1215

Fingar VH, Mang TS, Henderson BW (1988) Modification of photodynamic therapy induced hypoxia by Fluosol-DA (20%) and carbogen breathing in mice. Cancer Res 48:3350–3354

Fisher AMR, Murphree AL, Gomer CJ (1995) Clinical and preclinical photodynamic therapy. Laser Surg Med 17:2–31

Folkman J (1986) How is blood vessel growth regulated in normal and neoplastic tissue? GHA Clowes Memorial Award Lecture. Cancer Res 46:467–473

Folkman J (1990) What is the evidence that tumors are angiogenesis dependent? J Natl Cancer Inst 82:4–6

Foster TH, Murant RS, Bryant RG, Knox RS, Gibson SL, Hilf R (1991) Oxygen consumption and diffusion effects in photodynamic therapy. Radiat Res 126:296–303

Gerweck LE (1977) Modification of cell lethality at elevated temperatures. The pH effect. Radiat Res 70:224–235

Gerweck LE, Seetharaman K (1996) Cellular pH gradient in tumor versus normal tissue: potential exploitation for the treatment of cancer. Cancer Res 56:1194–1198

Gerweck LE, Gillette EL, Dewey WC (1974) Killing of Chinese hamster cells in vitro by heating under hypoxic or aerobic conditions. Eur J Cancer 10:691–693

Gerweck LE, Nygaard TG, Burlett M (1979) Response of cells to hyperthermia under acute and chronic hypoxic conditions. Cancer Res 39:966–972

Gonzales S, Arnfield MR, Meeker BE, Tulip J, Lakey WH, Chapman JD, McPhee MS (1986) Treatment of Dunning R3327-AT rat prostate tumors with photodynamic therapy in combination with misondiazole. Cancer Res 46:2858–2862

Griffiths JR (1991) Are cancer cells acidic? Br J Cancer 64:425–427

Grinstein S, Rotin D, Mason MJ (1989) Na^+/H^+ exchange and growth factor-induced cytosolic pH changes. Role in cellular proliferation. Biochim Biophys Acta 988:73–97

Gullino PM, Grantham FH, Smith SH, Haggerty AC (1965) Modification of the acid-base status of the internal milieu of tumours. J Natl Cancer Inst 34:857–869

Hall EJ, Roizen-Towle L (1975) Hypoxic sensitizers: radiobiological studies at the cellular level. Radiology 117:453–457

Henderson BW, Fingar VH (1987) Relationship of tumor hypoxia and response to photodynamic treatment in an experimental mouse tumor. Cancer Res 47:3110–3114

Hill SA, Pigott KH, Saunders MI, et al. (1996) Microregional blood flow fluctuations in murine and human tumours assessed using laser Doppler microprobes. Br J Cancer Suppl 74:260–263

Himeno T, Watanabe N, Yamauchi N, et al. (1990) Expression of endogenous TNF as a protective protein against the cytotoxicity of exogenous TNF. Cancer Res 50:4941–4945

Höckel M, Knoop C, Schlenger K, et al. (1993) Intratumoral pO_2 predicts survival in advanced cancer of the uterine cervix. Radiother Oncol 26:45–50

Ishizaka S, Kimoto M, Tsujii T (1992) Defect in generation of LAK cell activity under oxygen-limited conditions. Immunol Lett 32:209–214

Jähde E, Glusenkamp K-H, Klunder I, Hulser DF, Tietze LF, Rajewsky MF (1989) Hydrogen ion-mediated enhancement of cytotoxicity of bis-chloroethylating drugs in rat mammary carcinoma cells in vitro. Cancer Res 49:2965–2972

Jirtle RL (1988) Chemical modification of tumour blood flow. Int J Hyperthermia 4:355–371

Kallinowski F, Schlenger KH, Runkel S, Kloes M, Stohrer M, Okunieff P, Vaupel P (1989) Blood flow, metabolism, cellular microenvironment, and growth rate of human tumor xenografts. Cancer Res 49:3759–3764

Kung S-S, Goldberg ND, Dahl JL, Parks RE, Kline BE (1963) Potentiation of 5-Fluorouracil inhibition of flexner-jobling carcinoma by glucose. Science 141:627–628

Lynch EM, Sampson LE, Khalil AA, Horsman MR, Chaplin DJ (1995) Cytotoxic effect of tumour necrosis factor-alpha on sarcoma F cells at tumour relevant oxygen tension. Acta Oncol 34:423–427

Lyons JC, Kim GE, Song CW (1992) Modification of intracellular pH and thermosensitivity. Radiat Res 129:79–87

Madshus IH (1988) Regulation of intracellular pH in eukaryotic cells. Biochem J 250:1–8

Martin WMC, McNally NJ (1980) Cytotoxicity of adriamycin to tumour cells in vivo. Br J Cancer 42:881–889

Mikkelson RB, Wallach DFH (1982) Transmembrane ion gradients and thermochemotherapy. Proc Clin Biol Res 107:103–107

Minchinton AI, Durand RE, Chaplin DJ (1990) Intermittent blood flow in the KHT Sarcoma-flow cytometric studies using Hoechst 33342. Br J Cancer 62:195–200

Moan J, Sommer S (1985) Oxygen dependence of the photosensitising effect of hematoporphyrin derivative in NHIK 3025 cells. Cancer Res 45:1608–1610

Moolenaar WH, Tertoolen LGJ, De Laat SW (1984) The regulation of cytoplasmic pH in human fibroblasts. J Biol Chem 259:7563–7569

Moore BA, Palcic B, Skarsgard LD (1976) Radiosensitizing and toxic effects of the 2-nitromidazole Ro-07-0582 in hypoxic mammalian cells. Radiat Res 67:459–473

Mottram JC (1936) A factor of importance in the radiosensitivity of tumors. Br J Radiol 9:606–614

Moulder JE, Rockwell S (1987) Tumour hypoxia: its impact on cancer therapy. Cancer Metastasis Rev 5:313–341

Naldini A, Cesari S, Bocci V (1994) Effects of hypoxia on the cytotoxicity mediated by tumour necrosis factor. Lymphokine Cytokine 13:233–237

Newell K, Franchi A, Pouyssegur J, Tannock I (1993) Studies with glycolysis-deficient cells suggest that production of lactic acid is not the only cause of tumor acidity. Proc Natl Acad Sci USA 90:1127–1131

Olive PL (1989) Distribution, oxygenation and clonogenicity of macrophages in a murine tumour. Cancer Commun 1:93–100

Overgaard J (1994) Clinical evaluation of nitroimidazoles as modifiers of hypoxia in solid tumors. Oncol Res 6:509–518

Overgaard J, Horsman MR (1996) Modification of hypoxiainduced radioresistance in tumours by the use of oxygen and sensitisers. Semin Radiat Oncol 6:10–21

Overgaard J, Sand Hansen H, Overgaard M, Jorgensen K, Bastholt L, Berthelsen A, Pedersen M (1991) The Danish Head and Neck Cancer study group (DAHANCA) randomized trials with hypoxic radiosensitisers in carcinoma of the larynx and pharynx. In: Dewey WC, Edington M, Fry RJM (eds) Radiation research: a twentieth century perspective, vol II. Congress Proceedings ICRR, Toronto, pp 573–577

Park YMK, Anderson RL, Spitz DR, Hahn GM (1992) Hypoxia and resistance to hydrogen peroxide confer resistance to tumour necrosis factor in murine L929 cells. Radiat Res 131:162–168

Parkins CS, Chadwick JA, Chaplin DJ (1996) Inhibition of intracellular pH control and relationship to cytotoxicity of chlorambucil and vinblastine. Br J Cancer Suppl 74:75–77

Pass HI (1993) Photodynamic therapy in oncology: mechanisms and clinical use. J Natl Cancer Inst 85:443–456

Phillips RM, Hulbert PB, Bibby MC, Sleigh NR, Double JA (1992) In vitro activity of the novel indoloquinone EO9 and the influence of pH on cytotoxicity. Br J Cancer 65:359–364

Roizen-Towle L, Hall EJ (1978) Studies with bleomycin and misonidazole on aerated and hypoxic cells. Br J Cancer 37:254–260

Rojas A, Joiner MC, Hodgkiss RJ, Carl U, Kjellen E, Wilson GD (1992) Enhancement of tumour radiosensitivity and reduced hypoxia-dependent binding of a 2-nitroimidazole with normobaric oxygen and carbogen. A therapeutic comparison with skin and kidneys. Int J Radiat Oncol Biol Phys 23:361–366

Sampson LE, Chaplin DJ (1994) The influence of microenvironment on the cytotoxicity of TNF α in vitro. Int J Radiat Oncol Biol Phys 29:467–471

Scannel G, Waxman K, Kaeml GJ, Ioli G, Gatanga T, Yamamoto R, Granger GA (1993) Hypoxia induces a human macrophage cell line to release tumour necrosis factor α and its soluble receptors in vitro. J Surg Res 54:281–285

Siemann DW (1984) Modification of chemotherapy by nitroimidazoles. Int J Radiat Oncol Biol Phys 10:1585–1594

Siemann DW (1990) Enhancement of chemotherapy and nitroimidazole-induced chemopotentiation by the vasoactive drug hydralazine. Br J Cancer 62:348–353

Siemann DW, Chapman M, Beikirch A (1991) Effects of oxygenation and pH on tumor cell response to alkylating chemotherapy. Int J Radiat Oncol Biol Phys 20:287–289

Skarsgard LD, Skwarchuk MW, Vinczan A, Chaplin DJ (1993) The effect of pH on the aerobic and hypoxic cytotoxicity of SR 4233. Br J Cancer 68:681–683

Skarsgard LD, Acheson DK, Vinczan A, et al. (1995) Cytotoxic effect of RB 6145 in human tumour cell lines: dependence on hypoxia, extra- and intra-cellular pH and drug uptake. Br J Cancer 72:1479–1486

Smith E, Stratford IJ, Adams GE (1980) Cytotoxicity of adriamycin on aerobic and hypoxic Chinese hamster V79 cells in vitro. Br J Cancer 41:568–572

Sutherland RM, Eddy HA, Bareham B, Reich K, Vanantwerp D (1979) Resistance to adriamycin in multicellular spheroids. Int J Radiat Oncol Biol Phys 5:1225–1230

Suzuki M, Hori K, Abe I, Saito S, Sato H (1981) A new approach to cancer chemotherapy: selective enhancement of tumor blood flow with angiotensin II. J Natl Cancer Inst 67:663–669

Tannock IF (1968) The relation between cell proliferation and the vascular system in a transplanted mouse mammary tumour. Br J Cancer 22:258–273

Tannock IF (1970) Population kinetics of carcinoma cells, capillary endothelial cells and fibroblasts in a transplanted mouse mammary tumour. Cancer Res 30:2470–2474

Tannock IF, Rotin D (1989) Acidic pH in tumours and its potential for therapeutic exploitation. Cancer Res 49:4373–4384

Teicher BA, Holden SA (1987) A survey of the effect of adding Fluosol DA 20%/O₂ to treatment with various chemotherapeutic agents. Cancer Treat Rep 71:173–177

Teicher BA, Lazo JS, Sartorelli AC (1981) Classification of antineoplastic agents by their selevtive toxicities towards oxygenated and hypoxic tumor cells. Cancer Res 41:73–81

Thomlinson RH, Gray LH (1955) The histological structure of some human lung cancers and the possible implications for radiotherapy. Br J Cancer 9:539–549

Tozer GM, Shaffi KM, Prise VE, Bell KM (1996) Spatial heterogeneity of tumour blood flow modification induced by angiotensin II: relationship to receptor distribution. Int J Cancer 65:658–663

van Geel IPJ, Oppelaar H, Marijnissen JPA, Stewart FA (1996) Influence of fractionation and fluence rate in photodynamic therapy with photofrin or mTHPC. Radiat Res 145:602–609

Vanhaesebroeck B, Decoster E, Van Ostade X, Van Bladel S, Lenaerts A, Van Roy F, Fiers W (1992) Expression of an exogenous tumour necrosis factor (TNF) gene in TNF sensitive cell lines confers resistance to TNF mediated cell lysis. J Immunol 148:2785–2794

Varnes ME, Clay ME, Freeman K, Antunez AR, Oleinick NL (1992) Enhancement of photodynamic cell killing (with chloroaluminium phthalocyanine) by treatment of V79 cells with the ionophore nigericin. Cancer Res 50:1620–1625

Varnes ME, Bayne MT, Bright GR (1996) Reduction of intracellular pH is not the mechanism for the synergistic interaction between photodynamic therapy and nigericin. Photochem Photobiol 64:853–858

Vaupel P, Frinak S, Bicher HI (1981) Heterogeneous oxygen partial pressure and pH distribution in C3H mouse mammary adenocarcinoma. Cancer Res 41:2008–2013

Vaupel P, Kallinowski F, Okunieff P (1989) Blood flow, oxygen and nutrient supply, and metabolic microenvironment of human tumors: a review. Cancer Res 49:6449–6465

Volk T, Jähde E, Fortmeyer HP, Glusenkamp KH, Rajewsky MF (1993) pH in human tumour xenografts: effect of intravenous administration of glucose. Br J Cancer 68:492–500

Weibel ER (1984) The pathway for oxygen Structure and function in the mammalian respiratory system. Harvard University Press, Cambridge

Wike-Hooley JL, Haveman J, Reinhold HS (1984) The relevance of tumour pH to the treatment of malignant disease. Radiother Oncol 2:343–366

Wilson RE, Keng PC, Sutherland RM (1989) Drug resistance in Chinese hamster ovary cells during recovery from severe hypoxia. J Natl Cancer Inst 81:1235–1240

Wong CHW, Elwell JH, Oberley LW, Geoddel DV (1989) Manganous superoxide dismutase is essential for cellular resistance to cytotoxicity of tumour necrosis factor. Cell 58:923–931

Workman P, Stratford IJ (1993) The experimental development of bioreductive drugs and their role in cancer therapy. Cancer Metastasis Rev 12:73–82

Yamauchi N, Kuriyama H, Watanabe N, Neda H, Maeda M, Niitsu Y (1989) Intracellular hydroxyl radical production induced by recombinant tumour necrosis factor and its implication in the killing of tumour cells in vitro. Cancer Res 49:1671–1675

Zimmerman RJ, Chan A, Leadon SA (1989) Oxidative damage in murine tumour cells treated in vitro by recombinant human tumour necrosis factor. Cancer Res 49:1644–1648

Zyad A, Benard J, Tursz T, Clark R, Chouaib S (1994) Resistance to TNF alpha and Adriamycin in the human breast cancer MCF-7 cell line: relationship to MDR1, MnSOD and TNF gene expression. Cancer Res 54:825–831

14 Application of Nuclear Magnetic Resonance for Investigation of the Tumour Microenvironment

R.J. Maxwell

14.1 Introduction

In this chapter, consideration is given to nuclear magnetic resonance (NMR) as applied to experimental tumours in animals and in clinical research. Reference is made to in vitro studies which help to distinguish intrinsic properties of tumour cells from factors due to the microenvironment of solid tumours. NMR techniques can be used to provide anatomical, physiological and (bio)chemical information about tumours and, most importantly, it is often possible to access this information in a non-invasive way. The use of "straightforward" NMR imaging techniques is not discussed since this can be considered a relatively routine method for the detection of tumours in patients. Most of the contents could be labelled "NMR spectroscopy" while noting that techniques used for in vivo NMR spectroscopy have converged in recent years with those in NMR imaging, giving rise to "spectroscopic imaging", for example. In addition, the application of so-called functional imaging techniques in magnetic resonance is discussed since this is considered to be based on the oxygenation status of haemoglobin in the blood and therefore might be of considerable relevance to the tumour microenvironment. Overall, this review concentrates on biochemical and physiological information rather than purely structural or anatomical features of tumours.

14.1.1 The Basis of Nuclear Magnetic Resonance

The phrase "nuclear magnetic resonance" indicates that the method utilises a resonance phenomenon of certain atomic nuclei based on their magnetic properties. NMR is concerned with nuclei having a non-zero spin-quantum number and hence a magnetic moment. Commonly studied nuclei (e.g. ^{1}H, ^{13}C, ^{19}F and ^{31}P) have a spin-quantum number of 1/2, giving two possible spin states. Properties of important nuclei for in vivo studies are summarised in Table 14.1. The essence of the technique is that there is an energy difference between the two spin states when the nuclei experience an external magnetic field. "Resonance" can be brought about by application of a pulse of electro-magnetic radiation, typically in the radiofrequency range. This perturbation of the spin states of a population of nuclei can then be measured with a receiver tuned to the appropriate radiofrequency range. Each individual NMR method is characterised by its pulse sequence, typically consisting of an excitation pulse, an optional "preparation" period, the measurement period and an optional "relaxation" delay. The energy difference, although proportionate to the strength of the external magnetic field, is relatively small so that the technique is intrinsically insensitive. However, the essential requirements for making NMR measurements – the

R.J. Maxwell, PhD Århus University Hospitals, NMR Research Centre, Skejby Sygehus, DK-8200 Aarhus N, Denmark

Table 14.1. Summary of main NMR-detectable nuclei

Nucleus	Relative sensitivity	Natural abundance	Advantages	Disadvantages
^1H	100	99.98	Sensitivity; availability on clinical systems	Background (water, lipids); narrow frequency range
^2H	1.0	0.015	Short T_1; use of enriched compounds: e.g. D_2O uptake and other labelling experiments	Short T_2; low sensitivity; very narrow frequency range
^{13}C	1.6	1.1	Wide frequency range; low background; signal enhancement possible; use of enriched compounds (e.g. glucose, drugs)	Low sensitivity; cost of enrichment; need ^1H decoupling
^{19}F	83	100	Sensitivity; wide frequency range; no background; fluorinated anticancer drugs; fluorinated probes	Need fluorination
^{31}P	6.6	100	Wide frequency range; pH; "important" metabolites	Moderate sensitivity

provision of a powerful external magnetic field and the application and detection of radio frequency radiation – can be done non-invasively and, within certain limits, are non-harmful.

14.1.2
NMR Spectroscopy

The actual magnetic field experienced by nuclei (and hence their precise "resonance frequency") depends not only on the external magnetic field applied but also on the shielding effect of neighbouring electrons. The electron distribution around a nucleus depends on its chemical environment so that the resonance frequencies are sensitive to the chemical situation of the nucleus being studied. A plot of signal intensity versus resonance frequency constitutes the NMR spectrum. For a given nucleus, each chemical environment will have a characteristic resonance frequency which, after comparison with a reference and scaling to the applied magnetic field, is called its chemical shift. Note that although some chemical compounds have chemical shifts that are characteristic (i.e. constant) over a wide range of situations (e.g. the ^{31}P chemical shift of phosphocreatine), others can be affected by a variety of factors. Two forms of inorganic phosphate, HPO_4^{2-} and $H_2PO_4^-$, exist around physiological pH and have different ^{31}P chemical shifts. In fact, these two forms are in rapid exchange and only one signal is observed, with a frequency determined by their relative concentra-

tions: the chemical shift of inorganic phosphate is routinely used as a pH probe (MOON and RICHARDS 1973).

In addition to the electronic shielding reflected in the chemical shift, nuclei can also be sensitive to the magnetic properties of nuclei to which they are linked via chemical bonds. This *spin coupling* gives rise to a more complicated (multiplet) appearance in the spectrum. The consequence for in vivo spectra is usually that the signals appear to be attenuated and broadened (unless the individual multiplet components can be resolved). These effects can be eliminated for heteronuclear coupling (e.g. ^{13}C-^1H or ^{31}P-^1H) by applying radiofrequency irradiation at the frequency of one nucleus while the other nucleus is being observed. This is termed decoupling.

14.1.3
NMR Imaging

Traditionally, NMR imaging involves the detection of a single chemical form of a particular nucleus: in most cases (because of sensitivity and abundance) this is ^1H NMR of water. With NMR imaging, the applied magnetic field is no longer homogeneous but instead a magnetic field gradient is applied over the sample: therefore the water ^1H resonance frequency is spatially dependent. The spatial information about water distribution in the sample is encoded into frequency and phase information by, for example, application of appropriate magnetic field gradients during excitation, preparation and measurement periods of the pulse sequence. The image can be reconstructed from this (two- or three-dimensional) encoding by Fourier transformation.

14.1.4
Relaxation

The contrast seen in clinical ^1H NMR images is not just derived from variations in water concentration but is also affected by two relaxation processes, T1 and T2. T1 determines how quickly the excited spin states return to equilibrium. If T1 is long compared to the repetition time (TR) of the pulse sequence, then there will be partial saturation: this will cause signal loss in both imaging and spectroscopy but spatial variations in T1 will also contribute to image contrast. T2 determines how quickly signal is lost from the "transverse plane" (i.e. the plane in which it is detected). This typically results in signal loss during part of the preparation period (the echo time, TE) of imaging and some spectroscopy pulse sequences. If T2 is short, the lines in the NMR spectrum will be broad (overlapping and ultimately undetectable) and there will be signal loss during TE, again contributing to image contrast.

NMR relaxation is affected by molecular motion and by the presence of paramagnetic species. Slow molecular motions result in rapid T2 relaxation such that most signals may be too broad to be detected. In addition, the apparent T2 relaxation time (T2*) is also affected by magnetic field inhomogeneities, e.g. as a result of local variations in magnetic susceptibility. Paramagnetic ions can provide an efficient mechanism for T1 and T2 relaxation and can also affect T2* due to magnetic susceptibility effects. In the clinical setting, this is exploited by administration of Gd-DTPA, a chelate of gadolinium which is normally excluded from the brain, but which can cross into brain tumours and other lesions, providing additional image contrast. Although molecular oxygen is paramagnetic, the most important intrinsic paramagnetic relaxation in vivo is caused by deoxyhaemoglobin. Since oxyhaemoglobin is not paramagnetic, changes in haemoglobin oxygenation can affect the intensity of ^1H images. This is part of the basis of functional MR, since brain activity seems to result in localised improvement in oxygenation, the locations of this activity in response to a variety of stimuli can be studied.

Relaxation is often caused by dipolar interaction with neighbouring nuclei. Perturbation of the spin populations of these nuclei can change signal intensity: the nuclear Overhauser effect (NOE). The use of decoupling (to eliminate spin coupling) can provide suitable perturbation such that a signal enhancement due to NOE is also observed. It should be noted that the mechanisms, timings and power requirements for decoupling and NOE are quite different.

14.1.5
Sensitivity

The main limitations arise from the fundamental lack of sensitivity of NMR. The signal:noise or contrast:noise ratio of any measurement may be improved by signal averaging or by compromising spatial resolution. The use of specific pulse sequences for water suppression, spectral editing, spectroscopic localisation and metabolic imaging almost inevitably results in a lower signal:noise ratio than if simple pulse-observe measurements had been performed. In addition to any losses due to a reduction in the volume being sampled, there are also losses due to T_2 relaxation during any echo periods in the sequence, plus the effect of pulse or gradient imperfections. To minimise the duration of echo periods and to improve spatial resolution it is desirable to have the highest possible gradient strengths.

14.2
Tumour Energetics

14.2.1
Introduction

^{31}P was the first nucleus to be used to any extent for in vivo NMR spectroscopy measurements. The reasons for this probably include its relative sensitivity (better than ^{13}C or ^{15}N) and the lack of a dominant background signal (a long-standing problem with water and lipid signals in ^1H NMR). However, the real motivation for the earliest studies was the clear relevance of the information accessible with ^{31}P NMR to the biological questions of interest to the pioneering research groups (e.g. CHANCE et al. 1981; TAYLOR et al. 1983; GRIFFITHS et al. 1983). In particular, the study of bioenergetics is closely associated with the concentrations of "high-energy" phosphates (ATP and phosphocreatine, PCr), "low(er)-energy' phosphates (ADP and inorganic phosphate, P_i) and H^+ (as indicated by the pH-dependent chemical shift of P_i). In addition, information about phosphorylated intermediates in the glycolytic pathway (e.g. glucose-6-phosphate) might be available from the phosphomonoester region of the ^{31}P spectrum. Of course the main advantages of in vivo NMR are that the measurements are non-invasive and dynamic measurements can be performed on the same tissue during the course of its normal (or abnormal) activity. In the classic case of muscle metabolism, this allowed studies on muscle contraction either under control conditions or in combination with ischaemia.

^{1}H and ^{13}C NMR spectroscopy provide information about lactate concentration (MAXWELL et al. 1988; HE et al. 1995) and the kinetics of uptake and utilisation of energy sources (ARTEMOV et al. 1995).

The "bioenergetic status" has been derived from MRS measurements in slightly different ways, the most common and probably most reliable of which (for tumours) is the ratio β-NTP/P$_i$. The ratio of β-NTP to total signal in the ^{31}P spectrum (β-NTP/ΣP) and P$_i$/β-NTP are also used. Some authors have used PCr in their assessment of bioenergetic status and while this is entirely reasonable, it carries the risk of including signals from PCr present in surrounding normal tissue (especially muscle where [PCr] ≈ 24 mM); i.e. unless the localisation method is extremely effective and robust for relevant tumour sizes and shapes or the tumour site is distant from muscle, it is difficult to eliminate the possibility of

PCr contamination in the tumour spectrum. This problem is less severe for β-NTP because of the high PCr/NTP ratio in muscle. This alerts us to a further problem regarding a strict bioenergetic interpretation of in vivo ^{31}P spectra (i.e. can we assume that the metabolite concentrations or ratios reflect those inside individual cells?); since the spatial resolution is poor (typically 5–10 mm for animal studies and 20–50 mm for humans), the spectra are an average over a potentially heterogeneous tumour region (e.g. compared to the expected diffusion limit for oxygen from a tumour blood vessel of ca. 50–100 μm). This is also a problem in ^{1}H spectroscopy despite the better spatial resolution (1–5 mm in animal tumours). The resultant spectra should still be useful even though they reflect a weighted average over the various microenvironments within part (or all) of a tumour. We should not forget that, at best, this averaging over tumour cells occurs and, at worst, there is contamination from neighbouring tissue (e.g. muscle) and/or infiltrated non-tumour cells (e.g. macrophages).

TOZER and GRIFFITHS (1992) discussed the relationship between tumour oxygenation, cell death and cellular energy metabolism as measured by ^{31}P MRS. They considered that cells dying by apoptosis, and during mitosis (being a small fraction of the total and in any case being energised until phagocytosed), would be unlikely to affect the ^{31}P spectrum directly.

Table 14.2. NMR signals relevant to bioenergetics

NMR signal	Nucleus	Chemical shift (ppm)	Metabolite	Comments
P$_i$	^{31}P	3.5–5.5	Inorganic phosphate	pH-dependent; "low-energy" phosphate
PCr	^{31}P	0.0	Phosphocreatine	Chemical shift reference; "high-energy" phosphate; high concentration in muscle; may contaminate other tissues
γ-NTP	^{31}P	−2.5	γ-phosphate of ATP, GTP, UTP etc. β-phosphate of ADP, GDP, UTP etc.	
α-NTP	^{31}P	−7.57	α-phosphate of ATP, GTP, UTP etc. α-phosphate of ADP, GDP, UTP etc.	Chemical shift reference
β-NTP	^{31}P	−16.5	β-phosphate of ATP, GTP, UTP etc.	"High-energy" phosphate
tCr	^{1}H	3.05	Phosphocreatine Creatine	
Lactate	^{1}H	1.35	Lactate methyl group	
Glucose	^{13}C	93 96.5	α-1-[^{13}C]glucose β-1-[^{13}C]glucose	
Lactate	^{13}C	22	3-[^{13}C]lactate	Gives rate of lactate production from 1-[^{13}C]glucose

Since the membranes of histologically necrotic cells should be permeable to P_i and other low molecular weight compounds, they are unlikely to contribute significantly to the ^{31}P spectrum or indeed to lactate signals in the ^1H spectrum (i.e. they should diffuse into the interstitial fluid and then into the blood). However, as discussed in Sect. 14.4.2, it now seems that hypoxic and/or necrotic cells may be responsible for "mobile lipid signals" in ^1H spectra as a result of the formation of lipid droplets. It could also be the case that if the necrotic fraction becomes very high, the effective extracellular space is also high as in the murine C3H mammary carcinoma (KHALIL et al. 1995). For this tumour, a P_i compartment having a pH consistent with the pH_e (measured from 3-aminopropylphosphonate; GILLIES et al. 1994a) can often be detected in addition to a more alkaline (presumed to be intracellular) P_i compartment (McCoy et al. 1996b). Despite these reservations, it is reasonable to concur that the ^{31}P spectrum reflects the degree of hypoxia of the remaining viable cells, and the metabolic alterations required to sustain ATP synthesis as the oxygen supply diminishes. Table 14.2 summarises the main NMR signals of bioenergetic significance.

14.2.2
Relationship Between Oxygenation, Glucose Supply and Tumour Energetics

Extrapolation from, amongst other things, in vivo ^{31}P studies of muscle metabolism led to a common assumption that there would be a strong relationship between tumour bioenergetic status and oxygenation. The existence of such a relationship would suggest that parameters available non-invasively from ^{31}P NMR measurements could help to predict, in individual patients, the outcome of radiotherapy and other cancer treatments which are oxygen-dependent. However, some doubts about the importance of such a relationship in tumours could be gleaned from the early work of WARBURG (1930), who found that tumour cells tended to obtain much of their energy from glycolytic metabolism, even under aerobic conditions. REMPEL et al. (1996) have demonstrated a stable, fivefold amplification of the gene for hexokinase type II in a rapidly growing rat hepatoma line compared to normal hepatocytes with no rearrangement of the gene. Thus, hexokinase gene amplification provides one mechanism to explain the high glycolytic phenotype prevalent in tumour cells. This propensity towards glycolytic

metabolism, incidentally, led to a second assumption: that tumour cells would be acidic because of the consequent production of lactic acid. Although acidosis is observed by ^{31}P NMR measurements in muscle during acute episodes of exercise, especially when combined with ischaemia, tumours (both in animal models and in patients) have usually been found to have a neutral or even alkaline pH by ^{31}P NMR measurements based on the chemical shift of inorganic phosphate (the so-called pH_{MRS}, considered under most conditions to represent intracellular pH, pH_i). The reasons for this may be related to the difference between acute changes in energy demand and/or nutrient supply (which can actually reduce tumour pH_{MRS} in tumours), and to the ability of tumour cells to maintain a stable pH_i under chronic conditions (TOZER and GRIFFITHS 1992; STUBBS et al. 1994). The question of tumour pH is dealt with elsewhere in this volume so it will not be discussed at length here.

14.2.3
The Significance of ^{31}P NMR as an Indicator of Tumour Oxygenation

Two types of studies contribute to the question of the significance of ^{31}P NMR as an indicator of tumour oxygenation. Firstly, attempts have been made to correlate NMR-derived parameters with some other measure related to tumour oxygenation (e.g. pO_2 microelectrodes or radiation sensitivity). Secondly, acute changes in tumour oxygenation have been induced and the effect on the MR parameters has been monitored. Especially in the latter case, it is important to distinguish the effects of tumour blood flow changes on bioenergetic status from a pure influence of oxygenation. Although oxygen supply may indeed affect tumour energetics, such modulations also change the supply of other nutrients (glucose etc.) and the removal of metabolic end products (lactate etc.). Thus agents which primarily act by altering blood flow are discussed separately but it should be noted that agents which are expected to have a more direct effect on tumour oxygen supply can also influence tumour blood flow (e.g. carbon monoxide).

ROFSTAD et al. (1988) compared ^{31}P NMR data with intracapillary oxyhaemoglobin saturation over a range of tumour lines grown in mice. With increasing tumour volume, they observed a decrease in bioenergetic status (e.g. β-NTP) and in oxyhaemoglobin saturation in three of the tumour lines (KHT, RIF-1 and a human ovarian carcinoma xenograft, MLS). However, another human ovarian carcinoma (OW1) showed no change in either β-NTP or oxyhaemoglobin saturation with increasing tumour volume. They concluded that there was a direct relationship between tumour energy status (as indicated by ^{31}P NMR parameters) and oxygen supply conditions but that this relationship was not identical for all of the tumour lines. It is anticipated that this variability could be related to differences in tumour cell respiration rates and in the ability to tolerate hypoxia. Rasmussen et al. (1993) used ^{31}P NMR to study the metabolism of Ehrlich ascites tumour cells (wild-type and multidrug resistant) in vitro. Perfusion of the cells with azide, an inhibitor of mitochondrial respiration, had no effect on the NTP concentration, and caused no changes in glucose consumption and lactate production. Azide perfusion, combined with glucose depletion, caused a rapid drop in NTP, which was reversed after reperfusion with glucose. Kuin et al. (1994) studied the effect of the mitochondrial inhibitor meta-iodobenzylguanidine combined with moderate hyperglycaemia on pH_e (by microelectrode measurements), FDG uptake (by positron emission tomography) and pH_i and bioenergetic status (by ^{31}P NMR) of RIF-1 tumors in vivo. Despite a decrease in pH_e (to 6.2) and a two- to threefold increase in FDG uptake there was no change in pH_i or "bioenergetic status". Nordsmark et al. (1995), studying a murine mammary carcinoma implanted in the foot, found no significant effect of a period of carbon monoxide

breathing (660 ppm for 30 min) on NTP/P_i or pH_i despite the dramatic reduction in tumour oxygenation caused by this treatment (based on measurements of the radiobiological hypoxic fraction and by pO_2 microelectrodes). On the other hand, allowing the mice to breathe carbogen (95% O_2 plus 5% CO_2) did give rise to an increase in NTP/P_i in this tumour model and carbon monoxide breathing caused a redcution in β-NTP/P_i in SCCVII tumours (Fig. 14.1).

In contrast, some studies have demonstrated the importance of oxygen in determining tumour bioenergetic status. Kalra et al. (1994) studied the effect of a left-shifter of the oxygen-haemoglobin dissociation curve (BW 589C, which reduces the O_2 transport capacity) on ^{31}P spectra from the human colonic tumour xenograft HT-29 and from murine KHT and RIF-1 tumours implanted in mice. Doubling of $P_i/\Sigma P$ was observed 5–6 h after BW 589C for all three tumour types. However, although the left-shift due to BW 589C persists at 24 h, $P_i/\Sigma P$ returned to control levels in the RIF-1 and HT-29 tumours at this time. These results suggest that there was cellular metabolic adaptation to the reduction of oxygen delivery by BW 589C. This does not appear to involve 2,3-DPG as there was no significant alteration in tumour levels. The death of hypoxic cells may have contributed to the recovery of $P_i/\Sigma P$. Vaupel et al. (1994) studied the bioenergetic and metabolic status of murine FSaII tumours by ^{31}P MRS and by quantitative bioluminescence. β-NTP/P_i correlated with me-

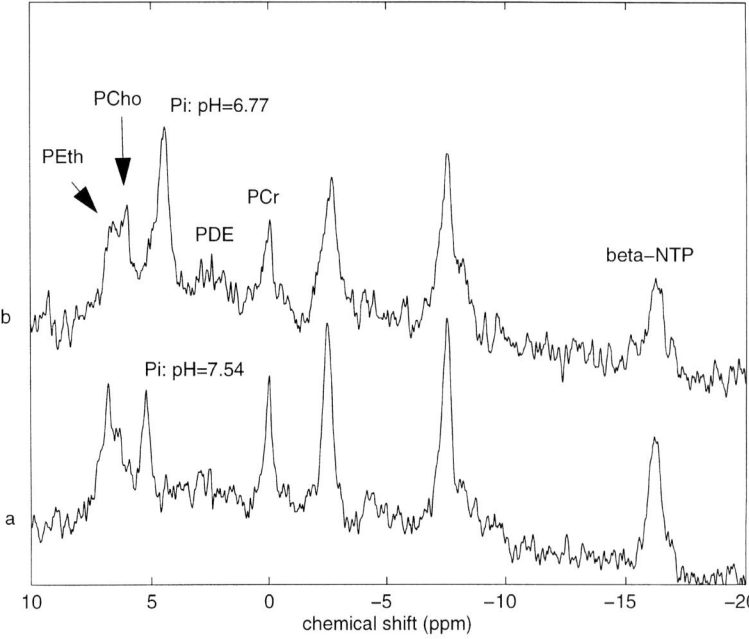

Fig. 14.1. ^{31}P NMR spectra (121 MHz, TR = 6 s, 160 averages) from an SCCVII murine tumour, (*a*) before and (*b*) 32–48 min after the start of carbon monoxide (660 ppm) breathing

dian tissue pO_2 (electrode measurement). pH declined during growth with intracellular acidosis being evident in tumours >350 mm^3. Elevated lactic acid corresponded to this acidification in small- and medium-sized tumours, whereas loss of ATP seems to be responsible for intracellular acidification (i.e. inability to regulate pH_i) in larger tumours (volume >350 mm^3). On average, median pO_2 values of ca. 10 mmHg represented a critical threshold for energy metabolism. At higher median pO_2, levels of ATP, phosphomonoester and total P_i were relatively constant and pH_i alkaline or neutral. In general, median $pO_2 < 10$ mmHg coincided with intracellular acidosis, ATP depletion, and elevated P_i.

The implication of these studies is that, provided there is an adequate supply of glucose (or perhaps of other nutrients), neither a pharmacological inhibition of mitochondrial respiration nor a dramatic reduction in oxygen delivery to the tumour necessarily has any effect on "tumour bioenergetic status" in the sense of the relative or absolute concentrations of P_i or NTP. However, it is possible that for a given tumour model (or for an individual tumour in a patient) there could be a range of oxygenation levels within which NTP or P_i or other ^{31}P NMR parameters are oxygen-dependent. In addition, a change in oxygen delivery may sometimes result in a change in glucose consumption but in other cases (e.g. where the tumour is already totally reliant on glycolytic metabolism) it might be unaffected by any further reduction in oxygenation.

14.2.4
Tumour Energetics and Blood Flow

LYNG et al. (1993) investigated bioenergetic status ((PCr + β-NTP)/Pi) in six human melanoma xenograft lines (in nude mice) by ^{31}P MRS and blood flow using ^{86}Rb uptake. Bioenergetic status, pH and blood supply per viable tumour cell decreased with increasing tumour volume for five of the cell lines. The magnitude of the falls in (PCr + β-NTP)/ΣP and in tumour pH correlated with the magnitude of the decrease in blood supply per viable tumour cell. Tumour pH decreased with decreasing tumour bioenergetic status, and the magnitude of this decrease was larger for the tumour lines showing a high blood supply than for those showing a low blood supply per viable tumour cell. No correlations across the tumour lines were found between tumour pH and tumour bioenergetic status or any other resonance ratio on the one hand and blood supply per

viable tumour cell on the other. Bioenergetic status and pH were correlated to blood supply per viable tumour cell within individual tumour lines. They concluded that the differences in the ^{31}P NMR spectrum between the tumour lines were probably caused by differences in their intrinsic biochemical properties rather than by the differences in blood supply per viable tumour cell and therefore that clinically useful parameters for prediction of tumour treatment can probably not be derived from a single ^{31}P spectrum.

14.2.5
Treatment-Induced Tumour Vascular Changes

Various types of tumour treatment result in changes to tumour blood supply. Tumour vasculature may or may not be the prime target for such treatments or even a significant influence on treatment outcome but these effects often contribute to tumour bioenergetic status, especially as reflected in the ^{31}P NMR spectrum. Dramatic reduction in bioenergetic status can occur after physical clamping of tumour blood supply, "pharmacological" clamping (with systemic vasodilators such as hydralazine), or treatments which result in blood vessel destruction (photodynamic therapy, high dose hyperthermia, flavone acetic acid). Improvement in tumour blood supply may occur after radiotherapy or chemotherapy.

14.2.5.1
Vasoactive Drugs

Several research groups have monitored the effect of *hydralazine* on blood flow reduction in murine tumours. This vasoactive drug has been shown to selectively lower blood flow to transplanted murine tumours and therefore to be potentially useful for enhancing drugs that act against hypoxic tumour cells (e.g. CHAPLIN and ACKER 1987) or enhance the effects of hyperthermia (e.g. HORSMAN et al. 1989). A consistent reduction in bioenergetic status (β-NTP/P_i or P_i/ΣP, and usually also in pH) has been reported in transplanted tumours (OKUNIEFF et al. 1988; BHUJWALLA et al. 1990; TOZER et al. 1990; BREMNER et al. 1994) but not in primary radiation-induced (FIELD et al. 1991) or spontaneous murine tumours (WOOD et al. 1992). It should be noted that some of the tumours studied by FIELD et al. (26%) and WOOD et al. (17%) did show a decline in bioenergetic status after hydralazine. Nevertheless, the general relevance of transplanted murine tumours for studies of agents that modify tumour vasculature was brought into question. More recently, NORDSMARK et al. (1996) reported a combined ^{31}P NMR and pO$_2$ microelectrode study of the effects of hydralazine on spontaneous mammary adenocarcinomas. All tumours showed an increase in the percentage of pO$_2$ values ≤5 mmHg after hydralazine and ten of 12 tumours showed a decrease in median pO$_2$. In addition, 75% of such tumours showed a decline in bioenergetic status (β-NTP/P$_i$) at 45 min after treatment. Analysis of the combined pO$_2$ and ^{31}P data in NORDSMARK et al. (1996) shows a very strong correlation between the relative change in β-NTP/P$_i$ (reflecting treatment-induced NTP hydrolysis) and the post-treatment pO$_2$ (% values ≤5 mmHg) in individual tumours but not between other combinations of oxygenation (e.g. initial pO$_2$ values) or bioenergetic (e.g. pre- or post-treatment β-NTP/P$_i$) parameters. The maximal reduction in bioenergetic status required a blood flow reduction sufficient to cause profound hypoxia (i.e. at least 95% pO$_2$ values ≤5 mmHg). These studies suggest that hydralazine can reduce tumour blood flow, oxygenation and bioenergetic status in some, but perhaps not all, spontaneous tumours in mice. BHUJWALLA et al. (1994) investigated the effect of combined blood flow reduction (due to hydralazine) and intratumoural glucose injection on RIF-1 tumour metabolism and pH. Tumour acidosis during the period of reduced

blood flow was not enhanced, relative to administration of hydralazine alone. Tumour NTP/P$_i$ decreased significantly within 20 min of hydralazine administration, whether or not glucose was injected, although NTP/P$_i$ values were slightly higher in tumours that received extra glucose. Lactate concentration was not significantly different in glucose-supplemented tumours, despite glucose concentrations that were four to five times higher. When the added glucose was labelled with ^{13}C, no correlation was detected between the pH and lactate production. Hydralazine-induced changes to the ^{1}H spectra of murine C3H mammary carcinomas are shown in Fig. 14.2.

The effects of the nitric oxide donor SIN-1 arginine (WOOD et al. 1993) and the nitric oxide synthase inhibitor NG-nitro-L-arginine (WOOD et al. 1994) on energy metabolism and on radiation sensitivity have been studied in murine tumours. In SCCVII tumours, SIN-1 reduced P$_i$/ΣP by 40%–50% within 10 min and increased X-ray sensitivity threefold, consistent with increased tumour oxygenation. In RIF-1 tumours NG-nitro-L-arginine increased P$_i$/ΣP by 250% at 45 min and increased radiation resistance three- to fivefold, consistent with the induction of tumour hypoxia. Both effects were dependent on NG-intro-L-arginine dose. These agents probably act by improving (SIN-1) or worsening (nitro-L-arginine) tumour blood flow through their actions on the nitric oxide system.

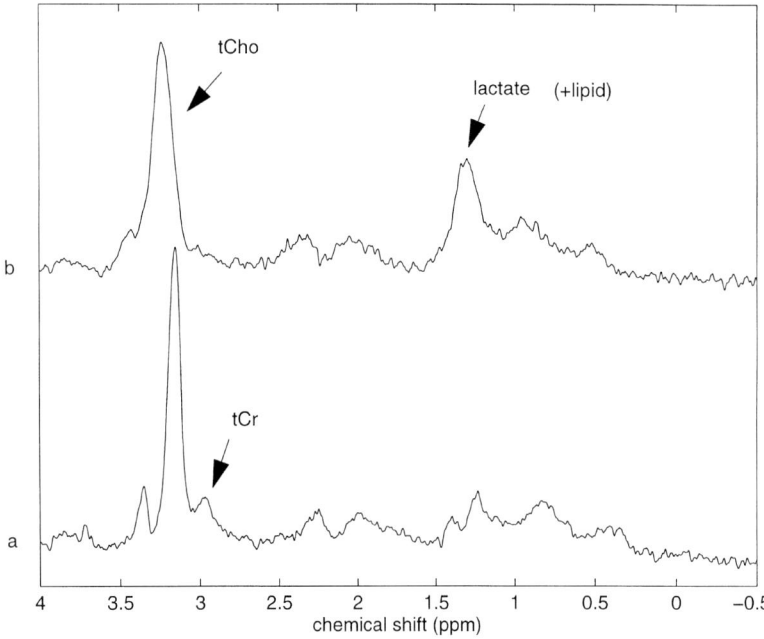

Fig. 14.2. ^{1}H NMR spectra (300 MHz, PRESS, TR = 2 s, TE = 270 ms, 512 averages) from a C3H murine mammary carcinoma, (*a*) before and (*b*) 32–48 min after hydralazine (5 mg/kg, i.p.)

14.2.5.2
Hyperthermia

Hyperthermia has been shown to be more effective under acidic conditions: it is not clear whether pH_i or pH_e is more important or what the relationship to other bioenergetic parameters is. However, since tumour vasculature and perhaps tumour cell membranes seem to be important targets, pH_e may be the most relevant prognostic parameter while changes in bioenergetic status and/or lipid metabolites may be expected to reflect treatment response. PRESCOTT et al. (1994) and SOSTMAN et al. (1994) investigated the effects of hyperthermia on human and canine patients with soft tissue sarcomas. A decrease in β-NTP/PME after the first hyperthermia treatment was associated with a greater chance of extensive necrosis in human tumours, but no significant relationships were observed between changes in tumor pH or phosphometabolite ratios and time to local failure in dogs. Post-treatment reduction in β-NTP/P_i correlated with thermal dose in dogs but not in humans. A relationship was also found between pretherapy pH and subsequent tumour necrosis in humans and pretherapy pH and the period of time elapsed until local failure in dogs for combined fractionated radiation therapy and hyperthermia. KIMURA et al. (1994) treated human lung tumour xenografts with radiotherapy (20 Gy) and/or hyperthermia (44°C, 10 min). There were significant decreases in β-NTP/P_i and pH_i 24 h after radiation plus hyperthermia but significant increases 6 days after the treatment relative to untreated controls and to pretreatment values. Tumour blood flow estimated by hydrogen clearance was reduced at 24 h and fully recovered to pretreatment level at 6 days. It seems likely that the treatment-induced blood flow changes were largely responsible for changes in tumour metabolism and were reflected by β-NTP/P_i and pH_i changes. VAUPEL et al. (1990) treated murine fibrosarcomas (FSaII) with hyperthermia (43.5°C for 15, 30 or 60 min) and observed a dose-dependent fall in β-NTP/P_i and pH_i. Similar effects were observed for mammary carcinomas (MCaIV) after exposure to the 30-min heating protocol.

14.2.5.3
Photodynamic Therapy

BREMMER et al. (1994) studied changes in energy metabolism occurring during and after treatment of murine RIF-1 tumours with photodynamic therapy.

The extent and duration of the increase in $P_i/\Sigma P$ were both light-dose dependent. Bioenergetic effects were greater if light was administered 1 h after the photosensitiser disulphonated phthalocyanine than for a 24-h delay, possibly due to differences in the distribution of photosensitiser within the tumour. Blood flow measurements (D_2O uptake and ^2H NMR) showed a 90% blood flow reduction within 10 min of the light treatment with a dose of 50 J. The data indicate that it is possible to observe very early changes in ^{31}P metabolism that can be attributed to direct cellular damage as opposed to the later changes indicative of overall tumour hypoxia caused by vascular damage.

14.2.5.4
Radiotherapy

MAHMOOD et al. (1994) studied the effects of radiation dose upon a hypoxic murine mammary carcinoma before and up to 9 days after irradiation (4–17 Gy). This is the same tumour model studied in a similar way after 32- or 65-Gy radiation doses (KOUTCHER et al. 1992). β-NTP/P_i increased after 8–65 Gy and decreased after 0- or 4-Gy doses. The energy status of the 8–65-Gy groups peaked between 1 and 4 days after irradiation. Hypoxic fraction measurements showed a significnat decrease in the 17-Gy and 32-Gy groups at 48h, coincident with the bioenergetic changes. PME/ΣP changes showed a strong dose dependence after irradiation. The downfield component of the PME peak (mostly PEth) increased relative to the upfield component, PCho. This dose-dependent ratio reached a maximum approximately 7 days post radiation. Changes in the levels of membrane phospholipid precursors may be related to alterations in cell proliferation or may be a result of radiation-induced membrane damage. This improvement in tumour energetics after radiotherapy is consistent with other reports, such as that of TOZER et al. (1989) who also showed an improvement in tumour blood flow. It seems likely that blood flow changes are the main explanation for the bioenergetic changes.

14.3
Tumour Oxygenation

Since ^{31}P MRS studies suggest tumour energy metabolism is at best an indirect measure of tumour oxygenation and that under some circumstances it appears to be decoupled from oxygen availability, we should consider some more direct approaches.

14.3.1
Fluorinated Hypoxia Probes

Nitroimidazoles have been developed for use as radiosensitisers and hypoxic cell cytotoxins because selective metabolism under hypoxic conditions results in compounds which are either activated by radiation or are toxic in their own right. Misonidazole was the lead compound in this series and a fluorinated analogue, Ro-07-0714, was monitored in tumours by ^{19}F NMR, initially because of its similarity (e.g. in terms of partition coefficient) to misonidazole. However, it was found that the fluorine signal was selectively retained in murine tumour types which are known to have high radiobiological hypoxic fractions, e.g. EMT-6 (MAXWELL et al. 1989). Similar results were obtained using a hexafluorinated analogue of misonidazole, CCI-103F (JIN et al. 1990; RALEIGH et al. 1991), and with a trifluorinated compound, SR-4554 (ABOAGYE et al. 1994). It appears that, following nitroreduction of these compounds under hypoxic conditions, one or more of the resultant metabolites (still containing fluorine label) are retained in the hypoxic tumour cells and can be detected by ^{19}F NMR even at times after administration (6–24 h) when the parent drug has been effectively cleared from the plasma and from normal, well-oxygenated tissue such as brain. Compounds of this type may prove useful as noninvasive probes of tumour hypoxia in both animal and human tumours but it should be noted that probe retention will depend not just on hypoxia but on the activity of enzymes with nitroreductase activity.

14.3.2
Perfluorocarbons

An alternative strategy for the evaluation of tumour oxygenation involves the use of perfluorocarbon compounds developed as artificial blood substitutes. The excessive leakiness of tumour vasculature allows these compounds to accumulate in the tumour interstitial space and because the ^{19}F relaxation times are dependent on oxygen tension (PARHAMI and FUNG 1983), they can be used as probes of tumour extracellular oxygenation (HEES and SOTAK 1993).

14.3.3
Blood-Oxygen-Level-Dependent Imaging

The development of functional magnetic resonance imaging is based on observations that MRI signal intensity in rat brain was increased by oxygen breathing (OGAWA et al. 1990) and that brain regions activated by visual stimulation also showed an increase in signal (KWONG et al. 1992). These effecs are mediated through an increase in blood haemoglobin saturation: since deoxyhaemoglobin is paramagnetic, it causes magnetic susceptibility effects which lower T_2^*, and hence lower image intensity such that improved blood oxygenation results in an increase in image intensity. The use of T_2^*-sensitive MR imaging techniques to monitor such changes has been termed "blood-oxygen-level-dependent" (BOLD) imaging. During brain activation, there is often a pardoxical increase in local blood oxygenation; it seems that the increased energy demand of activated regions results in an increase in blood supply but that oxygen consumption does not, at least initially, increase to the same extent. Depending on the pulse sequence and parameters used, the image intensity may also be increased by a direct effect of blood flow changes ("in-flow" effects).

BOLD imaging has been used to observe similar acute changes in animal tumours in response to the blood flow modifiers hydralazine (BHUJWALLA et al. 1993) or calcitonin gene-related peptide (CGRP; HOWE et al. 1993). The blood flow reduction caused by these agents was accompanied by a reduction in image intensity and, in the case of CGRP (which has a very short duration of action), by subsequent recovery. Improved tumour oxygenation can be achieved by changing the breathing gas and these changes are also reflected by an increase in image intensity. KARCZMAR et al. (1994) observed such effects in a rat mammary adenocarcinoma due to oxygen breathing while ROBINSON et al. (1995) observed very large increases for GH3 prolactinoma tumours due to carbogen breathing. These large changes in image intensity (perhaps ten times greater than observed in neurofunctional studies) may have been partly caused by blood flow changes, especially for carbogen breathing, and were much lower in some other tumour types (McCOY et al. 1996a). The implication is that the responsiveness of an individual tumour to changes in the breathing gas is highly variable but that its response can be monitored by BOLD imaging. Carbogen-induced changes to ^{1}H image intensity from murine C3H mammary carcinomas are shown in Fig. 14.3.

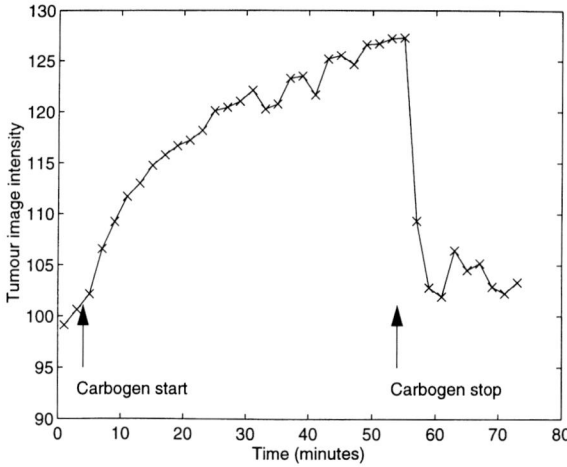

Fig. 14.3. Effect of carbogen breathing on the intensity of ^{1}H NMR images (300 MHz, gradient-recalled echo, TR = 200 ms, TE = 7 ms, flip angle = 30°) from a C3H murine mammary carcinoma

14.4
Lipid-Related Components

The main NMR signals of relevance to lipid metabolism are summarised in Table 14.3.

14.4.1
Phospholipid Metabolites

In his 1992 review of MRS studies of human tumours, NEGENDANK pointed out the typical metabolic characteristics of human cancers (from ^{1}H and ^{31}P spectra) as being high levels of phospholipid metabolites and a cellular pH that is more alkaline than normal. Neither of these conclusions fit in with what

was generally expected at the onset of (most of) these studies, i.e. the focus of attention was really concerned with disturbances in energy metabolism caused by poor tumour oxygenation. One good reason for overlooking (or at least postponing) in vivo investigation of (phospho) lipid metabolism was the difficulty in interpreting the spectroscopy data. From ^{31}P spectroscopy, the signals just downfield from inorganic phosphate (i.e. 6–7 ppm) are assigned to "phosphomonoesters", with potential contribution from several such compounds but, in tumours, principally arising from phosphorylcholine and phosphoethanolamine. Until recently, it has been difficult to separately resolve even these two components reliably, especially at low field. The use of ^{1}H decoupling removes the ^{31}P-^{1}H coupling contribution from the apparent linewidth of these signals and so reduces this problem (NEGENDANK et al. 1995). A similar problem exists in the "phosphodiester" region (i.e. 2.5–4 ppm) in which separate signals from glycerophosphorylcholine and glycerophosphorylethanolamine can be resolved at high field or with ^{1}H decoupling. The situation is hardly better for ^{1}H spectroscopy, the signal complex around 3.2 ppm referred to as Cho ("choline") or, better, as total choline (tCho), since it is actually a mixture of choline derivatives (the signal is from nine equivalent protons in the trimethylammonium headgroup of choline) and in tumours mostly consists of phosphorylcholine and glycerophosphorylcholine in addition to some contribution from choline itself and a coincident signal from taurine. Glycerophosphorylcholine (GPC) and glycerophosphorylethanolamine (GPE) can be considered as lipid synthesis intermediates, whereas phosphorylcholine (PCho) and phosphorylethanolamine (PEth)

Table 14.3. NMR signals relevant to lipid metabolism

NMR signal	Nucleus	Chemical shift (ppm)	Metabolite	Comments
PEth	^{31}P	ca. 6.7	Phosphorylethanolamine	pH-dependent
PCho	^{31}P	ca. 6.3	Phophorylcholine	pH-dependent
GPE	^{31}P	3.5	Glycerophosphorylethanolamine	
GPC	^{31}P	3.0	Glycerophosphorylcholine	
tCho	^{1}H	3.20 to 3.25	Glycerophosphorylcholine; phosphorylcholine; choline; (taurine)	
Lipid-CH$_2$	^{1}H	1.3	Lipid methylene groups	
Lipid-CH$_3$	^{1}H	0.9	Lipid methyl groups	
Lipid: HC $=$ CH	^{13}C	128		
Lipid-CH$_2$	^{13}C	23–35		

are considered to be lipid breakdown intermediates. However, PODO et al. (1996) point out that both biosynthetic and catabolic processes may contribute to PCho levels.

BHAKOO et al. (1996) showed (using cultured cells) that the tCho region of normal Schwann cell ^{1}H spectra was dominated by GPC whereas partial transformation increased PCho and full transformation further increased PCho such that it was the most intense of the tCho signals. Growth arrest resulted in reversion to high GPC and low PC. This is consistent with the findings of NEGENDANK: "In most ^{31}P studies, gliomas differ from normal brain (mostly glial cells) only in their reduced PDE and elevated pH. In ^{1}H MRS on the other hand, brain tumours including gliomas have elevated tCho ..." GILLIES et al. (1994b) found (for glioma cell lines in a bioreactor system) that the PEth/PCho ratio was low during cell growth (ca. 0.5 in log phase) and high (ca. 3–4) during stationary phase. FRYER et al. (1991) observed an increase in PEth/PCho during growth of multicellular spheroids, corresponding to a decrease in proliferative fraction. SMITH et al. (1993a,b) found a strong correlation between S-phase fraction and PCho content of an oestrogen-sesitive rat mammary carcinoma. Growth arrest induced by withdrawal of oestrogen led to lower PCho. LYNG et al. (1995) could not find any clear correlations between ^{31}P NMP peak ratios (including PME/PDE) and S-phase fraction or volume doubling time for a range of human melanoma xenograft lines grown in mice.

NEGENDANK (1992) noted that there was some inconsistency between ^{31}P and ^{1}H results for brain tumours: i.e. there was no regular observation of increased PME (^{31}P) despite an increase in tCho. Although there is a problem in interpreting this difference due to the use of larger voxel sizes for ^{31}P than for ^{1}H (i.e. greater risk of contamination by non-tumour tissue), a change in PEth/PCho need not be accompanied by a change in total PME signal. This could also account for the results of LYNG et al. (1995). The acquisition of better-resolved ^{31}P spectra (high field and/or ^{1}H decoupling) should help to resolve this issue.

BHUJWALLA et al. (1996) used ^{1}H MR spectroscopic imaging to study the effect of blood flow modifiers on murine tumour metabolism (RIF-1). Hydralazine has been shown to cause a dramatic blood flow reduction and a decline in bioenergetic status in this tumour model (BHUJWALLA et al. 1990) and also resulted in an increase in lactate. However, nicotinamide, which has been shown to increase blood flow in some murine tumours, did not give rise to any significant change in lactate but instead resulted in a fall in tCho signal. Extract studies showed that both GPC and PCho decreased after nicotinamide administration. No effects of nicotinamide on choline metabolites were observed in the ^{1}H spectra obtained from cultured RIF-1 cells. One explanation given for these results is that hypoxia leads to a breakdown of phospholipid membranes and a build-up of free fatty acids and GPC. Reduction of hypoxia by nicotinamide could therfore lower GPC levels (and also lower PCho since this is produced from GPC).

LI et al. (1996) obtained ^{1}H-decoupled ^{31}P spectra from soft tissue sarcomas and detected large PME signals (predominantly PEth rather than PCho) in 15 of 20 patients. GPC signals were detected in 4 of 20 cases but GPE was never observed. The PEth/PCho ratio (mean of 2.1) is similar to that in non-Hodgkin's lymphomas (NEGENDANK et al. 1995) and to solid RIF-1 tumours in mice but not to cultured RIF-1 cells (EVANOCHKO et al. 1984). It is likely that the differences seen in cell culture arise from the sensitivity to the concentrations of choline and ethanolamine in the culture medium (SHEDD et al. 1993). NEGENDANK et al. (1995) hypothesised that the high PEth/PCho ratio and very low GPE concentration resulted from sustained activation of phophatidylethanolamine-specific phospholipase D. LI et al. (1996) additionally found an unidentified PDE signal (at 2.2 ppm compared to PCr) in 7 of 20 soft tissue sarcomas, possibly arising from a mobile membrane component (since it was absent from chemically extracted tumour biopsies).

Despite several inconsistencies, a rough overall pattern of altered phospholipid metabolism is starting to become apparent. PCho is high in growth but may also be high in hypoxia, and low in growth arrest. GPC is low in growth, but high in hypoxia and growth arrest. PEth is low in growth (perhaps) and higher in growth arrest. In any case, tCho is high in tumours due to PCho (due to growth and/or hypoxia) and/or GPC (if there is hypoxia).

14.4.2
Mobile Lipids

In addition to the signals from intermediates of lipid metabolism, various groups have described the presence of "mobile lipid" signals in some tumours (large signals at 0.9 ppm and 1.3 ppm, especially in short TE spectra). The presence of such signals is associated with necrosis and therefore with a high grade (for brain tumours). The origin of these "mobile lipids" may well be from the formation of lipid droplets that has been shown to occur under hypoxic conditions. However, Freitas et al. (1995) have demonstrated histochemically the intracellular accumulation of neutral lipid droplets in perinecrotic tumour cells. It is not yet clear whether mobile lipid signals could be used as a marker of hypoxia. Kuesel et al. (1994) obtained ^{1}H MR spectra ex vivo from high grade astrocytomas and detected lipid signals at 5.33, 2.80, 2.04, 1.29 and 0.89 ppm. The intensities of most of these signals correlated with the degree of necrosis in the samples as determined by histology. They concluded that necrotic foci were also responsible for the lipid resonances detected in vivo at 1.3 and 0.9 ppm. Martinez-Perez et al. (1996) made similar observations in humans for high grade brain tumours from in vivo ^{1}H NMR spectra and found cytoplasmic and extracellular lipid droplets when the tumour biopsies were examined by electron microscopy.

Shungu et al. (1992), in a localised ^{1}H spectroscopy study of RIF-1 and EMT-6 tumours, found that while "adequate lipid suppression" could be achieved for RIF-1 tumours by outer volume suppression, this was not the case for EMT-6 tumours. Lipid selective MR imaging measurements showed that lipid was present only as an extratumoral rim around the RIF-1 tumours (therefore possible to eliminate from ^{1}H spectra using outer volume suppression) but was dispersed throughout the EMT-6 tumours. For that study, the authors were particularly interested in measuring lactate and the lipid was a contaminating signal, making it difficult to observe lactate signals. EMT-6 tumours have been shown to have a very high hypoxic fraction and RIF-1 generally to have a relatively low hypoxic fraction. Although there is as yet insufficient evidence to support this, it would be ironic indeed if lipid signals were found to be a better indicator of hypoxia than lactate!

Several hypotheses have been proposed to explain changes in lipid metabolism in tumours: (1) signal transduction as a consequence of oncogenic changes; (2) signal transduction as a consequence of altered growth factor/hormonal stimulation; (3) hypoxia-induced metabolic changes (i.e. adaptations to provide energy and/or synthetic intermediates under stress conditions); and (4) modulation by low pH_e. It is not clear which of these factors are most important but it is interesting to note that perturbations in tumour cell lipid metabolism may be determined by the state of the tumour microenvironment.

References

Aboagye EO, McCoy CL, Maxwell RJ, Tracy M, Workman P, Griffiths JR (1994) Assessment of tumour hypoxia by quantitative ^{19}F magnetic resonance spectroscopy. Proc Soc Magn Reson 3:1349

Artemov D, Bhujwalla ZM, Glickson JD (1995) In vivo selective measurment of {1-13C}-glucose metabolism in tumors by heteronuclear cross polarization. Magn Reson Med 33:151–155

Bhakoo KK, Land H, Williams S, Gadian D, Noble M (1996) Phosphorylcholine: a putative marker of cancer. Differential regulation of choline metabolites during oncogenic transformation. Anticancer Res 15:1690–1691

Bhujwalla ZM, Tozer GM, Field SB, Maxwell RJ, Griffiths JR (1990) The energy metabolism of RIF-1 tumours following hydralazine. Radiother Oncol 19:281–291

Bhujwalla ZM, Shungu DC, Glickson JD (1993) Susceptibility maps of tumours and their relationship to histopathological characteristics and flow. Proc Soc Magn Reson 12:142

Bhujwalla ZM, Shungu DC, Chatham JC, Wehrle JP, Glickson JD (1994) Glucose metabolism in RIF-1 tumors after reduction in blood flow: an in vivo 13C and 31P NMR study. Magn Reson Med 32:303–309

Bhujwalla ZM, Shungu DC, Glickson JD (1996) Effects of blood flow modifiers on tumor metabolism observed in vivo by proton magnetic resonance spectroscopic imaging. Magn Reson Med 36:204–211

Bremner JC, Bradley JK, Stratford IJ, Adams GE (1994) Magnetic resonance spectroscopic studies on "real-time" changes in RIF-1 tumour metabolism and blood flow during and after photodynamic therapy. Br J Cancer 69:1083–1087

Chance B, Eleff S, Leigh JS, Sokolow D, Sapega A (1983) Mitochondrial regulation of phosphocreatine/inorganic phosphate in exercising human muscle: a gated 31P-NMR study. Proc Natl Acad Sci USA 78:6714–6718

Chaplin DJ, Acker B (1987) The effect of hydralazine on the tumour cytotoxicity of the hypoxic cell cytotoxin RSU-1069: evidence for therapeutic gain. Int J Radiat Oncol Biol Phys 13:579–585

Evanochko WT, Sakai TT, Ng TC, et al. (1984) NMR study of in vivo RIF-1 tumors: analysis of perchloric acid extracts and identification of ^{1}H, ^{31}P and ^{13}C resonances. Biochim Biophys Acta 805:104–116

Field SB, Burney IA, Needham S, Maxwell RJ, Coggle JE, Griffiths JR (1991) Are transplanted tumours suitable as models for studies on vasculature? Int J Radiat Biol 60:255–260

Freitas I, Pontiggia P, Barni S (1995) Histochemical probes for the detection of hypoxic cells. Anticancer Res 10:611–620

Fryer JP, Schor PL, Jarrett KA, Neeman M, Sillerud LO (1991) Cellular energetics measured by phosphorus nuclear magnetic resonance spectroscopy are not correlated with chronic nutrient deficiency in multicellular tumor spheroids. Cancer Res 51:3831–3837

Gillies RJ, Barry JA, Ross BD (1994a) In vitro and in vivo 13C and 31P NMR analyses of phosphocholine metabolism in rat glioma cells. Magn Reson Med 32:310–318

Gillies RJ, Liu Z, Bhujwalla Z (1994b) 31P-MRS measurements of extracellular pH of tumors using 3-aminopropylphosphonate. Am J Physiol 267:195–203

Griffiths JR, Cady E, Edwards RH, McCready VR, Wilkie DR, Wiltshaw E (1983) ^{31}P NMR studies of human tumour in vivo. Lancet 25:1435–1436

He Q, Shungu DC, van Zijl PCM, Bhujwalla ZM, Glickson JD (1995) Single-scan in vivo lactate editing with complete lipid and water suppression by selective multiple-quantum-coherence transfer (SelMQC) with application to tumors. J Magn Reson B 106:203–211

Hees PS, Sotak CH (1993) Assessment of changes in murine tumour oxygenation in response to nicotinamide using ^{19}F NMR relaxometry of a perfluorocarbon emulsion. Magn Reson Med 29:303–310

Horsman MR, Christensen KL, Overgaard J (1989) Hydralazine-induced enhancement of hyperthermic damage in a C3H mammary carcinoma in vivo. Int J Hyperthermia 5:123–126

Howe FA, Robinson, SP, Griffiths JR (1993) Monitoring changes in oxygenation and blood perfusion of a subcutaneous tumour by functional magnetic resonance imaging. Proc Soc Magn Reson 12:642

Jin GY, Li S-J, Moulder JE, Raleigh JA (1990) Dynamic measurements of hexafluoromisonidazole (CCI-103F) retention in mouse tumours by 1H/19F magnetic resonance spectroscopy. Int J Radiat Biol 58:1025–1034

Kalra R, Bremner JC, Wood PJ, Sansom J, Counsell CJ, Stratford IJ, Adams GE (1994) 31P MRS to monitor the induction of tumour hypoxia by the modification of the oxygen affinity of haemoglobin using BW 589C. Int J Radiat Oncol Biol Phys 29:285–288

Karczmar GS, River JN, Vijayakumar S, Goldman Z, Lewis MZ (1994) Effects of hyperoxia on T2* and resonance frequency weighted magnetic resonance images of rodent tumors. NMR Biomed 7:3–11

Khalil AA, Horsman MR, Overgaard J (1995) The importance of determining necrotic fraction when studying the effect of tumour volume on tissue oxygenation. Acta Oncol 34:297–300

Kimura H, Itoh S, Kawamura Y, Nakatsugawa S, Ishii Y (1994) Metabolic alterations in implanted human tumors after combined radiation and hyperthermia therapy measured by in vivo 31P MRS. Magn Reson Imaging 12:109–119

Koutcher JA, Alfieri AA, Devitt ML, et al. (1992) Quantitative changes in tumor metabolism, partial pressure of oxygen and radiobiological oxygenation status post radiation. Cancer Res 52:4620–4627

Kuesel AC, Sutherland GR, Halliday W, Smith ICP (1994) 1H MRS of high grade astrocytomas: mobile lipid accumulation in necrotic tissue. NMR Biomed 7:149–156

Kuin A, Smets L, Volk T, et al. (1994) Reduction of intratumoral pH by the mitochondrial inhibitor m-iodobenzylguanidine and moderate hyperglycemia. Cancer Res 54:3785–3792

Kwong KK, Belliveau JW, Chesler DA, et al. (1992) Dynamic magnetic resonance imaging of human brain activity during primary sensory stimulation. Proc Natl Acad Sci USA 89:5675–5679

Li C-W, Kuesel AC, Padavic-Shaller KA, et al. (1996) Metabolic characterization of human soft tissue sarcomas in vivo and in vitro using proton-decoupled phosphorus magnetic resonance spectroscopy. Cancer Res 56:2964–2972

Lyng H, Olsen DR, Southon TE, Rofstad EK (1993) 31P-nuclear magnetic resonance spectroscopy in vivo of six human melanoma xenograft lines: tumour bioenergetic status and blood supply. Br J Cancer 68:1061–1070

Lyng H, Olsen DR, Petersen SB, Rofstad EK (1995) ^{31}P NMR spectroscopy studies of phospholipid metabolism in human melanoma xenograft lines differing in rate of tumour cell proliferation. NMR Biomed 8:65–71

Mahmood U, Alfieri AA, Thaler H, Cowburn D, Koutcher JA (1994) Radiation dose-dependent changes in tumor metabolism measured by 31P nuclear magnetic resonance spectroscopy. Cancer Res 54:4885–4891

Martinez-Perez I, Moreno A, Barba I, Capdevila A, Arus C (1996) Large lipid droplets observed by electron microscopy in 6 human brain tumors with lipid ^{1}H MRS in vivo patttern. Proc Int Soc Magn Reson Med 4:976

Maxwell RJ, Prysor-Jones RA, Jenkins JA, Griffiths JR (1988) Vasoactive intestinal peptide stimulates glycolysis in pituitary tumours; ^{1}H-NMR detection of lactate in vivo. Biochim Biophys Acta 968:86–90

Maxwell RJ, Workman P, Griffiths JR (1989) Demonstration of tumour-selective retention of fluorinated nitroimidazole probes by 19F magnetic resonance spectroscopy in vivo. Int J Radiat Oncol Biol Phys 16:925–929

McCoy CL, McIntyre DJO, Robinson SR, Aboagye EO, Griffiths JR (1996a) Magnetic resonance spectroscopy and imaging methods for measuring tumour and tissue oxygenation. Br J Cancer Suppl 74:226–231

McCoy CL, Maxwell RJ, Horsman MR, Overgaard J, Jorgensen HS, Griffiths JR (1996b) Measurement of intra- and extracellular pH of murine tumors pre- and post-therapy from a split peak in 31P spectra. Proc Int Soc Magn Reson Med 4:373

Moon RB, Richards JH (1973) Determination of intracellular pH by 31P magnetic resonance. J Biol Chem 248:7276–7278

Negendank W (1992) Studies of human tumours by MRS: a review. NMR Biomed 5:303–324

Negendank WG, Padavic-Shaller KA, Li C-W, et al. (1995) Metabolic characterization of human non-Hodgkin's lymphomas in vivo using proton-decoupled phosphorus magnetic resonance spectroscopy. Cancer Res 55:3286–3294

Nordsmark M, Grau C, Horsman MR, Jørgensen HS, Overgaard J (1995) Relationship between tumour oxygenation, bioenergetic status and radiobiological hypoxia in an experimental model. Acta Oncol 34:329–334

Nordsmark M, Maxwell RJ, Wood PJ, Stratford IJ, Adams GE, Overgaard J (1996) Effect of hydralazine in spontaneous tumours assessed by oxygen electrodes and 31P-magnetic resonance spectroscopy. Br J Cancer Suppl 74:232–235

Ogawa S, Lee T-M, Nayak AS, Glynn P (1990) Oxygenation-sensitive contrast in magnetic resonance image of rodent brain at high magnetic fields. Magn Reson Med 14:68–78

Okubo M (1993) Changes of 31P-MR spectroscopy in experimental tumors following radiotherapy combined with 5-fluorouracil compound (UFT). Tohoku J Exp Med 169:215–223

Okunieff P, Kallinowski F, Vaupel P, Neuringer LJ (1988) Effects of hydralazine-induced vasodilation on the energy metabolism of murine tumors studied by in vivo 31P-nuclear magnetic resonance spectroscopy. J Natl Cancer Inst 80:745–750

Parhami P, Fung BM (1983) Fluorine-19 relaxation study of perfluoro chemicals as oxygen carriers. J Phys Chem 87:1928–1931

Podo F, de Certaines JD, Bovee WMMJ (1996) Significance of magnetic resonance spectroscopy measurements in relation to biochemical processes and cellular control. In: Podo F, Bovee WMMJ, de Certaines JD, et al. (eds) Eurospin annual 1995–1996. Istituto Superiore di Sanita, Rome, pp 107–226

Prescott DM, Charles HC, Sostman HD, et al. (1994) Therapy monitoring in human and canine soft tissue sarcomas using magnetic resonance imaging and spectroscopy. Int J Radiat Oncol Biol Phys 28:415–423

Rajan SS, Wehrle JP, Li S, Steen RG, Glickson JD (1989) 31P NMR spectroscopic study of bioenergetic changes in radiation-induced fibrosarcoma-1 after radiation therapy. NMR Biomed 2:165–171

Raleigh JA, Franko AJ, Kelly DA, Trimble LA, Allen PS (1991) Development of an in vivo 19F magnetic resonance method for measuring oxygen deficiency in tumors. Magn Reson Med 22:451–466

Rasmussen J, Hansen LL, Friche E, Jaroszewski JW (1993) 31P and 13C NMR spectroscopic study of wild type and multidrug resistant Ehrlich ascites tumour cells. Oncol Res 5:119–126

Rempel A, Mathupala SP, Griffin CA, Hawkins AL, Pedersen PL (1996) Glucose catabolism in cancer cells: amplification of the gene encoding type II hexokinase. Cancer Res 56: 2468–2471

Robinson SP, Howe FA, Griffiths JR (1995) Non-invasive monitoring of carbogen-induced changes in tumour blood flow and oxygenation by functional magnetic resonance imaging. Int J Radiat Oncol Biol Phys 33:855–859

Rofstad EK, DeMuth P, Fenton BM, Sutherland RM (1988) ^{31}P Nuclear magnetic resonance spectroscopy studies of tumor energy metabolism and its relationship to intracapillary oxyhemoglobin saturation status and tumor hypoxia. Cancer Res 48:5440–5448

Shedd SF, Lutz NW, Hull WE (1993) The influence of medium formulation on phosphomonoester and UDP-hexose levels in cultured human colon tumor cells as observed by ^{31}P NMR spectroscopy. NMR Biomed 6:254–263

Shungu DC, Bhujwalla ZM, Wehrle JP, Glickson JD (1992) ^{1}H NMR spectroscopy of subcutaneous tumors: preliminary studies of effects of growth, chemotherapy and blood flow reduction. NMR Biomed 5:296–302

Smith TA, Baluch S, Titley JC, et al. (1993a) The effect of oestrogen ablation on the phospholipid metabolite content of primary and transplanted rat mammary tumours. NMR Biomed 6:209–214

Smith TA, Bush C, Jameson C, Titley JC, Leach MO, Wilman DE, McCready VR (1993b) Phospholipid metabolites, prognosis and proliferation in human breast carcinoma. NMR Biomed 6:318–323

Sostman HD, Prescott DM, Dewhirst MW, et al. (1994) MR imaging and spectroscopy for prognostic evaluation in soft-tissue sarcomas. Radiology 190:269–275

Stubbs M, Rodrigues L, Howe FA, Wang J, Jeong KS, Veech RL, Griffiths JR (1994) Metabolic consequences of a reversed pH gradient in rat tumors. Cancer Res 54:4011–4016

Taylor DJ, Bore PJ, Styles P, Gadian DG, Radda GK (1983) Bioenergetics of intact human muscle: a ^{31}P nuclear magnetic resonance study. Mol Biol Med 1:77–94

Tozer GM, Griffiths JR (1992) The contribution made by cell death and oxygenation to 31P MRS observations of tumour energy metabolism. NMR Biomed 5:279–289

Tozer GM, Bhujwalla ZM, Griffiths JR, Maxwell RJ (1989) Phosphorus-31 magnetic resonance spectroscopy and blood perfusion of the RIF-1 tumor following X-irradiation. Int J Radiat Oncol Biol Phys 16:155–164

Tozer GM, Maxwell RJ, Griffiths JR, Pham P (1990) Modification of the ^{31}P magnetic resonance spectra of a rat tumour using vasodilators and its relation to hypotension. Br J Cancer 62:553–560

Vaupel P, Okunieff P, Neuringer LJ (1990) In vivo ^{31}P-NMR spectroscopy of murine tumours before and after localized hyperthermia. Int J Hyperthermia 6:15–31

Vaupel P, Schaefer C, Okunieff P (1994) Intracellular acidosis in murine fibrosarcomas coincides with ATP depletion, hypoxia, and high levels of lactate and total P_i. NMR Biomed 7:128–136

Warburg O (1930) The metabolism of tumors. Constable, London

Wood PJ, Stratford IJ, Sansom JM, Cattanach BM, Quinney RM, Adams GE (1992) The response of spontaneous and transplantable murine tumours to vasoactive agents measured by ^{31}P magnetic resonance spectroscopy. Int J Radiat Oncol Biol Phys 22:473–476

Wood PJ, Stratford IJ, Adams GE, Szabo C, Thiemermann C, Vane JR (1993) Modification of energy metabolism and radiation response of a murine tumour by changes in nitric oxide availability. Biochem Biophys Res Commun 192: 505–510

Wood PJ, Sansom JM, Butler SA, et al. (1994) Induction of hypoxia in experimental murine tumors by the nitric oxide synthase inhibitor NG-nitro-L-arginine. Cancer Res 54: 6458–6463

15 The Potential Role of Positron Emission Tomography in Investigation of Microenvironment

C. LAUBENBACHER[1] and M. SCHWAIGER[2]

CONTENTS

15.1 Methodology of Positron Emission Tomography

Positron emission tomography (PET) represents a new imaging approach that permits measurement of physiologic and biochemical processes within various human organs. PET differs from conventional scintigraphic procedures, such as single photon emission tomography (SPECT), by its unique data acquisition which allows quantitative measurement of regional tissue radioactivity (HOFFMAN and PHELPS 1986) and employs radioisotopes of natural elements – ^{15}O, ^{11}C, and ^{13}N. These natural radioisotopes retain their normal biologic function and can be used to synthesize numerous positron-emitting radiopharmaceuticals. Due to the quantitative information regarding normal physiology provided by PET, it can detect pathophysiologic alterations that occur in disease states. In most diseases, biochemical abnormalities occur prior to anatomic changes. Thus, PET may detect these early biochemical changes before other imaging modalities.

[1,2] C. LAUBENBACHER, Dr. med. and M. SCHWAIGER, Prof. Dr. med., Klinik für Nuklearmedizin, Klinikum rechts der Isar der Technischen Universität München, Ismaninger Strasse 22, D-81675 Munich, Germany

To date, most research-related and clinical PET studies have been performed in neurology and cardiology (MAZZIOTTA and PHELPS 1986; SCHELBERT and SCHWAIGER 1986). However, it has become apparent that PET is also useful in oncologic research and in the clinical evaluation of cancer patients. In this chapter we will review the technical background of PET, address its possible application in the study of physiology and pathophysiology and summarize its application in oncology.

15.1.1 PET Imaging Principles and Technical Applications

Positron-emitting radionuclides have a nuclear imbalance characterized by an excess of protons. In order to restore stability to the nuclear structure, a proton is converted to a neutron and a positron is emitted. This energetic positron traverses a few millimeters of the tissue until it becomes thermalized by electrostatic interaction between the electrons and the atomic nuclide of the media. It then combines with a free electron to form a two-particle "atomlike" entity called positronium. The latter quickly decays by annihilation, the complete conversion of the positron and electron mass into energy, and generates a pair of photons which travel in nearly opposite directions (180° apart) with an energy of 511 KeV each (Fig. 15.1).

It is this unique characteristic of positron annihilation that is exploited for image formation. The opposed photons from positron decay can be detected by the use of pairs of colinearly aligned detectors. Photons interact with these detectors within a predefined time-window and are registered as radioactive events (coincidence counting). The detector pairs of a PET system are installed in a ring shaped pattern, allowing measurement of radioactivity along lines through the organ of interest at a series of angles and radial distances. This angular information is subsequently used to reconstruct tomo-

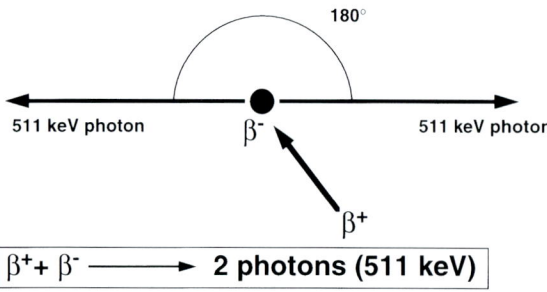

Fig. 15.1. Annihilation reation. During positron decay, a positively charged particle is emitted from the nucleus which interacts with an electron. During this annihilation reaction, a mass of postparticles is converted into energy in the form of two 511 keV photons

graphic images of regional radioactivity distribution. State-of-the-art PET devices consist of multiple closely packed rings of detectors that enable simultaneous imaging of several image planes.

To exploit fully the quantitative capabilities of PET, the images have to be corrected for photon attenuation (HOFFMAN and PHELPS 1986). Regional correction factors are derived from transmission images. These transmission scans are produced by placing a ring source around the body of the patient and measuring the absorption by the tissue. An example of a "CT-like" transmission scan with corresponding emission scan in a patient with lung cancer is shown in Fig. 15.2.

15.1.1.1
Resolution of PET

PET images are spatial maps of radioactivity distribution within tissue slices, analogous to autoradiograms obtained from selected tissue in animal experiments. The PET method, unlike autoradiogra-

phy, is noninvasive and may thus be used in clinical research, including longitudinal studies. A second difference between PET and tissue autoradiography is anatomic resolution. Typical film-autoradiographic methods for detection of ^{3}H and ^{14}C provide 50- to 100-μm resolution, allowing clear differentiation of regional tissue subtypes such as brain nuclei from surrounding fiber tracts. The spatial resolution of current PET instrumentation is approximately 4–10 mm. This resolution limitation results from the number and geometry of detectors in the instrument as well as the number of counts acquired from the image and their statistical precision. These factors vary between tomographs of different design as well as from study to study because of varying image acquisition times and tissue radioactivity levels. The ultimate theoretic limit of PET resolution, however, is the distance traveled by the positron in tissue before the annihilation reaction. Maximum tissue ranges vary according to initial positron energy (Table 15.1).

As a consequence of limited spatial resolution, small regions and thin structures such as micrometastases cannot be fully resolved. If anatomic structures smaller than the resolution are imaged, the "true" tracer concentration in such structures will be underestimated (HOFFMAN et al.

Table 15.1. Positron-emitting radionuclides

Nuclide	Half-life (min)	Energy (MeV)	Range (mm in water)
^{11}C	20.4	0.96	4.1
^{15}O	2.1	1.72	8.2
^{13}N	10.0	1.19	5.4
^{18}F	109.7	0.64	2.4
^{82}Rb	1.3	3.35	13.2

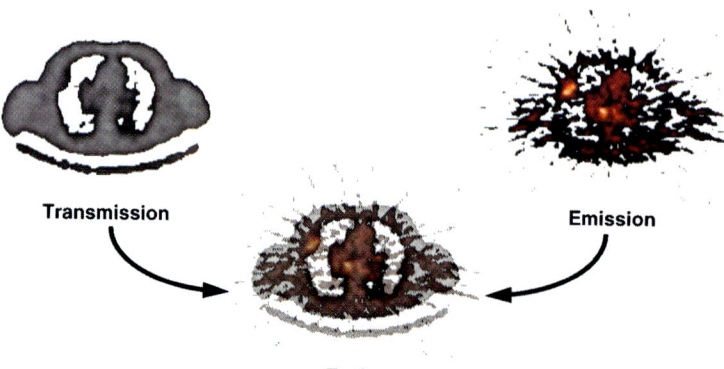

Fig. 15.2 Patient with lung cancer. Focal increased uptake of ^{18}F-deoxyglucose (FDG) in the tumor is clearly seen in the emission image. The fusion image, consisting of transmisson scan, which is necessary for attenuation correction, and emission scan, simplifies localization of the tumor

1979). Thus, partial volume effect limits the quantification of tracer in small structures. On the other hand, PET studies in large organs are not affected. Reconstructed PET data thus reflect average isotope concentrations in the imaged tissue. However, if the dimensions of a given organ of interest are defined by an imaging technique with high spatial resolution such as computed tomography (CT) or magnetic resonance imaging (MRI), correction factors can be applied to compensate for partial volume effect.

15.1.1.2
Tracer Kinetic Modeling

The primary strength of PET is the ability to use tracer kinetic models to yield measurements of physiologic processes in absolute units (HUANG and PHELPS 1986). To measure such processes after application of positron-labeled compounds, appropriate mathematical tracer kinetic models based on the principles of tracer and competitive enzyme kinetics are used. Models exist for the measurement of local rates of blood flow, membrane transport, metabolism, and receptor binding (HUANG and PHELPS 1986). Such models reduce complex biologic systems into a few compartments and attempt to estimate the rates of processes in those compartments. The most important factors in composing an adequate compartment model are: (1) choice of an optimal tracer, (2) construction of a compartment model (as simple as possible), and (3) quantification and validation of the model. An optimal tracer should measure only the process under investigation and there should be as few metabolic pathways as possible. The labeled radiopharmaceutical should retain its biologic and physiologic properties and should have a short biologic half-time in blood and in nontarget tissues for an optimal signal-to-noise ratio. Once a tracer kinetic model has been constructed and validated, the operation of the model can be incorporated into the PET computer system and used to convert tomographic images into specific rates of activity for physiologic processes. For example, PET scans with 18F-labeled deoxyglucose (FDG) and the appropriate kinetic model produce data that are converted to glucose metabolism per min and gram of tissue. Figure 15.3A shows an FDG image in a patient with two metastases of a breast carcinoma. By kinetic modeling and computer analysis, parametric images of the metabolic rate K_i, and for the blood space V_d, can be obtained (Fig. 15.3B,C; WONG and HICKS, 1994). For a detailed discussion of the principles and algorithms of specific tracer kinetic models, the reviews by Huang, Hawkins, and Koeppe are recommended (HAWKINS et al. 1992c; HUANG and PHELPS 1986; KOEPPE and HUTCHINS 1992).

15.2
Physiologic Parameters

The ability to identify structural alterations in cancer tissue has increased dramatically in the past decades with the development of noninvasive cross-sectional imaging techniques such as CT and MRI. Whereas significant advances have been made in morphologic imaging, less attention has been given to the development of in vivo methods of quantifying physiologic and biochemical tissue function. PET is a technique for obtaining physiologic information necessary not only to provide means of diagnosis of cancer based on altered tissue physiology, but also to serve as a tool for monitoring the effects of therapy on tissue metabolism. Correlative functional-anatomic imaging will permit the study of metabolic processes in the anatomic loci at which they occur and perhaps detect changes indicative of tumor response to therapy before alterations in structure occur. Therefore, PET studies of cancer have been employed to provide following information:

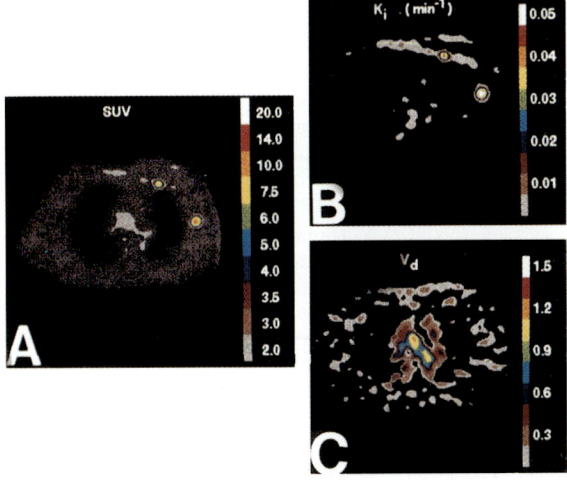

Fig. 15.3. A FDG image of a patient with two metastases of a breast carcinoma. By kinetic modelling and computer analysis, parametric images of the metabolic rate K_i (**B**) and for the blood space V_d (**C**) can be obtained. (Reprinted by permission of the Society of Nuclear Medicine from: Wong, et al. A clinically practical method to acquire parametric images of unidirectional metabolic rates and blood spaces. J Nucl Med 1994;35:1210. Figure 8)

Table 15.2. Radiopharmaceuticals used in PET studies of cancer

Process	Radiopharmaceutical	References
General biochemical and physiological processes		
Glucose metabolism (glycolysis)	[^{18}F]FDG	Numerous, e.g., HAWKINS et al. 1986 HUANG et al. 1980 PHELPS et al. 1979 REIVICH et al. 1979 SOKOLOFF et al. 1977 STRAUSS and CONTI 1991
Perfusion	[^{13}N]ammonia	CHEN et al. 1992 CHEN et al. 1991 GARDNER et al. 1993 KILLION et al. 1993 KUTEN et al. 1992 LORENZ et al. 1992 NITZSCHE et al. 1993 SCHELSTRAETE et al. 1982 SHIBATA et al. 1988
	^{82}Rb	BROOKS et al. 1984 DHAWAN et al. 1989 JARDEN et al. 1985 LAMMERTSMA et al. 1984 YEN et al. 1982
	[^{15}O]H$_2$O	HERHOLZ et al. 1988 ITO et al. 1982 KUBO et al. 1991 MARKHAM and SCHUSTER 1992 MINEURA et al. 1994 MINEURA et al. 1986 OGURO et al. 1993a OGURO et al. 1993b STRAUSS et al. 1989 TANIGUCHI et al. 1993 TYLER et al. 1987 VELAZQUEZ et al. 1991a
	[^{62}Cu]PTSM	MATHIAS et al. 1991 YOUNG et al. 1994
	^{68}Ga-MPO platelets	GOODWIN et al. 1993
	^{62}Cu-albumin conjugates	ANDERSON et al. 1993
Oxygen utilization	^{15}O$_2$	BEANEY et al. 1984 MINEURA et al. 1990 OGAWA et al. 1988 RHODES et al. 1983 SADATO et al. 1993
Amino acid uptake and protein synthesis	[^{11}C]methionine	Numerous, e.g., KIICHI et al. 1988 KUBOTA et al. 1988 KUBOTA et al. 1989a KUBOTA et al. 1989b MOSSKIN et al. 1987
	[^{11}C]leucine	HAWKINS et al. 1989 ISHIWATA et al. 1993a ISHIWATA et al. 1993b KEEN et al. 1989
	[^{13}N]glutamate	GELBARD et al. 1979 KNAPP et al. 1986 KNAPP et al. 1984 REIMAN et al. 1981 REIMAN et al. 1982

Table 15.2. *Continued*

Process	Radiopharmaceutical	References
Nucleic acid metabolism (DNA replication)	[^{11}C]thymidine	Higashi et al. 1994 Ishiwata et al. 1985 Kubota et al. 1991 Kubota et al. 1992a Mariat et al. 1988 Poupeye et al. 1989 Sato et al. 1992 Shields et al. 1990a Shields et al. 1990b van Eijkeren et al. 1992
Polyamine metabolism	[^{11}C]putrescine	Hiesiger et al. 1987 Hwang et al. 1989 Hwang et al. 1990
Glucosamine uptake	[^{18}F]fluoroacetyl-D-glucosamine	Fujiwara et al. 1990 Thorell et al. 1993
Lipoprotein metabolism	^{18}F-labeled *N*-lactitol-*S*-fluorophenacyl-cysteamine; [^{18}F]LCSH	Daugherty et al. 1992

Organ-specific or other specialized biochemical and physiological processes

BBB permeability/ vascular permeability	^{68}Ga-EDTA ^{68}Ga-transferrin	Hawkins et al. 1984 Kessler et al. 1984 Kaplan et al. 1992a Kaplan et al. 1992b Velazquez et al. 1991b
Receptor-specific ligands	[^{18}F]fluoroestradiol ^{18}F-norprogesterone ^{68}Ga-somatostatin analogue [^{124}I]MIBG	Katzenellenbogen 1980 Mintun et al. 1988 Van Brocklin et al. 1994 Dehdashti et al. 1991 Verhagen et al. 1991 Macke et al. 1993 Ott et al. 1992
Monoclonal antibodies	^{124}I-HMFG 1 (epithelial neoplasms) ^{124}I-3F8 (neuroblastoma, astrocytoma) ^{124}I-H17E2 [placental alkaline phosphatase (PLAP)] ^{124}I-BB5-G1 (parathyroid surface) ^{124}I-anti-CEA (colorectal carcinoma) ^{64}Cu-1A3 (colorectal carcinoma) ^{124}I-ICR 12 (anti-human c-erb B2 protooncogene) ^{124}I-MX35 / MH99 (ovarian cancer) ^{18}F-Mel-14 (gliomas)	Wilson et al. 1991 Daghighian et al. 1991 Larson et al. 1992 Miraldi 1989 Pentlow et al. 1991 Snook et al. 1990 Otsuka et al. 1991 Westera et al. 1991 Anderson et al. 1992 Bakir et al. 1992 Rubin et al. 1993 Vaidyanathan et al. 1992
Hypoxic cell markers	[^{18}F]fluoromisonidazole	Koh et al. 1991 Rasey et al. 1989 Rasey et al. 1990

Table 15.2. *Continued*

Process	Radiopharmaceutical	References
Chemotherapeutic agents	[^{18}F]fluorouridine (FUdR)	ABE et al. 1983
		ISHIWATA et al. 1991
		KUBOTA et al. 1992a
		SATO et al. 1992
		TSURUMI et al. 1990
	[^{18}F]5-fluorouracil	ABE et al. 1983
		DIMITRAKOPOULOU et al. 1993
		HOHENBERGER et al. 1993
		WOLF et al. 1990
	[^{11}C]nitrosoureas (BCNU, SarCNU)	MITSUKI et al. 1991
	[^{18}F]tamoxifen	YANG et al. 1990
	[^{13}N]cisplatin	GINOS et al. 1987
Iodine metabolism	[^{124}I]iodine ion	FLOWER et al. 1990
		FLOWER et al. 1989
		LAMBRECHT et al. 1988
Skeletal metabolic (osteoblastic) activity	[^{18}F]fluoride ion	BLAU et al. 1972
		CHARKES et al. 1978
		HAWKINS et al. 1992a
		HOH et al. 1993
		REEVE et al. 1988
		TSE et al. 1994b
		WOOTEN and DOVE 1986
	[^{86}Y]yttrium citrate	HERZOG et al. 1993
Reticuloendothelial system	^{68}Ga-HSA microspheres	OKADA et al. 1990
		WESTERBERG and LANGSTROM 1994

1. Quantification of tumor perfusion, assessment of pH
2. Evaluation of tumor energy metabolism (substrate metabolism)
3. Quantification of anabolic metabolism (DNA replication, protein synthesis)
4. Characterization of cell membrane function (receptors, antibodies)
5. Characterization of specific cell functions (expression of proteins)
6. Tracing of radiolabeled cytostatic agents (pharmacokinetics)

Various procedures have been used to label more than 200 biologic substrates and drugs with positron-emitting isotopes (BARRIO 1986; FOWLER and WOLF 1986). These labeled compounds constitute a large potential resource for development of PET bioassay methods. A list of some of the processes and corresponding positron-emitting radiopharmaceuticals used in cancer studies is given in Table 15.2. The list is intended to illustrate the most commonly used agents in oncologic studies.

15.2.1
Quantification of Tumor Perfusion and pH

15.2.1.1
Tumor Perfusion

Noninvasive evaluation of regional tissue blood flow is important because many diseases affect the delivery of oxygen and other nutrients necessary for energy production. A number of invasive techniques, used in animal experiments, cannot be applied easily to the clinical situation. The PET scan noninvasively measures regional tracer concentration and permits the quantification of tissue perfusion in milliliters per minute per gram of organ (HUANG et al. 1985; MARTIN and RAICHLE 1983; SHAH et al. 1985).

In experimental studies, with the possibility of injecting the radiopharmaceutical intraarterially, microaggregated albumin microspheres labeled with ^{68}Ga or ^{11}C can be used to accurately assess organ blood flow (BELLER et al. 1979). Instead of albumin, a "chemical microsphere" ([^{62}Cu]PTSM, copper(II) pyruvaldehyde bis-(N4-thiosemicarbazone)) that does not require injection into the left ventricle and can be labeled with the generator-produced [62]Cu is used for assessment of regional blood flow in heart (SCHWAIGER and MUZIK 1991) and in brain (FUJIBAYASHI et al. 1993). In oncology, MATHIAS et al. (1991) investigated its feasibility for assessing tumor blood flow in hamsters with colorectal carcinoma cell implants and found good correlation of PET data with those derived from microsphere measurements. Since the mechanism of PTSM trapping is linked to glutathione levels in tissue and tumor tissue glutathione levels might vary, its utility for assessing tumor blood flow has to be determined in larger experimental studies and probably for each individual tumor entity.

In clinical studies, two other types of intravenous flow tracers are available for the assessment of regional blood flow. Examples of the *first type* are [82]Rb and [^{13}N]ammonia. These tracers are extracted from plasma and trapped in tissue in proportion to blood flow. The tissue retention fraction of both tracers is inversely related to flow, limiting flow estimates at high flow rates. The rapid plasma extraction and slow tissue clearance are advantageous and allow clear distinction of tissue activity from vascular activity. [82]Rb has a short half-life of only 76 s, allowing repeated measurements of blood flow at brief intervals. [^{13}N]ammonia is converted to [^{13}N]glutamine by glutamine synthetase reaction (BERGMAN et al. 1980). The use of [^{13}N]ammonia in oncology is novel and few experimental studies in animals have been done. CHEN et al. (1991, 1992) investigated the feasibility of dynamic PET examinations in determining the hepatic arterial blood flow in dogs and normal volunteers (CHEN et al. 1991), finding a close relationship between PET-derived data and those obtained by the gold standard (microsphere blood flow measurement). SHIBATA et al. (1988) examined the blood supply of hepatocellular carcinomas and metastatic liver tumors in 23 patients. He found a strong correlation between the demonstration of rich tumor vessels by hepatic angiography and by [^{13}N]ammonia uptake of the tumor. In seven cases with metastatic liver tumor, the accumulation of [^{13}N]ammonia was lower than in normal liver throughout the scan. The results suggested that blood flow of liver tumors can be assessed by dynamic PET. Therefore, this method may provide information for the characterization of tumors (differentiation of hemangiomas) as well as for the evaluation of treatment (differences in tumor blood supply before and after therapeutic interventions). Furthermore, the determination of renal blood flow

by PET was first evaluated in animals (CHEN et al. 1992). Especially renal function is of great interest in monitoring therapy effects. [^{13}N]ammonia PET was used in the evaluation of acute radiation effects on renal blood flow in dogs (KILLION et al. 1993; KUTEN et al. 1992) before examinations in humans (LORENZ et al. 1992; NITZSCHE et al. 1993) were performed. These studies demonstrated that the determination of absolute blood flow measurements by [^{13}N]ammonia PET correlates well with experimentally obtained flow data by microsphere technique. Thus, it seems to be possible to noninvasively evaluate blood flow and its alteration during therapy in humans.

The second type of flow tracers are *inert diffusible substances*, such as [^{15}O]water and [^{11}C]butanol (RAICHLE et al. 1976). These tracers accumulate and clear from organs as a function of flow. [^{15}O]water is the most widely used inert flow tracer, and it is avidly extracted by tissue, with a first-pass extraction fraction close to 100%. Animal studies on heart tissue have shown that this high extraction is maintained as a function of flow, indicating that the distribution of [^{15}O]water is limited by flow rather than diffusion (BERGMAN et al. 1984). In addition to this favorable physiologic property, [^{15}O]water has a short physical half-life (2 min), which allows repetitive flow measurements in short intervals. However, the high concentration of [^{15}O]water in the vascular space may impair determination of tissue tracer concentration, especially in organs with large vascular volume such as the heart. In contrast the brain has a small vascular space; therefore, [^{15}O]water is the method of choice for regional blood flow determination in brain studies (RAICHLE et al. 1981). With the use of dynamic PET scanning, blood and tissue activity can be differentiated, and quantification of regional blood flow can be achieved with reasonable accuracy. Very few studies in oncology have been performed with [^{15}O]water to determine blood flow. KAHN et al. (1994) demonstrated in young normal adults that PET with [^{15}O]water can assess bone marrow blood flow prior to therapeutic interventions. The perfusion of tumors and metastases can be defined by this method as shown by HOHENBERGER et al. (1993) in colorectal liver metastases.

It has been shown that PET can quantitatively assess tumor blood flow. Flow measurements may be useful in defining heterogeneity of perfusion and provide measurements of potential drug delivery. Due to the short half-life of PET tracers, flow agents can be combined with other tracers such as FDG, allowing for combined evaluation of flow and substrate metabolism.

15.2.1.2
Noninvasive Determination of pH by PET

Changes of intracellular pH can dramatically affect cell metabolism. The noninvasive PET method to determine pH noninvasively is based on the biodistribution of weak acids or bases. Neutral molecules pass through the capillary membrane more easily than charged ions. Under the assumption that the capillary membrane is pervious to neutral molecules and impermeable to charged ions, the concentration of the neutral molecules will be the same for all tissue compartments during steady state, while the concentration of the charged particles depends on pH (Fig. 15.4). CO_2 and 5,5-dimethyl-2,4-oxyzolidin-dion (DMO) have been investigated as weak acids in normal brain tissue (ROTTENBERG et al. 1984) as well as in brain tissue after infarction (SYROTA et al. 1985). Correlation of PET-derived pH with data obtained by direct measurements was weak, indicating that the assumption of a selective capillary membrane, made as a theoretical basis, may be invalid. Additionally, CO_2 is continuously exhaled without a steady state phase, thus requiring serial imaging for correction for metabolites. Up to now, the two tracers, CO_2 and DMO, have only been used in brain, since an experimental validation of this method in tumors is still missing.

15.2.2
Energy Metabolism (Substrate Metabolism)

15.2.2.1
Evaluation of Oxygen Consumption and Hypoxia

Several factors are known to influence the probability of tumor control after radiation. These include tumor oxygen tension distribution, glutathione content, intrinsic radiation sensitivity, etc. It is well established that cells are more sensitive to radiation in the presence of oxygen than in its absence (see review in VAUPEL 1990). As a result, tumor hypoxia represents an important factor in radiation therapy. Hypoxic radioresistance has been demonstrated in many animal tumors and has been directly demonstrated in a variety of human tumors (MOLLS et al. 1994; VAUPEL et al. 1990). In vivo demonstration of hypoxia requires measurements with oxygen electrodes. However, this technique has limited clinical application due to its invasiveness. PET with $^{15}O_2$ or $[^{18}F]$fluoromisonidazole might provide means of noninvasively demonstrating tumor hypoxia.

The steady state (autoradiographic) method using $^{15}O_2$ inhalation and PET is a simple and practical way of imaging blood flow and oxygen metabolism (SADATO et al. 1993). OGAWA (OGAWA et al. 1988) demonstrated by this method that, after radio-chemotherapy of human gliomas, glucose consumption and blood volume decreased in tumor tissue, while in normal gray matter, blood flow, blood volume, and oxygen consumption did not show any significant changes. Oxygen extraction, glucose consumption, and glucose extraction, however, decreased significantly. Besides a variety of central nervous tumors and meningiomas (MINEURA et al. 1990), there have also been studies in peripheral tumors such as breast carcinomas. In these tumors, BEANEY found elevated regional blood flow, oxygen utilization, and blood volume (BEANEY et al. 1984). A disadvantage of this technique is that it often leads to underestimation of the oxygen extraction fraction due to tissue heterogeneity as shown in a simulation study by Lammertsma (LAMMERTSMA and JONES 1992). However, due to the noninvasiveness of PET measurements, this method may at least be used for

Determination of pH by PET

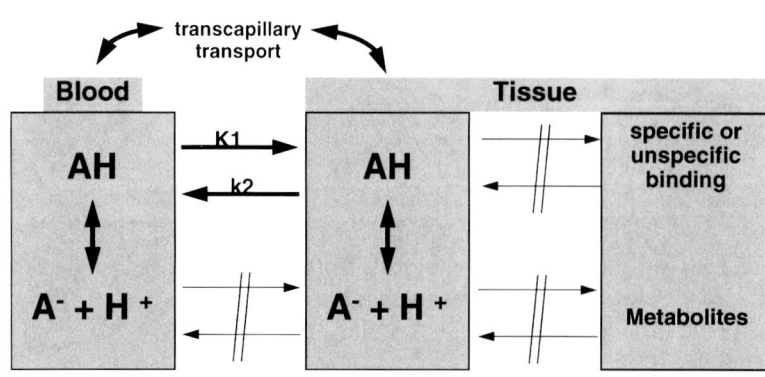

Fig. 15.4. Compartment model for the determination of pH by PET. Under the assumption that the capillary wall is totally permeable to neutral molecules, totally impermeable to charged ions, and that there is no further specific or unspecific binding or metabolites, the distribution of charged ions is influenced only by pH (K1, k2: transport constants)

Fig. 15.5. Mechanism of intracellular binding of nitroimidazoles (e.g., FMISO). Reductase enzymes convert the parent compound into a nitro radical anion which is bound intracellularly. In the presence of oxygen, the parent compound is generated. (Modified from MARTIN et al. 1989)

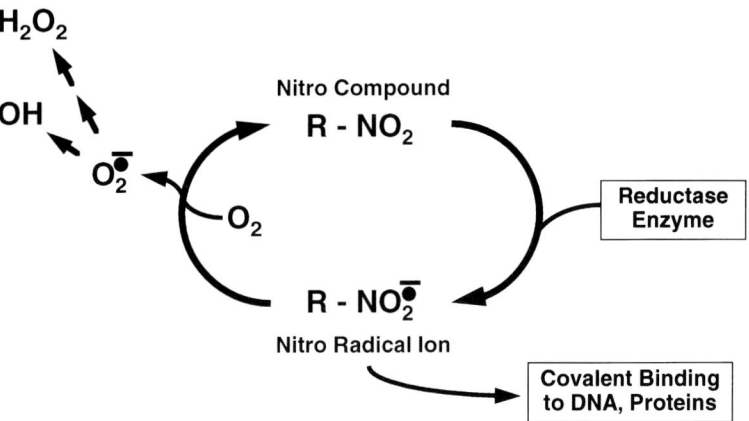

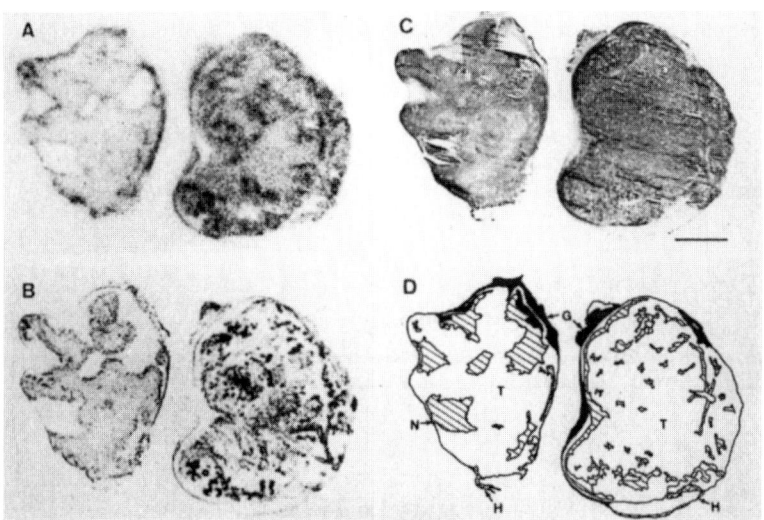

Fig. 15.6A–D. Autoradiograms with FDG (**A**), [³H]thymidine (**B**), a photomicrograph of the specimen (**C**), and the illustration of the micrograph (**D**) of xenotransplanted FM3A (*left*) and MH134 (*right*) tumors. *T*, tumor cells; *G*, granulation tissue; *N*, necrosis; *H*, host normal tissue (mouse). (Reprinted by permission of the Society of Nuclear Medicine from: Kubota, et al. Active and passive mechanisms of fluorine-18 fluorodeoxyglucose uptake by proliferating and prenecrotic cancer cells in vivo: a microautoradiographic study. J Nucl Med 1994;35:1068. Figure 1)

assessment of changes in oxygen metabolism during and after therapeutic interventions.

Fluoromisonidazole (FMISO) is a fluorinated analogue of the radiosensitizer misonidazole. Although unproven, the favored mechanism of drug binding involves reduction of the nitro group to a reactive anion radical. The nitro anion radical is a metabolite which, when reduced by a second electron, binds to intracellular molecules. In the presence of oxygen, the favored reaction is reformation of the less reactive parent compound in a futile cycle. The parent compound is freely diffusible and clears from tissue as a function of time (Fig. 15.5; MARTIN et al. 1989). Necrotic tissue is believed to fail in binding these compounds because enzymatic capacity is lacking

(MARTIN et al. 1992). When oxygen is reduced, the parent (FMISO) is quickly regenerated and metabolites do not accumulate (KOH et al. 1991). After evaluation in vitro by RASEY and her group (1990) in tumor cells, they and KOH et al. (1991) investigated a variety of human tumors, demonstrating a close correlation between the direct measurements by oxygen electrodes and PET-derived data. Up to the present, the overall results with ¹⁸F-misonidazole are controversial, since the exact uptake mechanism is not yet known. Nevertheless, this technique might provide a non-invasive tool for selecting patients likely to benefit from radiosensitizers in future if further studies can clarify the meaning of the [18F]misonidazole signal.

15.2.2.2
Glucose Metabolism

The most commonly used metabolic tracer for PET imaging is ^{18}F-deoxyglucose (FDG; IDO et al. 1978). Initially, ^{14}C-labeled deoxyglucose (DG) was employed by SOKOLOFF et al. in animal experiments using autoradiography (Fig. 6) to develop a tracer kinetic model method for determining cerebral glucose utilization (SOKOLOFF et al. 1977). Deoxyglucose is transported into tissue and phosphorylated in a manner identical to glucose. However, DG-6-phosphate is trapped in the cells and accumulates in proportion to exogenous glucose utilization (Fig. 15.7). Thus, the use of deoxyglucose represents an approach with a substrate analogue aimed at isolating a segment (transport and phosphorylation) of a complex reaction sequence and permitting accurate estimates of the rate of these processes. The labeling of deoxyglucose with ^{18}F provided a positron-emitting form of this glucose for its application in human subjects.

One of the major advantages of PET is the ability to measure substrate metabolism quantitatively. A three-compartment tracer kinetic model for FDG has been developed with rate constants for transport as well as phosphorylation, allowing the quantitative assessment of regional glucose utilization (MARSHALL et al. 1983). These metabolic measurements using FDG have been validated in the brain and heart, demonstrating the possibility of in vivo quantification of glucose metabolism in these organs by PET (KRIVOKAPICH et al. 1982; RATIB et al. 1982).

FDG uptake in tissue is influenced by plasma glucose levels and the relationship of glucose and FDG with respect to transport and phosphorylation. In the above-mentioned model, both plasma glucose levels and a correction factor (lumped constant) for the different biochemical behavior of FDG are included. This constant corrects differences in K_m values of FDG and glucose for transport (glucose-transporter) and phosphorylation (hexokinase reaction) in tissue. The lumped constant has been defined for heart and brain tissue, but no data are available for tumor tissue. Since transport proteins for glucose may vary from tissue to tissue, the application of the lumped constant defined in brain or heart tissue may lead to erroneous estimates of glucose utilization. Further studies are necessary to validate the FDG method for the quantification of tumor glucose utilization rate in tumor tissue using the established tracer kinetic model. Simplification of the FDG method using the Patlak approach yields a FDG tissue incorporation rate corrected for plasma glucose levels (PATLAK et al. 1983). The advantage of this method is the reduction of FDG kinetics in blood and tissue to a linear function that can be determined on a pixel-by-pixel basis from dynamic PET images. Such data can be used to generate parametric images (Fig. 3B,C) representing regional glucose utilization in milligrams per minute per gram of tissue. However, calculation of glucose utilization based on Patlak data also requires the lumped constant. A detailed discussion of the FDG tracer kinetic model is provided in reviews (HAWKINS et al. 1992b; HUANG and PHELPS 1986).

^{18}F Deoxyglucose

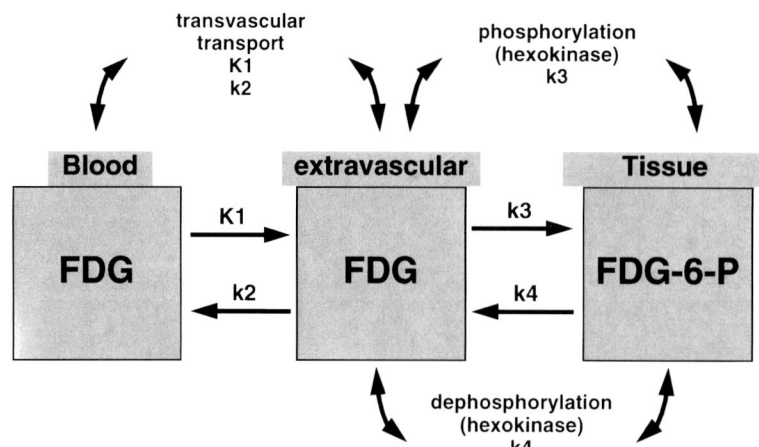

Fig. 15.7. Compartment model for ^{18}F-deoxyglucose (*FDG*). This model describes the transvascular transport (*K1, k2*) as well as the phosphorylation (*k3, k4*) of FDG in tissue

A "semiquantitative" method for describing tissue FDG uptake has been introduced by Strauss et al. The standardized FDG uptake value (SUV) relates ^{18}F tissue concentration at 40–60 min after tracer injection to injected FDG dose and body weight. This simple parameter is widely used to quantitate FDG uptake in tumors:

$$SUV = \frac{\text{tissue concentration } (nCi/g)}{\text{injected dose } (nCi)/\text{body weight } (g)}$$

Calculation of SUV does not take into consideration the competition of "cold" glucose and FDG for transport and phosphorylation. This limitation applies specifically to patients with diabetes mellitus and high plasma glucose levels. Previous studies have demonstrated a relationship between SUV values and plasma glucose levels (LANGEN et al. 1993; WAHL et al. 1992). Furthermore, normalization of FDG uptake relation to body weight may not reflect the varying relative contribution of glucose-utilizing cells to overall tracer body retention. However, the simple determination of FDG uptake by SUV technique may be sufficient for clinical determination of empirical values distinguishing benign from malign tissue as shown in several studies (see review in STRAUSS and CONTI 1991).

15.2.2.2.1
CLINICAL APPLICATION OF [^{18}F]FDG PET
Increased glycolysis is considered to be the most important and specific metabolic characteristic of cancer cells (WARBURG 1925). Therefore, noninvasive determination of glucose metabolism by [^{18}F]FDG PET in tumor tissue has been proposed for following indications:

1. Detection and staging of cancer
2. Noninvasive assessment of the grade of differentiation
3. Noninvasive assessment of proliferative activity
4. Detection of recurrent tumors
5. Therapy monitoring

Table 15.3. Sensitivity and specificity of [^{18}F]FDG PET

Tumor	n	FDG PET		Control			References
		Sens. (%)	Spec. (%)	Sens. (%)	Spec. (%)	Method	
Proven diagnostic accuracy							
Brain tumor	39	94	80	n.g.	n.g.	Biopsy	Numerous, e.g., MACAPINLAC 1994
Lung cancer	88	97	89	52	69	CT,	Numerous, e.g.,
				48	64	MR	COLEMAN 1994
Breast cancer	321	96	96	100	100	Biopsy	Numerous, e.g., ADLER 1994
Head and neck cancer							Numerous, e.g.,
Primary tumor	30	97	n.g.	77	n.g.	MR	REGE et al. 1994
Lymph node metastases	32	94	n.g.	91	n.g.	MR	REGE et al. 1994
Pancreatic cancer	40	93	85	n.g.	n.g.	Biopsy	BARES et al. 1994
Promising first results							
Ovarian cancer							
Primary tumor	177	87	94	70	42	CT	HUBNER et al. 1994
Recurrent tumor	24	93	100	73	38	CT	HUBNER et al. 1994
Lymphoma	49	100	100	n.g.	n.g.	CT	WAHL 1994
	12	86	80	n.g.	n.g.	Biopsy	UYGUR et al. 1994
Bone metastases	20	90	100	n.g.	n.g.	Bone scan	TSE et al. 1994a
Melanoma	15	100	100	n.g.	n.g.	n.g.	WAHL 1994
Rectal carcinoma							
Primary tumor	24	100	85	70	43	CT	GUPTA 1994
Metastases	16	90	66	60	100	CT	GUPTA 1994

CT, Computer tomography; MR, magnetic resonance imaging; n.g., not given.

Detection and Staging of Cancer by [¹⁸F]FDG PET. Because of promising experimental and clinical results, [¹⁸F]FDG PET is beginning to mature as an important clinical diagnostic modality in oncology. Numerous studies, including those on nearly every human tumor, are currently under way to investigate the diagnostic accuracy of [¹⁸F]FDG PET imaging in detection and staging of cancer. Preliminary results demonstrate that [¹⁸F]FDG PET yields high sensitivity and specificity in the detection of cancer in selected patient groups (see Table 15.3). However, nontumoral accumulation of [¹⁸F]FDG in abscesses and other inflammatory processes has been reported (JONES et al. 1994; KUBOTA et al. 1994), limiting the specificity of the [¹⁸F]FDG signal for tumor tissue. A detailed discussion of clinical results is offered by reviews by COLEMAN, STRAUSS, and HAWKINS (COLEMAN 1991; HAWKINS et al. 1991; STRAUSS and CONTI 1991). An example of increased [¹⁸F]FDG uptake in lung cancer is given in Fig. 15.2

and a whole-body scan of a patient with sarcoidosis is shown in Fig. 15.8.

Although the first clinical results are very promising, most studies comprise small patient populations. Additionally, there are few prospective studies which compare the diagnostic accuracy of [¹⁸F]FDG PET and morphologically orientated diagnostic procedures in large nonselected patient groups. Such studies are necessary to define the cost/benefit ratio of PET in the diagnostic workup of patients with suspected or proven cancer.

Noninvasive Assessment of the Grade of Differentiation. It is well established that the grade of differentiation has major impact on prognosis and influences the choice of therapy. Several groups investigated the relationship between [¹⁸F]FDG uptake and the grade of differentiation in primary tumors and metastases. Contradicting results have been described. In glioma studies, the glucose utilization rate showed a positive correlation to the pathologic grade of the glioma (DI CHIRO et al. 1982) and to patient survival time (PATRONAS et al. 1985). ALAVI et al. (1988) reported that prognosis of patients with primary brain tumors with high [¹⁸F]FDG uptake was worse than with bradytrophic brain tumors. Among the high-grade glioma patients, PET results allowed division into a group with better prognosis (bradytrophic, 78% 1-year survival) and a group with poor prognosis (tachytrophic, 29% 1-year survival). These results suggest that glucose-metabolic studies of brain tumors may provide an independent measure of aggressiveness and supplement histopathologic grading. In musculoskeletal tumors as well (ADLER et al. 1990, 1991; KERN et al. 1988), a significant correlation between [¹⁸F]FDG uptake and grade of malignancy has been reported.

However, other authors could not confirm the correlation between the rate of glucose utilization and the grade of glioma (TYLER et al. 1987) nor observe such a relationship in many other tumors (LESKINEN et al. 1991b; MINN et al. 1988a). Since glucose uptake reflects overall substrate metabolism, the relationship between [¹⁸F]FDG uptake and tumor grade may be complex and vary from tumor to tumor.

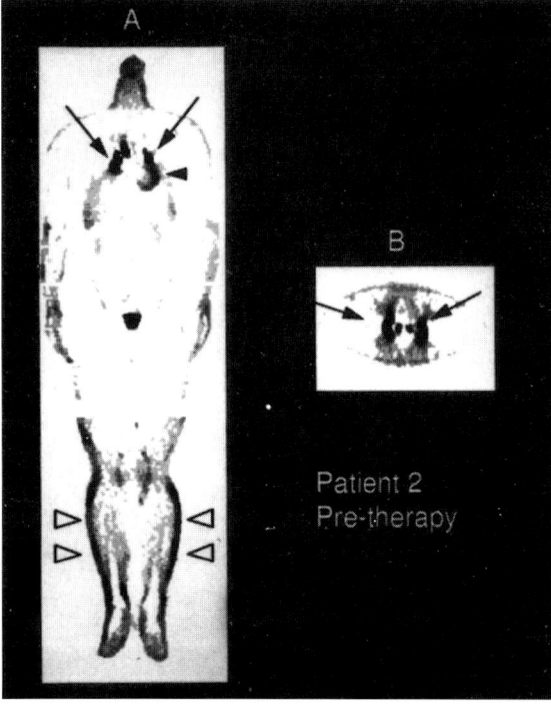

Fig. 15.8A,B. Whole-body FDG scan of a patient with sarcoidosis: coronal (**A**) and transaxial (**B**) slice through the lung hilum showing intense FDG uptake by hilar lymph nodes (*arrow*) as well as uptake into erythema nodosum on the lower limbs (*open arrowhead*). The heart is indicated by the *closed arrowhead*. (Reprinted by permission of the Society of Nuclear Medicine from: Lewis, et al. Uptake of fluorine-18-fluorodeoxy-glucose in sarcoidosis. J Nucl Med 1994;35:1648. Figure 2)

Noninvasive Assessment of Proliferative Activity. Considerable efforts have been devoted to elucidating the role of an increased glycolytic rate as a marker of cell proliferation. Recently, new cell biological techniques of flow cytometry have been introduced by which DNA content of cells in suspension can be analyzed and the percentage of proliferating cells calculated from the DNA histogram. Immuno-histochemical labeling with the antibody Ki-67 also provides an estimation of the growth fraction of the tumor. Both techniques require tissue samples, making these procedures invasive and thus not suitable for routine clinical use. PET would avoid this invasiveness, allowing for repeated intraindividual measurements. However, similar, contradictory results as to the assessment of the histologic differentiation of tumors have been found for the evaluation of proliferative activity by [^{18}F]FDG PET.

In animal studies, Sweeney et al. have shown that the glycolytic rates correlated with the rates of tumor growth in a series of Morris minimum deviation hepatomas (SWEENEY et al. 1963). In rat brain tumors, the high glucose utilization area correlated well with the distribution of BUdR-positive nuclei (WATANABE et al. 1989), indicating a positive correlation of [^{18}F]FDG uptake with proliferative activity.

There are data from patients with head and neck cancers that demonstrate a clear relationship between the proliferative rates of cells in the biopsied samples as measured by flow cytometry (GRIFFETH et al. 1992; MINN et al. 1988a) and the intensity of [^{18}F]FDG uptake. A recent report by Haberkorn also suggested that, at least in a subset of head and neck cancer patients, there is a relationship between the proliferative rate of the tumor and the extent of tumor [^{18}F]FDG uptake. Rapidly proliferating tumors displayed increased levels of [^{18}F]FDG uptake (HABERKORN et al. 1991b). Additionally, the Ki-67 labeling index in head and neck tumor showed good correlation to [^{18}F]FDG uptake (OKADA et al. 1992). Similarly, an inverse correlation between glucose utilization and tumor doubling time was found in meningiomas (DI CHIRO et al. 1987). In contrast to these results, experimental data indicated that the increased glucose transport seen in malignant cells was independent of tumor growth rate and transformation-specific (WEBER et al. 1984).

Another indicator of the nonspecificity of [^{18}F]FDG accumulation in tumor cells is that it also accumulates in intratumoral macrophages and granulation tissue (KUBOTA et al. 1992c, 1994) and represents neutrophilic activity as shown in rabbits with experimental pneumonia (JONES et al. 1994). HIGASHI et al. (1993) demonstrated that tumor [^{18}F]FDG uptake is an excellent indicator of the total number of viable tumor cells, but it does not reflect the growth rate of the tumor cells, as studied in HTB77IP3 ovarian carcinoma cells. In an editorial to this report, KUBOTA et al. (1993a) reviewed currently available data and suggested a model for [^{18}F]FDG and thymidine uptake in tumor tissue. In this model, tumor tissue is characterized as a mixture of several cell types. Tumor tissue can be divided (a) into neoplastic tissue, consisting of proliferating cells, viable but nonproliferating cells, and necrotic tissue, and (b) into non-neoplastic tissue, consisting of macrophages, neutrophils, stroma, fibroblasts, and scar tissue. Except necrotic and scar tissues, all cell types accumulate [^{18}F]FDG. Since S-phase-proliferating neoplastic tissue and fibroblasts are only a subset of viable cells, the resulting [^{18}F]FDG signal only partially represents the proliferative activity of the tumor (Fig. 15.9).

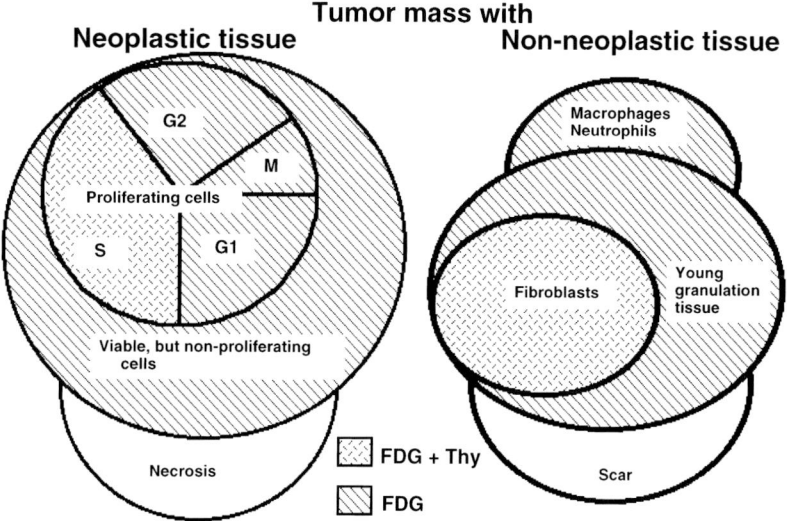

Fig. 15.9. Comparison of FDG and thymidine (*Thy*) accumulation in various cellular elements in a tumor. (Modified from KUBOTA et al. 1993)

Detection of Recurrent Tumors. The differentiation between recurrent tumor and scar is often a major clinical problem in the follow-up of tumor patients after initial treatment. Most experience with PET has been gained in brain tumors and colorectal tumors. As early as 1982, Patronas was able to differentiate radiation necrosis from recurrent brain tumors by PET (PATRONAS et al. 1982). To date, it is generally accepted that [^{18}F]FDG PET is able to distinguish between therapy-induced necrosis or fibrosis and persistent or recurrent brain tumors (COLEMAN et al. 1991; DI CHIRO et al. 1988). For extracranial tumors, the results are controversial. Strauss et al. reported a high diagnostic accuracy in assessing recurrent colorectal cancer (STRAUSS et al. 1989). Although there are no data on the nature of the posttreatment [^{18}F]FDG-PET signal, it has been speculated that

some of the residual [^{18}F]FDG uptake after radiation may be due to an inflammatory response to radiation (HABERKORN et al. 1991a). In summary, available data indicate that the interpretation of an increased uptake of [^{18}F]FDG after therapy is strongly dependent on the time interval between the end of therapy and the PET study. Short-term studies may have high false-positive results due to inflammatory processes, while high [^{18}F]FDG uptake observed a long time after therapy (at least 6–8 weeks) should be indicative of tumor recurrence.

Therapy Monitoring. Changes of regional metabolic rate for glucose after therapeutic interventions have been investigated with [^{18}F]FDG in a variety of human tumors. The underlying hypothesis is that biochemical changes will occur earlier than a detectable

Table 15.4. Therapy monitoring with [^{18}F]FDG

Tumor	n	Intervention	References
Cell			
MCF		CT	HABERKORN et al. 1992
Premonocytic line		CT	SLOSMAN et al. 1993
Animal			
Murine lung tumor		CT	MINN et al. 1992
Rat, AH109A		RT	KUBOTA et al. 1991
Rat, rhabdomyosarcoma		RT	DAEMEN et al. 1992
Murine breast cancer, rat hepatoma		RT	ABE et al. 1986
Human			
Brain tumor	11	CT	OKADA et al. 1991
Brain tumor	10	CT	LANGEN et al. 1989
Brain tumor	4	RT	ROZENTAL et al. 1991a
Brain tumor	9	Various	TOMURA et al. 1990
Glioma	2	RT	GRIEBEL et al. 1991
Glioma	14	RT + CT	ROZENTAL et al. 1991b
Glioma	6	CT	ROZENTAL et al. 1993
Glioma	13	RT + CT	OGAWA et al. 1988
Pituitary microadenoma	4	CT	FRANCAVILLA et al. 1991
Head and neck	19	RT	CHAIKEN et al. 1993
Head and neck	18	CT	HABERKORN et al. 1993
Head and neck	25	RT	GREVEN et al. 1994
Head and neck	11	RT	REGE et al. 1993
Head and neck	19	RT	MINN et al. 1988b
Breast cancer	10	RT + CT	MINN and SOINI 1989
Breast cancer	11	CT	WAHL et al. 1993
Lung cancer	5	RT + CT	ABE et al. 1990
Lung cancer	18	RT + CT	STRAUSS and CONTI 1991
Liver tumor	35	RT + CT	OKAZUMI et al. 1992
Liver tumor	17	CT	NAGATA et al. 1990
Liver tumor	8	CT	SMITH et al. 1992
Recurrent colorectal cancer	44	RT	HABERKORN et al. 1991a
Various	26	Various	ICHIYA et al. 1991

CT, Chemotherapy; RT, radiotherapy.

decrease in tumor size. In tumor-bearing animals, a dose-related decrease of [^{18}F]FDG uptake has been observed, at least for irradiated tumors (ABE et al. 1986; DAEMEN et al. 1992). Only limited data on the response in human neoplasms are available. Although there is a tendency towards a more marked decrease of [^{18}F]FDG uptake in radiosensitive than in radioresistant tumors, a decrease in [^{18}F]FDG uptake has also been recorded in tumors with a poor clinical response (ICHIYA et al. 1991; MINN et al. 1988b). In contrast, other groups found that, independent of other intervening treatment procedures, irradiation always results in a decreased [^{18}F]FDG uptake (FRANCAVILLA et al. 1991). Likewise, differing results were obtained in the assessment of response to chemotherapy. Changes in the uptake of [^{18}F]FDG have been seen which appeared to depend on the schedule of drug administration and the interval between therapy and PET studies (LANGEN et al. 1989; ROZENTAL et al. 1991b). In summary, timing of the [^{18}F]FDG scan after therapy seems to play a critical role in assessing therapeutic efficacy (HABERKORN et al. 1991a).

Table 15.4 lists oncologic studies with [^{18}F]FDG PET before and after therapy. The range of tumor types is broad, yet the number of patients small. In addition, the treatment regimes vary both within and between studies. To date, only few data are available to define the clinical role of [^{18}F]FDG PET studies in therapy monitoring. Future investigations have to define the timing of PET studies following therapy and evaluate the feasibility of [^{18}F]FDG imaging to assess the effect of various cytotoxic or cytostatic interventions. It is expected that metabolic imaging with pharmaceuticals such as [^{18}F]FDG may provide an important endpoint in the early definition of therapy effect in oncology.

15.2.3
Anabolic Metabolism

15.2.3.1
Protein Synthesis/Amino Acid Uptake

It is generally accepted that malignant cells demonstrate an increased level of amino acid uptake and protein metabolism (BUSCH et al. 1959). Therefore, PET using labeled amino acids has prominent clinical applications in the field of oncology, since it offers the opportunity of quantitative in vivo measurement of amino acid uptake and metabolism (protein synthesis rates, PSR) in tissues, provided

appropriate kinetic models are developed. As early as the late 1970s and early 1980s, ^{11}C- or ^{13}N-labeled amino acids have been proposed for studying amino acid metabolism (HENZE et al. 1982). A theoretical limit to this method was proposed, since the short half-life of these radionuclides as compared to the "slow" rate of amino acid metabolism may limit the accuracy of this technique. However, the amino acid turnover in tumors is high, allowing demonstration of these pathophysiologic conditions with marked changes in protein synthesis by PET.

To date, a great number of positron-emitting amino acids has been synthesized and used for animal and clinical studies. The amino acids most widely used for the evaluation of amino acid uptake and metabolism are methionine, tyrosine, leucine, and phenylalanine, which will be described below in detail.

15.2.3.1.1
METHIONINE

Methionine, an essential sulfur-containing amino acid, is the most important supplier of methyl groups; it reacts with ATP to adenosine-methionine, whose methyl group is used for many biosynthetic pathways. It also serves as supplier of sulfur for protein anabolism. The main metabolic pathway of L-[^{11}C]methyl-methione is protein incorporation (ISHIWATA et al. 1988a). The influence of minor metabolic pathways on measuring PSR by PET is still under discussion (JONES et al. 1985; LUNDQVIST et al. 1985). For instance, in brain, L-methionine is a precursor of S-adenosyl-L-methionine (BALDESARINI and KOPIN 1986), which plays an important role in biochemical transmethylation processes (ANDREOLI et al. 1978; USDIN et al. 1979).

Since the labeling of L-methyl-methionine with ^{11}C is simple and reliable and the radiochemical yield high, the majority of PET studies with amino acids have been performed with L-[^{11}C]methyl-methionine since the first report of COMAR et al. in 1976.

Clinically, this amino acid has been applied to a variety of tumors including brain tumors (ISHIWATA et al. 1993a), gliomas (OGAWA et al. 1993), pituitary adenomas (BERGSTROM et al. 1991), prolactinomas (DAEMEN et al. 1991b), breast cancer (HUOVINEN et al. 1993; LESKINEN et al. 1991a), head and neck cancer (LESKINEN et al. 1992), lung cancer (FUJIWARA et al. 1989; KUBOTA et al. 1992b), and noncentral lymphomas (LESKINEN et al. 1991b), demonstrating the potential of this agent to map the distribution of viable tumor cells, both in the brain and in tumors outside the CNS. Nearly all of these studies have been performed to image and detect tumors and metastases, demonstrating high sensitivity and specificity of this method (see overview in HAWKINS et al. 1992c). However, to date there are no prospective studies comparing the usefulness of methionine-PET and conventional imaging methods in tumor detection.

A few studies have evaluated the use of [^{11}C]methionine in therapy monitoring during radiation (KUBOTA et al. 1989b), demonstrating a sharp decrease in methionine uptake during therapeutic intervention. These preliminary results indicate that [^{11}C]methionine uptake correlates well with the loss of viable tumor cells. The possibility of differentiation between residual masses and recurrent lung cancer in 21 patients was demonstrated by KUBOTA et al. (1993b), while SAWATAISHI et al. (1992) and OGAWA et al. (1994) reported similar results in patients with primary brain lymphomas.

To date, there have been only few studies to compare the sensitivity and specificity of [^{18}F]FDG and [^{11}C]methionine PET imaging. In 24 patients with suspected lung cancer, no significant differences were observed between the two tracers, suggesting that PET studies using either methionine (Met) or [^{18}F]FDG may be useful in the differential diagnosis of lung tumors (KUBOTA et al. 1990). In 15 patients clinically suspected of having recurrent brain tumor or radiation injury, both tracers were also used. PET with [^{11}C]methionine clearly delineated the extent of recurrent brain tumor as focal areas of increased accumulation of [^{11}C]methionine and was useful in early detection of recurrent brain tumor. In these patients, PET with [^{18}F]FDG also proved useful in the evaluation of recurrent tumors. Low [^{18}F]FDG uptake was observed in all patients with radiation injury, but was also found in one patient with recurrent malignant brain tumor. Overall, the authors of this study suggested the combined use of Met-PET and [^{18}F]FDG PET to improve the accuracy of differentiation between recurrent brain tumor with low uptake of [^{18}F]FDG and radiation injury (OGAWA et al. 1991). In 14 patients with head and neck cancer, 19 of 21 malignant lesions that could be evaluated were visible with both tracers. Tracer uptake was measured as standardized uptake values (SUV) and as K_i values according to PATLAK et al. The mean SUV was similar in both tracers, whereas the K_i values in [^{11}C]methionine studies were always higher than in [^{18}F]FDG studies. Both [^{18}F]FDG and [^{11}C]methionine were effective in PET imaging of head and neck cancer (LINDHOLM et al. 1993). Leskinen compared the usefulness of both tracers in 14 patients with non-Hodgkin's lymphomas, demonstrating a significantly higher uptake rate for [^{11}C]methionine than for [^{18}F]FDG. Especially intermediate- and low-grade lymphomas had poor [^{18}F]FDG uptake. [^{11}C]methionine seemed to be preferable in detecting tumors, while [^{18}F]FDG was superior to [^{11}C]methionine in distinguishing the high-grade malignant lymphomas from the other grades (LESKINEN et al. 1991b).

15.2.3.1.2
TYROSINE

Tyrosine is an aromatic amino acid and an important precursor for the synthesis of many proteins (ISHIWATA et al. 1988b). By labeling tyrosine with ^{11}C, a PET-suitable tracer was found. As for all other labeled amino acids, due to the missing validation of kinetic models and metabolic pathways, it is not clear to date whether PET with [^{11}C]tyrosine measures PSR adequately or whether its uptake primarily reflects amino acid uptake systems. Preliminary studies on the biodistribution of ^{11}C-labeled tyrosine have been performed in Walker 256 carcinosarcoma-bearing rats (DAEMEN et al. 1991a) and in mice bearing mammary carcinomas (ISHIWATA et al. 1993b), demonstrating high tumor uptake. In rat rhabdomyosarcomas (DAEMEN et al. 1992), a decrease in tracer uptake after radiation therapy has been shown. Clinically, this tracer has already been used for detection of ocular melanomas (VAN LANGEVELDE et al. 1988) and prolactinomas (DAEMEN et al. 1991b), confirming the high tracer uptake demonstrated in animal studies. Because of the relatively slow PSR in the primate brain, COENEN et al. (1989) proposed that L-2-[^{18}F]fluorotyrosine with a longer-lived radionuclide would be a more suitable tracer to study PSR. Brain uptake of [^{18}F]tyrosine increased with time and the tracer was incorporated into cerebral proteins without significant catabolism. These properties could be rather promising for measuring PSR and contrast with the properties of the parent amino acid tyrosine. In contrast, brain uptake of [^{14}C]tyrosine decreased over time, probably because of decarboxylation superimposed on increasing radioactivity due to protein incorporation (ISHIWATA et al. 1988b).

15.2.3.1.3
LEUCINE

Leucine is an essential amino acid which plays an important role in intermediate protein metabolism.

In the brain, radiolabeled leucine has been used to measure protein synthesis using tracer kinetic modeling (PHELPS et al. 1984). ISHIWATA et al. (1993b) investigated brain and tumor uptake of L-[³H]methyl-methionine ([³H]Met), L-1-[¹⁴C]leucine [¹⁴C]Leu) and L-2-(¹⁸F)fluorotyrosine ([¹⁸F]Tyr) and their incorporation into the acid-precipitable fraction (APF) in mice bearing mammary carcinomas in order to compare the potential of positron-emitting analogues for measuring PSR in these tissues. ¹¹C-labeled leucine had the fastest incorporation rate. Ishiwata proposed that, due to the short half-life of the positron-emitting radionuclides, this amino acid might be the most suitable for measuring PSR by PET.

15.2.3.1.4
PHENYLALANINE

Phenylalanine, an essential amino acid, is metabolized in vivo to oxaloacetic acid and acetyl-CoA. Besides the wide use of [¹¹C]L-3,4-dihydroxyphenylalanine (L-DOPA) in studies in neurology, especially in Parkinson's patients, it has also been used to detect melanoma in an animal tumor model (VAN LANGEVELDE et al. 1988). A fluorinated phenylalanine derivate (L-2-[¹⁸F]fluorophenylalanine) has been tested in normal volunteers and patients with brain tumors to assess the metabolic rate of phenylalanine in these tissues (MIURA et al. 1992). Interestingly, with another ¹⁸F-labeled phenylalanine compound, 4-borono-D-L-phenylalanine (BPA; ISHIWATA et al. 1991b), which is a potential target for cancer treatment with boron neutron capture therapy, initial biodistribution determinations in mice bearing mammary carcinoma and melanomas have been undertaken (ISHIWATA et al. 1991a, 1992). These preliminary results confirmed that the ¹⁰B concentrations in tissue, which are crucial for this kind of radiation therapy, can be assessed in vivo.

In summary, there is general agreement on the clinical usefulness of PET with labeled amino acids in tumor visualization and staging, but for quantitative measurement of PSR, several problems regarding the above-mentioned metabolic pathways are unresolved. An experimentally validated compartment model for quantitation of PSR has not yet been defined. For therapy monitoring, the same basic problems as for [¹⁸F]FDG exist. Especially for evaluation of tumor response to therapy, in which quantitative measurements are important, further investigations with radiolabeled amino acids are required.

15.2.3.2
DNA Replication

DNA replication is a direct measure of cell proliferation. Cell proliferation is an important prognostic factor in malignant disease (MEYER et al. 1983) and may have therapeutic implications (SHACKNEY et al. 1978), both in terms of diagnostic utility and in predicting and monitoring responses to treatment.

Nucleosides, such as thymidine or uridine, are components of DNA. PET with such labeled tracers has the potential for noninvasively measuring DNA synthesis. In order to validate this potential, one needs to construct accurate biochemical models that reflect the metabolism of such tracers, including their uptake and degradation as well as their incorporation into DNA (SHIELDS et al. 1990b).

15.2.3.2.1
THYMIDINE

Thymidine (TdR), a nucleoside, is used in DNA synthesis and is taken up by dividing cells. PET with [¹¹C]TdR provides an opportunity to measure the rate of uptake of TdR noninvasively in vivo. The uptake of TdR can be regarded as an index of cell proliferation (HUGHES et al. 1958; LEA et al. 1966).

In oncology, [¹¹C]TdR has already been used for visualization of head and neck tumors (VAN EIJKEREN et al. 1992), non-Hodgkin's lymphoma (MARIAT et al. 1988), and gliomas (SATO et al. 1992). Early promising results in monitoring the response to radiotherapy have been shown in a human adenocarcinoma cell line (HIGASHI et al. 1994) and in a rat AH109A tumor model (KUBOTA et al. 1991). However, there are still some basic problems in the evaluation of this tracer. The metabolic pathway found in proliferating tissues of mice and dogs demonstrated different amounts of metabolites (SHIELDS et al. 1990b). Moreover, the same group reported that the ¹¹C-labeled metabolites CO_2/HCO_3^- are not eliminated from the body rapidly enough for their contribution to PET images to be neglected (SHIELDS et al. 1992). The use of [¹¹C]TdR for tumor imaging has been demonstrated, but the metabolic pathways have to be determined in detail for quantitative measurement of DNA synthesis by tracer kinetic modeling.

A new therapy regimen is "suicide-gene-therapy." The most widely used system is based on the principle of transducting herpes simplex virus type 1 TdR kinase (HSV1-TK) genes into the genome of targeted cells. Under normal circumstances, this gene is blocked by a regulatory sequence. Upon infection or alteration of such a targeted cell by a promoter (for example, oncogenes, HIV), the regulatory sequence switches to high-level expression of HSV1-TK. The second step is the application of aciclovir or ganciclovir, nucleoside analogues, that become toxic after phosphorylation by HSV1-TK. Thus, high levels of toxic, phosphorylated compounds accumulate in infected or transformed cells and cause cell destruction. Promising first results have been demonstrated in HIV-infected CD4 cells (CARUSO and KLATZMANN 1992) and in murine CT26 adenocarcinoma (PLAUTZ et al. 1991). A possible future application of PET with [^{11}C]TdR (or an analog substrate of thymidin-kinase) may be the monitoring of this "suicide-gene-therapy" noninvasively in vivo.

15.2.3.2.2
FLUOROURIDINE

The same is true for uridine, another nucleoside consisting of uracil and ribose. The metabolic fate of the fluorine-labeled analogue 2'-deoxy-5-[^{18}F]fluorouridine (FdUrd), a PET tracer of nucleic acid metabolism in tumors, was investigated in murine FM3A mammary carcinoma and in humans (ISHIWATA et al. 1991). These early results indicate, that [^{18}F]FdUrd is a promising tracer of nucleic acid metabolism for evaluating the proliferative potential of tumors (SATO et al. 1992).

15.2.4
Characterization of Cell Membrane Function

Cell membrane properties can be characterized by determination of receptors and antigens located on the cell surface. In conventional nuclear medicine, with single-photon emitters, both methods have been widely evaluated. An example of a SPECT receptor tracer is the ^{111}In-labeled somatostatin analogue pentatreotide (Octreo-Scan) which binds specifically to somatostatin receptors, primarily located on GEP tumors. Additionally, many antibodies have been labeled with ^{123}I for radioimmunodiagnosis and with ^{131}I for radioimmunotherapy.

By labeling these compounds with positron emitters and by the unique advantage of PET, the absolute quantification of radioactivity in tissue, it should be possible to quantify noninvasively the distribution of receptors and antigens in humans.

15.2.4.1
Receptor Imaging

Radioactive-labeled receptor agonists or antagonists are commonly used for the in vitro assay of receptor systems. The use of positron-emitting receptor ligands allows the in vivo measurement of receptor systems, and a considerable number of receptor ligands have been synthesized (Table 15.5). It is feasible to localize cerebral dopamine, opiate, serotonin, benzodiazepine (BDZ), and muscarinic cholinergic receptors using PET. Peripheral receptor systems studied by PET include the alpha- and beta- adrenergic receptors as well as the muscarinic cholinergic receptors (ROBINSON et al. 1987; SYROTA et al. 1983). Most of these studies were performed in the heart and have not been extended to other organs.

Initially, the relative distribution of ligand binding was explored and related to the in vitro studies

Table 15.5. Receptor ligands imaged in the human brain by PET

Receptor type	Ligand	Reference
Dopamine, D$_2$	[^{18}F]Spiperone	PERLMUTTER et al. 1987
	[^{11}C]Methylspiperone	WAGNER et al. 1983
	[^{18}F]Methylspiperone	ARNETT et al. 1986
	[^{76}Br]Bromospiperone	BARON et al. 1986
	[^{11}C]Pimozide	BARON et al. 1985
	[^{11}C]Raclopride	FARDE et al. 1986
Opiate	[^{11}C]Carfentanil	FROST et al. 1985
Cholinergic, muscarinic	[^{11}C]Scopolamine	FREY et al. 1987
Benzodiazepine	[^{11}C]Ro 15-1788	PERSSON et al. 1985
	[^{11}C]Suriclone	
Serotonin, S$_2$	[^{11}C]Methylspiperone	ROBINSON et al. 1987
	[^{11}C]2-Br-LSD	

employing autoradiography (YOUNG et al. 1986). These more qualitative studies allow the definition of the anatomic location of receptor binding as well as the comparison of tracer uptake in different organs. However, the quantification of binding sites by PET should be possible, allowing the noninvasive assessment of receptor densities (FREY et al. 1985).

Applications of in vivo receptor ligand binding methods are potentially diverse, ranging from studies of disease or therapy-related alterations in the number of binding sites to dynamic tests of synaptic function. The latter is based on measurements of changes in free receptor sites caused by altered levels of endogenous neurotransmitter. Thus, either physiologic or pharmacologic challenges that alter presynaptic activity or modify binding of the endogenous transmitter may indirectly produce changes in radioligand binding to receptor sites. Because of the short half-lives of positron-emitting radioisotopes, ligand studies can be combined with PET studies of flow or other tissue function. The combination of tracers of presynaptic function such as ^{18}F-labeled dopamine or ^{18}F-labeled norepinephrine analogues with postsynaptic ligand studies seems promising; it allows characterization of neurotransmission and differentiation of the effect of disease on pre- and postsynaptic sites (WIELAND et al. 1989, 1990). All of the above-mentioned receptors have been studied widely in neurology and cardiology.

In oncology, as early as 1980 Katzenellenbogen demonstrated the possibility that receptor-binding radiopharmaceuticals may be suitable for imaging breast tumors and studied estrogen–receptor interactions and selectivity of tissue uptake of halogenated estrogen analogues (KATZENELLENBOGEN et al. 1980). A few years later, as many as 13 patients with primary breast tumors had been studied with 16 alpha-[^{18}F]fluoroestradiol-17 beta (MINTUN et al. 1988). There was excellent correlation between uptake within the primary tumor as measured by PET and the tumor estrogen–receptor concentration measured in vitro after excision. This technique may provide an in vivo method of assessing estrogen receptors in primary and metastatic breast cancers and thus may guide management of this disease with antiestrogen chemotherapy. Recently, seven alpha-methyl-substituted estrogens have been labeled with ^{18}F as potential breast tumor imaging agents (VAN BROCKLIN et al. 1994).

A positron-emitting progestin compound, 21-[^{18}F]fluoro-16 alpha-ethyl-19-norprogesterone (FENP), has recently been evaluated by Verhagen (VERHAGEN et al. 1991) and Dehdashti (DEHDASHTI et al. 1991) for detection of progesteron-positive breast carcinomas. The latter group has already used this tracer in eight patients, but yielded only low sensitivity due to high background activity. Additionally, there was no correlation between tracer-uptake and progesterone-receptor levels, indicating that FENP is not a suitable agent for imaging progesterone receptors in humans.

In summary, due to complicated labeling techniques and unknown metabolic pathways, few studies with labeled receptors have been performed in oncology up to the present. For the future, more basic research is required to extend the use of receptor studies from cardiology and neurology to oncology.

15.2.4.2
Antibody Imaging

Radioactive-labeled monoclonal antibodies have been used to identify the location and extent of several tumors in experimental and clinical studies. It is not clear at present if the kinetics of antibody binding and the clearance of the antibody from blood allow imaging with relatively short-lived radioisotopes employed for PET imaging.

In oncology, PET is potentially useful for the quantitative imaging of radiolabeled antibodies, leading in turn to improved dosimetry in radioimmunotherapy. ^{124}I is a positron-emitting nuclide with appropriate chemical properties and half-life (4.2 days) for such studies, since the radiolabeling of antibodies with iodine is well understood and the half-life permits measurements over several days. Some preliminary studies with ^{124}I-labeled antibodies are listed in Table 15.2. Other PET radionuclides with feasible physical properties are ^{64}Cu and probably ^{18}F. With these tracers, first experience was gathered with the ^{64}Cu-labeled antibody 1A3 (colorectal carcinoma) and with ^{18}F-Mel-14 (gliomas; VAIDYANATHAN et al. 1992).

As for antibodies labeled with single-photon emitters, the basic problem of antibody imaging is the specificity of the antigens on the surface of tumor cells and the availability of the antibody in the tumor. With further basic investigations, these problems may be overcome. For the first time, a noninvasive quantitative assessment of the biodistribution of labeled antibodies by PET appears to be possible.

15.2.5
Characterization of Specific Cell Functions

Another field of possible use of PET in the future is the quantitative assessment of some specific cell functions. Some of the systems studied up to the present are reticuloendothelial function, vascular permeability, the polyamine/monoamine system, and bone metabolism.

Overall, all of the tracers used, which are listed in Table 15.2, are new and have been used only in selected cases. Whether these compounds will continue to play a role in oncology has to be investigated in further studies.

15.2.6
Tracing of Radiolabeled Cytostatic Agents (Chemotherapeutics)

Chemotherapeutic agents labeled with positron emitters are promising as tumor imaging agents. Additionally, they present an opportunity to perform in vivo numerical pharmacokinetic studies on a level not possible before the advent of PET technology. Although in vivo results in humans with malignancies are relatively limited with these agents, the potential of imaging the distribution of chemotherapeutic agents and evaluating the kinetics of these agents in vivo is promising both as a diagnostic modality and as a method of optimizing treatment protocols. Up to the present, several drugs commonly used in oncology have been labeled with positron-emitting radionuclides.

5-Fluorouracil (5-FU) is the most common cytostatic agent used for chemotherapy in patients with colorectal tumors. The biodistribution of the fluorinated PET tracer ($[^{18}F]$5-fluorouracil) has been studied in animals (Abe et al. 1983) and in patients with liver metastases from colorectal carcinoma (Dimitrakopoulou et al. 1993), indicating notable intra- and interindividual variation in uptake. The results of the later group suggest that FU retention in different metastases mainly depends on early FU uptake into tumor cells (Dimitrakopoulou et al. 1993). The same group compared perfusion, assessed by $[^{15}O]$water, and $[^{18}F]$5-FU uptake of tumors and normal liver tissue after intravenous and intraarterial injection (Hohenberger et al. 1993). The perfusion of normal liver tissue was similar after intravenous and intraarterial assessment. Metastases were found to be poorly perfused compared to normal liver tissue after i.v. examinations. By

intraarterial injection, the perfusion in metastases increased slightly. However, single metastases showed up to ten times greater perfusion with the i.a. injection route than with the i.v. one. Additionally, lesions with no change or lower perfusion were also observed. Generally, accumulation of $[^{18}F]$5-FU in metastases after i.v. infusion was less than after i.a. injection. Correlation of i.v. perfusion and uptake was moderate ($r = 0.54$, $p = 0.0001$). The authors conclude that perfusion as measured by PET with $[^{15}O]$water does not generally predict uptake of $[^{18}F]$5-FU in colorectal liver metastases. To measure the bio-availability of 5-FU in single metastases, use of PET and $[^{18}F]$5-FU seems to be the most accurate method. It would allow identification of individual patients with considerably greater accumulation of 5-FU following i.a. administration. Thus, patients, who should profit from a crossover to intrahepatic chemotherapy can be selected by PET.

The pharmacokinetics of cisplatin have been assessed by labeling this drug with ^{13}N ($[^{13}N]$cisplatin). Similar to $[^{18}F]$5-FU, it has been shown that $[^{13}N]$cisplatin is suitable for comparison of drug delivery by i.a. versus i.v. chemotherapy in malignant brain tumors (Ginos et al. 1987). Again, large intra- and interindividual differences have been found. Recently, a fluorinated analogue of tamoxifen (fluorotamoxifen) has been synthesized for studying the delivery of this anti-estrogen (Yang et al. 1994). Thus, patients can be selected who should profit from tamoxifen therapy.

An example of the basic evaluation of new chemotherapeutic drugs in vivo is provided by Mitsuki et al. (1991) who compared the pharmacokinetics of the ^{11}C-labeled, well-established drug BCNU and the newly synthesized SarCNU in gliomas. The distribution of these drugs, as measured by PET, indicates that SarCNU has a higher tumor-to-brain ratio than BCNU. Additionally, the authors could demonstrate by simultaneous use of ^{68}Ga-EDTA as an indicator of blood-brain barrier integrity that the new nitrosourea, SarCNU, most likely enters brain tissue by a different mechanism than BCNU, which enters by diffusion.

Besides the possibility of labeling cytostatic drugs, it also possible to label any other interesting drug or new pharmaceutical in medicine. For example, Livni synthesized fluorinated fluconazol, an antimycotic drug, and studied its biodistribution in rabbits (Livni et al. 1992).

In summary, experience in the use of PET-labeled drugs is currently limited. However, first results demonstrate that this technique is suitable for selecting individual patients who will profit from certain therapy regimes and certain drugs and for testing the biodistribution of new drugs in vivo quantitatively.

15.3
Current and Future Role of PET in Oncology

The multimodality approach to cancer treatment includes surgery, radiotherapy, chemotherapy, hyperthermia, and immunotherapy. Most of the data on which we base therapeutic strategies have been derived from in vitro studies or animal tumor models. More information is required on the physiology of in vivo human tumors and their response to therapy.

The particular advantages of the PET technique are its higher sensitivity, higher resolution, and higher quality of image compared to that found in conventional nuclear medicine. Another advantage is the possible use of a wide range of "biologic" tracers due to "natural" radionuclides which do not alter the biologic and physiologic properties of tracers. Additionally, PET allows the regional tissue concentration of a positron-emitting radionuclide to be measured in absolute units noninvasively in vivo if valid tracer models can be formulated that accurately describe the fate of an administered biologic tracer. To date, it has become possible to measure a

variety of physiologic and pathophysiologic parameters as listed in Table 15.2. For exact quantitative determination, generally accepted tracer models exist for the assessment of regional tissue blood flow, tissue blood volume, and oxygen utilization. It is possible to determine these parameters in absolute units as shown in Table 15.6. Other tracer kinetic models, such as the model for regional glucose utilization and for amino acid uptake and protein synthesis, are still open to criticism. In these studies, in which tracer models are thought to be less than complete (e.g., due to insufficient biological data), only a semiquantitative or qualitative assessment of the pathophysiologic state may be possible.

To date, most human PET cancer studies have been performed with [^{18}F]FDG or [^{11}C]methionine for "pure" tumor imaging. It is already generally accepted that these tracers are good imaging agents for tumors, yielding high sensitivity and specificity (Table 15.3). However, more specific radiopharmaceuticals are required if other features of tumor metabolism are to be observed. In near future, further labeled amino acids may prove to be good tracers for quantitative measurements of tumor proliferation, and the role of other tracers, such as ^{18}F-labeled misonidazole for imaging of hypoxia, may increase.

Recently, a new type of in vivo probe has been introduced in nuclear medicine: antisense oligodeoxynucleotide. It has been widely used for the

Table 15.6. Physiologic parameters that have been quantitatively studied with PET

Physiologic parameters	Units
Accepted tracer model	
Regional tissue blood flow	ml / 100 ml min
	(ml / g min)
Regional tissue blood volume	ml / 100 ml
	(ml / g)
Regional oxygen utilization	ml O$_2$ / 100 ml min
	(ml O$_2$ / g min)
Tracer model still open to criticism	
Regional glucose utilization	mg glucose / 100 ml min
	(mg glucose / g min)
Amino acid uptake/protein synthesis	nmol / 100 ml min
	(nmol / g min)
Only applicable to brain with intact BBB	
Blood-brain barrier integrity	
Regional pH	
Cerebral hematocrit	
No generally accepted tracer model	
Drug uptake	
Receptor studies	
Antibody studies	

regulation of gene expression in cell culture, plant engineering, and the design of chemotherapeutic agents. In principle, an antisense oligonucleotide can be used to target any known sense RNA. Antisense reagents are believed to block the expression of targeted proteins by hybridizing with their mRNA and preventing its translation. This Watson-Crick base pair formation confers an extremely high degree of specificity to deoxynucleic acid structure. Calculations estimate that an oligonucleotide of 15 to 17 nucleotides in length would have a unique sequence compared to the entire human genome. In principle, this specificity has been the driving force behind cur-

rent attempts to develop oligodeoxynucleotides for therapy and diagnosis of human disease (PIWNICA-WORMS 1995). Studies with ^{32}P- or ^{3}H-labeled molecules have already demonstrated that it is possible to determine this sort of "receptor-mediated" uptake in vitro by radioactive tracers (Lu et al. 1994). Other preliminary results demonstrated the possibility of noninvasively imaging c-myc oncogene messenger RNA with ^{111}In-labeled antisense probes in a mammary tumor-bearing mouse model (DEWANJEE et al. 1994). It is reasonable to combine the high specificity of these agents with the high sensitivity of PET to evaluate biodistribution and effects in human.

Table 15.7. Selection of some induced effects on genes where the in vivo effect of gene therapy could probably be measured noninvasively by PET

Species/cell	Effector/inducer	Primary effect	Secondary effect	Possible PET measurement	References
Murine myeloid cell line 32D	IL-3 deprivation	c-myc + or-nithine-decar-boxylase down-regulation	Accumulation of cells in G1-phase, eventually cell death	Ornithine-decarboxylase activity	ASKEW et al. 1991
Mouse	Peroxisome proliferator methylclofenapate	Increase of cytochrome P-450 4a-10 RNA in liver		?	BELL et al. 1993
Mouse	Decrease of intracellular glutathione by diamide	Upregulation of glutathione S transferase Ya gene expression	Induction of chloramphenicol-acetyltransferase activity	Glutathione level	BERGELSON et al. 1994
Mouse, neuroblastoma cell line (N2a)	Alpha-difluoro-methyl ornithine (suicide inhibitor of ornithine decarboxylase)	Reduction of polyamine content		Polyamine metabolism	CHEN 1991
Bacteriophage T3	Adenosyl-methionine-hydrolase-encoding gene	Lowers S-adenosyl-L. methionine level		Methionine metabolism	COLLIER et al. 1994
Mouse macrophage cell line	Lipopolysaccharide	Activation of the inducible nitric oxid synthase gene (i-NOS)	Increase of i-NOS, TNF-alpha, IFN-beta	Lipopolysaccharide metabolism	FUJIHARA et al. 1994
Mouse T-cell	Antisense c-myc oligonucleotide	Inhibition of cyclin-dependent kinase-2	Blocks T cells in the G1 phase	Radiolabeled oligonucleotide	KIM et al. 1994
Rat, VSMC	Interferon-gamma	Induction of NO synthase and generation of nitric oxide	Inhibition of pro-liferation of vascular smooth muscle cells	NO	NUNOKAWA et al. 1994
Ehrlich carcinoma cells	Ornithine	Induction of or- nithine decarboxylase gene	Increase in ODC activity	Ornithine metabolism	SANCHEZ et al. 1992
Human gingival fibroblasts	Interleukin-1	Induction of metalloproteinase genes	Increased expression of collagenase (MMP-1), stromelysin (MMP-3), plasminogen activator (PA)	?	TEWARI et al. 1994

Additionally, it seems possible to measure non-invasively the in vivo effects of new therapy regimes such as the "suicide gene therapy." Table 15.7 gives a selection of some induced effects on genes. All of these studies have been performed in cell cultures or animal models. If these therapy regimes are used in humans, it should be possible to study the in vivo effects by PET.

References

Abe Y, Fukuda H, Ishiwata K, et al. (1983) Studies on 18F-labeled pyrimidines. Tumor uptakes of 18F-5-fluorouracil, 18F-5-fluorouridine, and 18F-5-fluorodeoxyuridine in animals. Eur J Nucl Med 8:258–261

Abe Y, Matsuzawa T, Fujiwara T (1986) Assessment of radiotherapeutic effects on experimental tumors using 18F-2-fluoro-2-deoxy-D-glucose. Eur J Nucl Med 12:325–328

Abe Y, Matsuzawa T, Fujiwara T, et al. (1990) Clinical assessment of therapeutic effects on cancer using 18F-2-fluoro-2-deoxy-D-glucose and positron emission tomography: preliminary study of lung cancer. Int J Radiat Oncol Biol Phys 19:1005–1010

Adler L (1994) Cost effectiveness of PET in breast cancer. In: ICP, The Institute for Clinical PET. Proceedings of the Sixth Annual International PET Conference, Washington, DC

Adler LP, Blair HF, Williams RP, et al. (1990) Grading liposarcomas with PET using [18F]FDG. J Comput Assist Tomogr 14:960–962

Adler LP, Blair HF, Makley JT, et al. (1991) Noninvasive grading of musculoskeletal tumors using PET. J Nucl Med 32:1508–1512

Alavi JB, Alavi A, Chawluk J, et al. (1988) Positron emission tomography in patients with glioma. A predictor of prognosis. Cancer 62:1074–1078

Anderson CJ, Connett JM, Schwarz SW, et al. (1992) Copper-64-labeled antibodies for PET imaging. J Nucl Med 33:1685–1691

Anderson CJ, Rocque PA, Weinheimer CJ, Welch MJ (1993) Evaluation of copper-labeled bifunctional chelate-albumin conjugates for blood pool imaging. Nucl Med Biol 20:461–467

Andreoli VM, Agnoli A, Fazio C (1978) Transmethylations and the central nervous system. Springer, Berlin Heidelberg New York, pp 112–149 (Monographieu aus dem gresamtgebicte der Psychiatrie, vol 18)

Arnett CD, Wolf AP, Shiue CY, et al. (1986) Improved delineation of human dopamine receptors using (18F)-N-methylspiroperidol and PET. J Nucl Med 27:1878–1885

Askew DS, Ashmun RA, Simmons BC, Cleveland JL (1991) Constitutive c-myc expression in an IL-3-dependent myeloid cell line suppresses cell cycle arrest and accelerates apoptosis. Oncogene 6:1915–1922

Bakir MA, Eccles S, Babich JW, et al. (1992) c-erbB2 protein overexpression in breast cancer as a target for PET using iodine-124-labeled monoclonal antibodies. J Nucl Med 33:2154–2160

Baldesarini RJ, Kopin IJ (1986) S-Adenosyl-methionine in brain and other tissues. J Neurochem 13:769–777

Bares R, Klever P, Hauptmann S, et al. (1994) F-18 fluorodeoxyglucose PET in vivo evaluation of pancreatic glucose metabolism for detection of pancreatic cancer. Radiology 192:79–86

Baron JC, Comar D, Zarifian E, et al. (1985) Dopaminergic receptor sites in human brain: positron emission tomography. Neurology 35:16–24

Baron JC, Maziere B, Loch C, et al. (1986) Loss of striatal (76Br) bromospiperone binding sites demonstrated by positron tomography in progressive supranuclear palsy. J Cereb Blood Flow Metab 6:131–135

Barrio JR (1986) Biochemical principles in radio-pharmaceutical design and utilization. In: Phelps ME, Mazziotta JC, Schelbert HR (eds) Positron emission tomography and autoradiography: principles and applications for the brain and heart. Raven, New York, p 451

Beaney RP, Lammertsma AA, Jones T, et al. (1984) Positron emission tomography for in-vivo measurement of regional blood flow, oxygen utilization, and blood volume in patients with breast carcinoma. Lancet 1:131–134

Bell DR, Plant NJ, Rider CG, et al. (1993) Species-specific induction of cytochrome P-450 4A RNAs: PCR cloning of partial guinea-pig, human and mouse CYP4A cDNAs. Biochem J 294:173–180

Beller GA, Alten WJ, Cochavi S, Hnatowich D, Brownell GI (1979) Assessment of regional myocardial perfusion by positron emission tomography after intracoronary administration of Ga-68 labeled albumin microspheres. J Comput Assist Tomogr 3:447–454

Bergelson S, Pinkus R, Daniel V (1994) Intracellular glutathione levels regulate Fos/Jun induction and activation of glutathione S-transferase gene expression. Cancer Res 54:36–40

Bergman SR, Hack S, Tewson T, Welch MJ, Sobel RE (1980) The dependence of accumulation of N-13 NH3 by myocardium on metabolic factors and its implications for the quantitative assessment of perfusion. Circulation 61:34–41

Bergman SR, Fox KAA, Rand AL, et al. (1984) Quantification of regional myocardial blood flow in vivo with H2O-15. Circulation 70:724–731

Bergstrom M, Muhr C, Lundberg PO, Langstrom B (1991) PET as a tool in the clinical evaluation of pituitary adenomas. J Nucl Med 32:610–615

Blau M, Ganatra R, Bender MA (1972) F-18 fluoride for bone imaging. Semin Nucl Med 2:31–37

Brooks DJ, Kensett MJ, Lammertsma AA, et al. (1984) Quantitative measurement of blood brain barrier permeability using rubidium-82 and positron emission tomography. J Cereb Blood Flow Metab 4:535–545

Busch H, Davis JR, Honig GR, Anderson DC, Nair PV, Nyhan WL (1959) The uptake of a variety of amino-acids into nuclear proteins of tumors and other tissues. Cancer Res 19:1030–1039

Caruso M, Klatzmann D (1992) Selective killing of CD4+ cells harboring a human immunodeficiency virus-inducible suicide gene prevents viral spread in an infected cell population. Proc Natl Acad Sci USA 89:182–186

Chaiken L, Rege S, Hoh C, et al. (1993) Positron emission tomography with Fluorodeoxyglucose to evaluate tumor response and control after radiation therapy. Int J Radiat Oncol Biol Phys 27:455–464

Charkes ND, Makler PTJ, Philips C (1978) Studies of skeletal tracer kinetics. I. Digital computer solution of a five-compartment model of (18F)fluoride kinetics in humans. J Nucl Med 19:1301–1309

Chen BC, Huang SC, Germano G, et al. (1991) Noninvasive quantification of hepatic arterial blood flow with nitrogen-

13-ammonia and dynamic positron emission tomography. J Nucl Med 32:2199–2206

Chen BC, Germano G, Huang SC, et al. (1992) A new noninvasive quantification of renal blood flow with N-13 ammonia, dynamic positron emission tomography, and a two-compartment model. J Am Soc Nephrol 3:1295–1306

Chen ZP, Chen KY (1991) Differentiation of a mouse neuroblastoma variant cell line whose ornithine decarboxylase gene has been amplified. Biochim Biophys Acta 1133:1–8

Coenen HH, Kling P, Stöcklin G (1989) Cerebral metabolism of L-(2-18F)fluorotyrosine, a new PET tracer of protein synthesis. J Nucl Med 30:1367–1372

Coleman RE (1991) Single photon emission computed tomography and positron emission tomography in cancer imaging. Cancer 67:1261–1270

Coleman RE (1994) Clinical applications in lung cancer. In: ICP, The Institute for Clinical PET. Proceedings of the Sixth Annual International PET Conference, Washington, DC

Coleman RE, Hoffman JM, Hanson MW, Sostman HD, Schold SC (1991) Clinical application of PET for the evaluation of brain tumors. J Nucl Med 32:616–622

Collier GB, Mattson TL, Connaughton JF, Chirikjian JG (1994) A novel Tn10 tetracycline regulon system controlling expression of the bacteriophage T3 gene encoding S-adenosyl-L-methionine hydrolase. Gene 148:75–80

Comar D, Cartron JC, Maziere M, Marazano C (1976) Labelling and metabolism of methionine-methyl-11C. Eur J Nucl Med 1976:11–14

Daemen BJ, Elsinga PH, Ishiwata K, Paans AM, Vaalburg W (1991a) A comparative PET study using different 11C-labelled amino acids in Walker 256 carcinosarcoma-bearing rats. Int J Radiat Appl Instrum [B] 18:197–204

Daemen BJ, Zwertbroek R, Elsinga PH, Paans AM, Doorenbos H, Vaalburg W (1991b) PET studies with L-[1-11C]tyrosine, L-[methyl-11C]methionine and 18F-fluorodeoxyglucose in prolactinomas in relation to bromocriptine treatment. Eur J Nucl Med 18:453–460

Daemen BJ, Elsinga PH, Paans AM, Wieringa AR, Konings AW, Vaalburg W (1992) Radiation-induced inhibition of tumor growth as monitored by PET using L-[1-11C]tyrosine and fluorine-18-fluorodeoxyglucose. J Nucl Med 33:373–379

Daghighian F, Pentlow KS, Larson SM, et al. (1991) In vivo kinetics of radiolabeled antibody: PET studies of I-124 labeled 3F8 MAB in human glioma (abstr). J Nucl Med 32:1021–1022

Daugherty A, Kilbourn MR, Dence CS, Sobel BE, Thorpe SR (1992) Quantitative assessment of lipoprotein metabolism by positron emission tomography with an 18F-containing residualizing label. Int J Radiat Appl Instrum [B] 19:411–416

Dehdashti F, McGuire AH, Van Brocklin H, et al. (1991) Assessment of 21-[18F]fluoro-16-alpha-ethyl-19-norprogesterone as a positron-emitting radiopharmaceutical for the detection of progestin receptors in human breast carcinomas. J Nucl Med 32:1532–1537

Dewanjee MK, Ghafouripour AK, Kapadvanjwala M, et al. (1994) Noninvasive imaging of c-myc oncogene messenger RNA with indium-111-antisense probes in a mammary tumor-bearing mouse model. J Nucl Med 35:1054–1063

Dhawan V, Poltorak A, Moeller JR, et al. (1989) Positron emission tomographic measurement of blood-to-brain and blood-to-tumour transport of 82Rb. I. Error analysis and computer simulations. Phys Med Biol 34:1773–8174

Di Chiro G, De LaPaz R, Brooks R (1982) Glucose utilization of cerebral gliomas measured by (^{18}F) fluorodeoxyglucose and positron emission tomography. Neurology 32:1323–1329

Di Chiro G, Hatazawa J, Katz DA, Rizzoli HV, De Michele DJ (1987) Glucose utilization by intracranial meningiomas as an index of tumor aggressivity and probability of recurrence: a PET study. Radiology 164:521–526

Di Chiro G, Oldfield E, Wright DC, et al. (1988) Cerebral necrosis after radiotherapy and/or intraarterial chemotherapy for brain tumors: PET and neuropathologic studies. Am J Roentgenol 150:189–197

Dimitrakopoulou A, Strauss LG, Clorius JH, et al. (1993) Studies with positron emission tomography after systemic administration of fluorine-18-uracil in patients with liver metastases from colorectal carcinoma. J Nucl Med 34:1075–1081

Farde L, Hall H, Ehrin E, Sedvall G (1986) Quantitative analysis of D2 dopamine receptor binding in the living human brain by PET. Science 231:258–264

Flower MA, Schlesinger T, Hinton PJ, et al. (1989) Radiation dose assessment in radioiodine therapy. 2. Practical implementation using quantitative scanning and PET, with initial results on thyroid carcinoma. Radiother Oncol 15:345–357

Flower MA, Irvine AT, Ott RJ, et al. (1990) Thyroid imaging using positron emission tomography – a comparison with ultrasound imaging and conventional scintigraphy in thyrotoxicosis. Br J Radiol 63:325–330

Fowler JS, Wolf AP (1986) Positron emitter-labeled compounds: priorities and problems. In: Phelps ME, Mazziotta JC, Schelbert HR (eds) Positron emission tomography and autoradiography: principles and applications for the brain and heart. Raven, New York, p 391

Francavilla TL, Miletich RS, De Michele D, et al. (1991) Positron emission tomography of pituitary macroadenomas: hormone production and effects of therapies. Neurosurgery 28:826–833

Frey K, Koeppe RA, Jewett DM, et al. (1987) The in vivo distribution of (11C) scopolamine in human brain determined by positron emission tomography (abstr). Soc Neurosci 13:1658–1663

Frey KA, Hichwa RD, Ehrenkaufer RLE, Agranoff BW (1985) Quantitative in vivo receptor binding. III. Tracer kinetic modeling of muscarinic cholinergic receptor binding. Proc Natl Acad Sci USA 82:6711–6715

Frost JJ, Wagner HN, Dannals RF, et al. (1985) Imaging opiate receptors in the human brain by positron tomography. J Comput Assist Tomogr 9:231–237

Fujibayashi Y, Wada K, Taniuchi H, Yonekura Y, Konishi J, Yokoyama A (1993) Mitochondria-selective reduction of 62Cu-pyruvaldehyde bis(N4-methylthiosemicarbazone) (62Cu-PTSM) in the murine brain: a novel radiopharmaceutical for brain positron emission tomography (PET) imaging. Biol Pharm Bull 16:146–149

Fujihara M, Ito N, Pace JL, Watanabe Y, Russell SW, Suzuki T (1994) Role of endogenous interferon-beta in lipopolysaccharide-triggered activation of the inducible nitric-oxide synthase gene in a mouse macrophage cell line, J774. J Biol Chem 269:12773–12778

Fujiwara T, Matsuzawa T, Kubota K, et al. (1989) Relationship between histologic type of primary lung cancer and carbon-11-L-methionine uptake with positron emission tomography. J Nucl Med 30:33–37

Fujiwara T, Kubota K, Sato T, et al. (1990) N-[18F]Fluoroacetyl-D-glucosamine: a potential agent for cancer diagnosis. J Nucl Med 31:1654–1658

Gardner SF, Lazarus HM, Bednarczyk EM, et al. (1993) High-dose cyclophosphamide-induced myocardial damage during BMT: assessment by positron emission tomography. Bone Marrow Transplant 12:139–144

Gelbard AS, Benua RS, Laughlin JS, et al. (1979) Quantitative scanning of osteogenic sarcoma with nitrogen-13-labeled L-glutamate. J Nucl Med 20:782–784

Ginos JZ, Cooper AJL, Dhawan V, et al. (1987) (13N)Cisplatin PET to assess pharmacokinetics of intra-arterial versus intravenous chemotherapy for malignant brain tumors. J Nucl Med 28:1844–1852

Goodwin DA, Lang EV, Atwood JE, et al. (1993) Viability and biodistribution of 68Ga MPO-labelled human platelets. Nucl Med Commun 14:1023–1029

Greven KM, Williams D, Keyes JJ, et al. (1994) Positron emission tomography of patients with head and neck carcinoma before and after high dose irradiation. Cancer 74:1355–1359

Griebel M, Friedman HS, Halperin EC, et al. (1991) Reversible neurotoxicity following hyperfractionated radiation therapy of brain stem glioma. Med Pediatr Oncol 19:182–186

Griffeth LK, Dehdashti F, McGuire AH, et al. (1992) PET evaluation of soft-tissue masses with fluorine-18 fluoro-2-deoxy-D-glucose. Radiology 182:185–194

Gupta NC (1994). Clinical applications in colorectal cancer. In: ICP, The Institute for Clinical PET. Proceedings of the Sixth Annual International PET Conference, Washington, DC

Haberkorn U, Strauss LG, Dimitrakopoulou A, et al. (1991a) PET studies of fluorodeoxyglucose metabolism in patients with recurrent colorectal tumors receiving radiotherapy. J Nucl Med 32:1485–1490

Haberkorn U, Strauss LG, Reisser C, et al. (1991b) Glucose uptake, perfusion, and cell proliferation in head and neck tumors: relation of positron emission tomography to flow cytometry. J Nucl Med 32:1548–1555

Haberkorn U, Reinhardt M, Strauss LG, et al. (1992) Metabolic design of combination therapy: use of enhanced fluorodeoxyglucose uptake caused by chemotherapy. J Nucl Med 33:1981–1987

Haberkorn U, Strauss LG, Dimitrakopoulou A, et al. (1993) Fluorodeoxyglucose imaging of advanced head and neck cancer after chemotherapy. J Nucl Med 34:12–17

Hawkins RA, Phelps ME, Huang SC, et al. (1984) A kinetic evaluation of blood brain barrier permeability in human brain tumors with 68Ga EDTA and positron computed tomography. J Cereb Blood Flow Metab 4:515–527

Hawkins RA, Phelps ME, Huang SC (1986) Effects of temporal sampling, glucose metabolic rates and disruptions of the blood brain barrier (BBB) on the FDG model with and without a vascular compartment: studies in human brain tumors with PET. J Cereb Blood Flow Metab 6:170–183

Hawkins RA, Huang SC, Barrio JR, et al. (1989) Estimation of local cerebral protein synthesis rates with L-(1-11C)leucine and PET: methods, model and results in animals and humans. J Cereb Blood Flow Metab 9:446–460

Hawkins RA, Hoh C, Dahlbom M, et al. (1991) PET cancer evaluations with FDG. J Nucl Med 32:1555–1558

Hawkins RA, Choi Y, Huang SC, et al. (1992a) Evaluation of the skeletal kinetics of fluorine-18-fluoride ion with PET. J Nucl Med 33:633–642

Hawkins RA, Choi Y, Huang SC, Messa C, Hoh CK, Phelps ME (1992b) Quantitating tumor glucose metabolism with FDG and PET. J Nucl Med 33:339–344

Hawkins RA, Hoh C, Glaspy J, et al. (1992c) The role of positron emission tomography in oncology and other whole-body applications. Semin Nucl Med 22:268–284

Henze E, Schelbert HR, Barrio JR (1982) Evaluation of myocardial metabolism with N-13 and C-11 labeled amino acids and positron computed tomography. J Nucl Med 23:671–677

Herholz K, Ziffling P, Staffen W, et al. (1988) Uncoupling of hexose transport and phosphorylation in human gliomas demonstrated by PET. Eur J Cancer Clin Oncol 24:1139–1150

Herzog H, Rosch F, Stocklin G, Lueders C, Qaim SM, Feinendegen LE (1993) Measurement of pharmacokinetics of yttrium-86 radiopharmaceuticals with PET and radiation dose calculation of analogous yttrium-90 radiotherapeutics. J Nucl Med 34:2222–2226

Hiesiger E, Follwer JS, Wolf AP, et al. (1987) Serial PET studies of human cerebral malignancy with (1-11C)putrescine and (1-11C)2-deoxy-D-glucose. J Nucl Med 28:1251–1261

Higashi K, Clavo AC, Wahl RL (1993) Does FDG uptake measure proliferative activity of human cancer cells? In vitro comparison with DNA flow cytometry and tritiated thymidine uptake. J Nucl Med 34:414–419

Higashi K, Clavo AC, Wahl RL (1994) In vitro assessment of 2-fluoro-2-deoxy-D-glucose, L-methionine and thymidine as agents to monitor the early response of a human adenocarcinoma cell line to radiotherapy. J Nucl Med 34:773–779

Hoffman EJ, Phelps ME (1986) Positron emission tomography: principles and quantification. In: Phelps M, Mazziotta J, Schelbert H (eds) Positron emission tomography and autoradiography: principles and applications for the brain and heart. Raven, New York, p 237

Hoffman EJ, Huang SC, Phelps ME (1979) Quantitation in positron emission computed tomography. 1. Effect of object size. J Comput Assist Tomogr 3:299–305

Hoh CK, Hawkins RA, Dahlbom M, et al. (1993) Whole body skeletal imaging with [18F]fluoride ion and PET. J Comput Assist Tomogr 17:34–41

Hohenberger P, Strauss LG, Lehner B, Frohmuller S, Dimitrakopoulou A, Schlag P (1993) Perfusion of colorectal liver metastases and uptake of fluorouracil assessed by H2(15)O and [18F]uracil positron emission tomography (PET). Eur J Cancer 29:1682–1686

Huang SC, Phelps ME (1986) Principles of tracer kinetic modeling in positron emission tomography and autoradiography. In: Phelps ME, Mazziotta JC, Schelbert HR (eds) Positron emission tomography and autoradiography: principles and applications for the brain and heart. Raven, New York, p 287

Huang SC, Phelps ME, Hoffman EJ, Sideris K, Selin CJ, Kuhl DE (1980) Non-invasive determination of local cerebral metabolic rate of glucose in man. Am J Physiol 238:E69

Huang SC, Schwaiger M, Carson RE, Carson J, Phelps ME, Schelbert HR (1985) Quantitative measurement of myocardial blood flow with oxygen-15 water and positron computed tomography: an assessment of potential and problems. J Nucl Med 25:616–624

Hubner KF, Smith GT, Hunter K, et al. (1994) Assessment of primary and recurrent cancer of the ovary using F-18-FDG PET. In: ICP, The Institute for Clinical PET. Proceedings of the Sixth Annual International PET Conference, Washington, DC

Hughes WL, Bond VP, Brecher G, et al. (1958) Cellular proliferation in the mouse revealed by autoradiography with tritiated thymidine. Proc Natl Acad Sci USA 44:476–483

Huovinen R, Leskinen KS, Nagren K, Lehikoinen P, Ruotsalainen U, Teras M (1993) Carbon-11-methionine and PET in evaluation of treatment response of breast cancer. Br J Cancer 67:787–791

Hwang DR, Lang LX, Mathias CJ, Kadmon D, Welch MJ (1989) N-3-[18F]fluoropropylputrescine as potential PET imaging agent for prostate and prostate derived tumors. J Nucl Med 30:1205–1210

Hwang DR, Mathias CJ, Welch MJ, McGuire AH, Kadmon D (1990) Imaging prostate derived tumors with PET and N-(3-[18F]fluoropropyl)putrescine. Int J Radiat Appl Instrum [B] 17:525–532

Ichiya Y, Kuwabara Y, Otsuka M, et al. (1991) Assessment of response to cancer therapy using fluorine-18-fluorodeoxyglucose and positron emission tomography. J Nucl Med 32:1655–1660

Ido T, Wan CN, Casella V, et al. (1978) Labeled 2-deoxy-D-glucose analogs: 18F-labeled 2-deoxy-2-fluoro-D-glucose. J Labelled Compd Radiopharm 24:174–178

Ishiwata K, Ido T, Abe Y, Matsuzawa T, Murakami M (1985) Studies on 18F-labeled pyrimidines. III. Biochemical investigation of 18F-labeled pyrimidines and comparison with 3H-deoxythymidine in tumor-bearing rats and mice. Eur J Nucl Med 10:39–44

Ishiwata K, Vaalburg W, Elsinga PH, Paans AM, Woldring MG (1988a) Comparison of L-[1-11C]methionine and L-methyl-[11C]methionine for measuring in vivo protein synthesis rates with PET. J Nucl Med 29:1419–1427

Ishiwata K, Vaalburg W, Elsinga PH, Paans AM, Woldring MG (1988b) Metabolic studies with L-[1-14C]tyrosine for the investigation of a kinetic model to measure protein synthesis rates with PET. J Nucl Med 29:524–529

Ishiwata K, Ido T, Kawamura M, Kubota K, Ichihashi M, Mishima Y (1991a) 4-Borono-2-[18F]fluoro-D,L-phenylalanine as a target compound for boron neutron capture therapy: tumor imaging potential with positron emission tomography. Int J Radiat Appl Instrum [B] 18:745–751

Ishiwata K, Ido T, Mejia AA, Ichihashi M, Mishima Y (1991b) Synthesis and radiation dosimetry of 4-borono-2-[18F]fluoro-D,L-phenylalanine: a target compound for PET and boron neutron capture therapy. Int J Radiat Appl Instrum [A] 42:325–328

Ishiwata K, Sato K, Kameyama M, Yoshimoto T, Ido T (1991c) Metabolic fates of 2′-deoxy-5-[18F]fluorouridine in tumor-bearing mice and human plasma. Int J Radiat Appl Instrum [B] 18:539–545

Ishiwata K, Shiono M, Kubota K, et al. (1992) A unique in vivo assessment of 4-[10B]borono-L-phenylalanine in tumour tissues for boron neutron capture therapy of malignant melanomas using positron emission tomography and 4-borono-2-[18F]fluoro-L-phenylalanine. Melanoma Res 2:171–179

Ishiwata K, Kubota K, Murakami M, et al. (1993a) Re-evaluation of amino acid PET studies: can the protein synthesis rates in brain and tumor tissues be measured in vivo? J Nucl Med 34:1936–1943

Ishiwata K, Kubota K, Murakami M, Kubota R, Senda M (1993b) A comparative study on protein incorporation of L-[methyl-3H]methionine, L-[1-14C]leucine and L-2-[18F]fluorotyrosine in tumor bearing mice. Nucl Med Biol 20:895–899

Ito M, Lammertsma AA, Wise RJS, et al. (1982) Measurement of regional cerebral blood flow and oxygen utilization in patients with cerebral tumors using O-15 and positron emission tomography: analytical techniques and preliminary results. Neuroradiology 23:63–74

Jarden JO, Dhawan V, Poltorak A, et al. (1985) Positron emission tomographic measurement of blood-to-brain and blood-to-tumor transport of 82Rb: The effect of dexamethasone and whole-brain radiation therapy. Ann Neurol 18:636–646

Jones HA, Clark RJ, Rhodes CG, Schofield JB, Krausz T, Haslett C (1994) In vivo measurement of neutrophil activity in experimental lung inflammation. Am J Respir Crit Care Med 149:1635–1639

Jones RM, Cramer S, Sargent T, Budinger TF (1985) Brain protein synthesis rates measured in vivo using methionine and leucine. J Nucl Med 1985:168–169

Kahn D, Weiner GJ, Ben HS, et al. (1994) Positron emission tomographic measurement of bone marrow blood flow to the pelvis and lumbar vertebrae in young normal adults. Blood 83:958–963

Kaplan JD, Calandrino FS, Schuster DP (1992a) Effect of smoking on pulmonary vascular permeability. A positron emission tomography study. Am Rev Respir Dis 145:712–715

Kaplan JD, Trulock EP, Anderson DJ, Schuster DP (1992b) Pulmonary vascular permeability in interstitial lung disease. A positron emission tomographic study. Am Rev Respir Dis 145:1495–1498

Katzenellenbogen JA, Carlson KE, Heiman DF, et al. (1980) Receptor-binding radiopharmaceuticals for imaging breast tumors: estrogen-receptor interactions and selectivity of tissue uptake of halogenated estrogen analogs. J Nucl Med 21:550–558

Keen RE, Barrio JR, Huang SC, et al. (1989) In-vivo protein synthesis rates with leucyltransfer RNA used as precursor pool: determination of biochemical parameters to structure tracer kinectic models for positron emission tomography. J Cereb Blood Flow Metab 9:429–445

Kern KA, Brunetti A, Norton JA, et al. (1988) Metabolic imaging of human extremity musculoskeletal tumors by PET. J Nucl Med 29:181–186

Kessler RM, Goble JC, Bird JH, et al. (1984) Measurement of blood-brain barrier permeability with positron emission tomography and (68Ga)EDTA. J Cereb Blood Flow Metab 4:323–328

Kiichi I, Vaalburg W, Elsinga PH, et al. (1988) Comparison of L-(1–11C)methionine and L-methyl-(11C)methionine for measuring in vivo protein synthesis rates with PET. J Nucl Med 29:1419–1427

Killion D, Nitzsche E, Choi Y, Schelbert H, Rosenthal JT (1993) Positron emission tomography: a new method for determination of renal function. J Urol 150:1064–1068

Kim YH, Buchholz MA, Chrest FJ, Nordin AA (1994) Up-regulation of c-myc induces the gene expression of the murine homologues of p34cdc2 and cyclin-dependent kinase-2 in T lymphocytes. J Immunol 152:4328–4335

Knapp WH, Helus F, Sinn HJ, et al. (1984) N-13-L-Glutamate uptake in malignancy: its relationship to blood flow. J Nucl Med 25:989–997

Knapp WH, Helus F, Layer K, et al. (1986) Nitrogen-13-glutamate uptake and perfusion in Walker 256 carcinosarcoma before and after single-dose-irradiation. J Nucl Med 27:1604–1610

Koeppe RA, Hutchins GD (1992) Instrumentation for positron emission tomography: tomographs and data processing and display systems. Semin Nucl Med 22:162–181

Koh WJ, Rasey JS, Evans ML, et al. (1991) Imaging of hypoxia in human tumors with (F-18)fluoromisonidazole. Int J Radiat Oncol Biol Phys 22:199–212

Krivokapich J, Huang SC, Phelps ME, et al. (1982) Estimation of rabbit myocardial metabolic rate for glucose using fluorodeoxyglucose. Am J Physiol 243:H884

Kubo S, Yamamoto K, Magata Y, et al. (1991) Assessment of pancreatic blood flow with positron emission tomography and oxygen-15 water. Ann Nucl Med 5:133–138

Kubota K, Matsuzawa T, Fujiwara T, et al. (1988) Differential diagnosis of solitary pulmonary nodules with positron emission tomography using (11C)L-methionine. J Comput Assist Tomogr 12:794–796

Kubota K, Matsuzawa T, Fujiwara T, et al. (1989a) Comparison of C-11 methionine and F-18 fluorodeoxyglucose for the differential diagnosis of lung tumor (abstr). J Nucl Med 30:788–789

Kubota K, Matsuzawa T, Takahashi T, et al. (1989b) Rapid and sensitive response of carbon-11-L-methionine tumor uptake to irradiation. J Nucl Med 30:2012–2016

Kubota K, Matsuzawa T, Fujiwara T, et al. (1990) Differential diagnosis of lung tumor with positron emission tomography: a prospective study. J Nucl Med 31:1927–1932

Kubota K, Ishiwata K, Kubota R, et al. (1991) Tracer feasibility for monitoring tumor radiotherapy: a quadruple tracer study with fluorine-18-fluorodeoxyglucose or fluorine-18-fluorodeoxyuridine, L-[methyl-14C]methionine, [6-3H]thymidine, and gallium-67. J Nucl Med 32:2118–2123

Kubota K, Ishiwata K, Yamada S, et al. (1992a) Tumor radiotherapy monitoring with radioscintigraphy tracers: a comparative study with multiple-tracer technique. Tohoku J Exp Med 168:437–439

Kubota K, Yamada K, Yoshioka S, Yamada S, Ito M, Ido T (1992b) Differential diagnosis of idiopathic fibrosis from malignant lymphadenopathy with PET and F-18 fluorodeoxyglucose. Clin Nucl Med 17:361–363

Kubota R, Yamada S, Kubota K, Ishiwata K, Tamahashi N, Ido T (1992c) Intratumoral distribution of fluorine-18-fluorodeoxyglucose in vivo: high accumulation in macrophages and granulation tissues studied by microautoradiography. J Nucl Med 33:1972–1980

Kubota K, Kubota R, Yamada S (1993a) FDG accumulation in tumor tissue. J Nucl Med 34:419–421

Kubota K, Yamada S, Ishiwata K, et al. (1993b) Evaluation of the treatment response of lung cancer with positron emission tomography and L-[methyl-11C]methionine: a preliminary study. Eur J Nucl Med 20:495–501

Kubota R, Kubota K, Yamada S, Tada M, Ido T, Tamahashi N (1994) Microautoradiographic study for the differentiation of intratumoral macrophages, granulation tissues and cancer cells by the dynamics of fluorine-18-fluorodeoxyglucose uptake. J Nucl Med 35:104–112

Kuten A, Roval HD, Griffeth LK, et al. (1992) Positron emission tomography in the study of acute radiation effects on renal blood flow in dogs. Int Urol Nephrol 24:527–529

Lambrecht RM, Woodhouse N, Phillips R, et al. (1988) Investigational study of iodine-124 with a positron camera. Am J Physiol Imaging 3:197–200

Lammertsma AA, Jones T (1992) Low oxygen extraction fraction in tumours measured with the oxygen-15 steady state technique: effect of tissue heterogeneity. Br J Radiol 65:697–700

Lammertsma AA, Brooks DJ, Frackowiak RSJ, et al. (1984) A method to quantitate the fractional extraction of rubidium-82 across the blood brain barrier using positron emission tomography. J Cereb Blood Flow Metab 4:523–534

Langen KJ, Roosen N, Kuwert T, et al. (1989) Early effects of intra-arterial chemotherapy in patients with brain tumours studied with PET: preliminary results. Nucl Med Commun 10:779–790

Langen KJ, Braun U, Kops ER, et al. (1993) The influence of plasma glucose levels on fluorine-18-fluorodeoxyglucose uptake in bronchial carcinomas. J Nucl Med 34:355–359

Larson SM, Pentlow KS, Volkow ND, et al. (1992) PET scanning of iodine-124-3F9 as an approach to tumor dosimetry during treatment planning for radioimmunotherapy in a child with neuroblastoma. J Nucl Med 33:2020–2023

Lea MA, Morris HP, Weber G (1966) Comparative biochemistry of hepatomas. VI. Thymidine incorporation into DNA as a measure of hepatoma growth rate. Cancer Res 26:465–469

Leskinen KS, Nagren K, Lehikoinen P, Ruotsalainen U, Joensuu H (1991a) Uptake of 11C-methionine in breast cancer studied by PET. An association with the size of S-phase fraction. Br J Cancer 64:1121–1124

Leskinen KS, Ruotsalainen U, Nagren K, Teras M, Joensuu H (1991b) Uptake of carbon-11-methionine and fluorodeoxyglucose in non-Hodgkin's lymphoma: a PET study. J Nucl Med 32:1211–1218

Leskinen KS, Nagren K, Lehikoinen P, Ruotsalainen U, Teras M, Joensuu H (1992) Carbon-11-methionine and PET is an effective method to image head and neck cancer. J Nucl Med 33:691–695

Lindholm P, Leskinen KS, Minn H, et al. (1993) Comparison of fluorine-18-fluorodeoxyglucose and carbon-11-methionine in head and neck cancer. J Nucl Med 34:1711–1716

Livni E, Fischman AJ, Ray S, et al. (1992) Synthesis of 18F-labeled fluconazole and positron emission tomography studies in rabbits. Int J Radiat Appl Instrum [B] 19:191–199

Lorenz CH, Powers TA, Partain CL (1992) Quantitative imaging of renal blood flow and function. Invest Radiol 27:S109–114

Lu XM, Fischman AJ, Jyawook SL, Hendricks K, Tompkins RG, Yarmush ML (1994) Antisense DNA deliver In vivo: liver targeting by receptor-mediated uptake. J Nucl Med 35:269–275

Lundqvist H, Stalnacke CG, Langstrom B, Jones B (1985) Labeled metabolites in plasma after intravenous administration of (11CH3)-L-methionine. In: Greitz T et al. (eds) The metabolism of the human brain studied with positron emission tomography. Raven, New York, pp 233–240

Macapinlac HA, Larson SM, Blasberg RG, et al. (1994) Clinical indications for FDG PET evaluation of intracranial mass lesions. In: ICP, The Institute for Clinical PET. Proceedings of the Sixth Annual International PET Conference. Washington, DC

Macke HR, Smith JP, Maina T, et al. (1993) New octreotide derivatives for in vivo targeting of somatostatin receptor-positive tumors for single photon emission computed tomography (SPECT) and positron emission tomography (PET). Horm Metab Res Suppl 27:12–17

Mariat P, Ferrant A, Labar D, et al. (1988) In vivo measurement of carbon-11 thymidine uptake in Non-Hodgkin's lymphoma using positron emission tomography. J Nucl Med 29:1633–1637

Markham J, Schuster DP (1992) Effects of nonideal input functions on PET measurements of pulmonary blood flow. J Appl Physiol 72:2495–2500

Marshall RC, Huang SC, Nash WW, Phelps ME (1983) Investigation of the 18-fluorodeoxyglucose tracer kinetic model to accurately measure the myocardial metabolic rate for glucose during ischemia: preliminary notes. J Nucl Med 24:1060–1067

Martin GV, Caldwell JH, Rasey JS, Grunbaum Z, Cerqueira M, Krohn KA (1989) Enhanced binding of the hypoxic cell marker (F-18) fluoromisonidazole in ischemic myocardium. J Nucl Med 30:194–201

Martin GV, Caldwell JH, Graham MM, et al. (1992) Noninvasive detection of hypoxic myocardium using fluorine-18-fluoromisonidazole and positron emission tomography. J Nucl Med 33:2202–2208

Martin WR, Raichle ME (1983) Cerebellar blood flow and metabolism in cerebral hemisphere infarction. Ann Neurol 14:168–173

Mathias CJ, Welch MJ, Perry DJ, et al. (1991) Investigation of copper-PTSM as a PET tracer for tumor blood flow. Int J Radiat Appl Instrum [B] 18:807–811

Mazziotta JC, Phelps ME (1986) Positron emission tomography studies of the brain. In: Phelps M, Mazziotta J, Schelbert H (eds) Positron emission tomography and autoradiography: principles and applications for the brain and heart. Raven, New York, p 493

Meyer JS, Friedman E, McCrate MM, et al. (1983) Prediction of early course of breast carcinoma by thymidine labeling. Cancer 51:1879–1886

Mineura K, Yasuda T, Kowada M (1986) Positron emission tomographic evaluation of histological malignancy in gliomas using oxygen-15 and fluorine-18-fluorodeoxyglucose. Neurol Res 8:164–168

Mineura K, Sasajima T, Kowada M, Shishido F, Uemura K (1990) Positron emission tomography (PET) study in patients with meningiomas. No To Shinkei 42:145–151

Mineura K, Sasajima T, Kowada M, et al. (1994) Perfusion and metabolism in predicting the survival of patients with cerebral gliomas. Cancer 73:2386–2394

Minn H, Soini I (1989) [^{18}F]Fluorodeoxyglucose scintigraphy in diagnosis and follow up of treatment in advanced breast cancer. Eur J Nucl Med 15:61–66

Minn H, Joensuu H, Ahonen A, Klemi P (1988a) Fluorodeoxyglucose imaging: a method to assess the proliferative activity of human cancer in vivo. Comparison with DNA flow cytometry in head and neck tumors. Cancer 61:1776–1781

Minn H, Paul R, Ahonen A (1988b) Evaluation of treatment response to radiotherapy in head and neck cancer with fluorine-18 fluorodeoxyglucose. J Nucl Med 29:1521–1525

Minn H, Kangas L, Kellokumpu LP, et al. (1992) Uptake of 2-fluoro-2-deoxy-D-[U-14C]-glucose during chemotherapy in murine Lewis lung tumor. Int J Radiat Appl Instrum [B] 19:55–63

Mintun MA, Welch MJ, Siegel BA, et al. (1988) Breast cancer: PET imaging of estrogen receptors. Radiology 169:45–52

Miraldi F (1989) Monoclonal antibodies and neuroblastoma. Semin Nucl Med 19:282–294

Mitsuki S, Diksic M, Conway T, Yamamoto YL, Villemure JG, Feindel W (1991) Pharmacokinetics of 11C-labelled BCNU and SarCNU in gliomas studied by PET. J Neurooncol 10:47–55

Miura S, Murakami M, Kanno I, Iida H, Uemura K (1992) Phenylalanine transport in the living human brain by a dynamic PET of L-[2-18F]-fluorophenylalanine. Nippon Rinsho 50:1457–1460

Molls M, Feldmann HJ, Fuller J (1994) Oxygenation of locally advanced recurrent rectal cancer, soft tissue sarcoma and breast cancer. Adv Exp Med Biol 345:459–463

Mosskin M, von Holst H, Bergstrom M, et al. (1987) Positron emission tomography with 11C-methionine and computed tomography of intracranial tumours compared with histopathologic examination of multiple biopsies. Acta Radiol 28:673–681

Nagata Y, Yamamoto K, Hiraoka M, et al. (1990) Monitoring liver tumor therapy with [18F]FDG positron emission tomography. J Comput Assist Tomogr 14:370–374

Nitzsche EU, Choi Y, Killion D et al. (1993) Quantification and parametric imaging of renal cortical blood flow in vivo based on Patlak graphical analysis. Kidney Int 44:985–996

Nunokawa Y, Ishida N, Tanaka S (1994) Promoter analysis of human inducible nitric oxide synthase gene associated with cardiovascular homeostasis. Biochem Biophys Res Commun 200:802–807

Ogawa T, Uemura K, Shishido F, et al. (1988) Changes of cerebral blood flow, and oxygen and glucose metabolism following radiochemotherapy of gliomas: a PET study. J Comput Assist Tomogr 12:290–297

Ogawa T, Kanno I, Shishido F, et al. (1991) Clinical value of PET with 18F-fluorodeoxyglucose and L-methyl-11C-methionine for diagnosis of recurrent brain tumor and radiation injury. Acta Radiol 32:197–202

Ogawa T, Shishido F, Kanno I, et al. (1993) Cerebral glioma: evaluation with methionine PET. Radiology 186:45–53

Ogawa T, Kanno I, Hatazawa J, et al. (1994) Methionine PET for follow-up of radiation therapy of primary lymphoma of the brain. Radiographics 14:101–110

Oguro A, Taniguchi H, Koyama H, et al. (1993a) Quantification of human splenic blood flow (quantitative measurement of splenic blood flow with H2(15)O and a dynamic state method). Ann Nucl Med 7:245–250

Oguro A, Taniguchi H, Koyama H, et al. (1993b) Relationship between liver function and splenic blood flow (quantitative measurement of splenic blood flow with H2(15)O and a dynamic state method). Ann Nucl Med 7:251–255

Okada J, Yoshikawa K, Imazeki K, et al. (1991) Change of cerebral glucose metabolism by antineoplastic drug. Am J Physiol Imaging 6:162–166

Okada J, Yoshikawa K, Itami M, et al. (1992) Positron emission tomography using fluorine-18-fluorodeoxyglucose in malignant lymphoma: a comparison with proliferative activity. J Nucl Med 33:325–329

Okada S, Ohto M, Kuniyasu Y, Higashi S, Arimizu N, Uematsu S (1990) Estimation of the reticuloendothelial function by positron emission computed tomography (PET) study in chronic liver disease. Nippon Shokakibyo Gakkai Zasshi 87:90–99

Okazumi S, Isono K, Enomoto K, et al. (1992) Evaluation of liver tumors using fluorine-18-fluorodeoxyglucose PET: characterization of tumor and assessment of effect of treatment. J Nucl Med 33:333–339

Otsuka FL, Welch MJ, Kilbourn MR, Dence CS, Dilley WG, Wells SJ (1991) Antibody fragments labeled with fluorine-18 and gallium-68: in vivo comparison with indium-111 and iodine-125-labeled fragments. Int J Radiat Appl Instrum [B] 18:813–816

Ott RJ, Tait D, Flower MA, Babich JW, Lambrecht RM (1992) Treatment planning for 131I-mIBG radiotherapy of neural crest tumours using 124I-mIBG positron emission tomography. Br J Radiol 65:787–791

Patlak OC, Blasberg RG, Fenstermacher JD (1983) Graphical evaluation of blood-to-brain transfer constants from multiple-time uptake data. J Cereb Blood Flow Metab 3:1–7

Patronas NJ, Di Chiro G, Brooks RA, et al. (1982) Work in progress: (18F)fluorodeoxyglucose and positron emission tomography in the evaluation of radiation necrosis of the brain. Radiology 144:885–889

Patronas NJ, Di Chiro G, Kufta C, et al. (1985) Prediction of survival in glioma patients by means of positron emission tomography. J Neurosurg 62:816–822

Pentlow KS, Graham MC, Lambrecht RM, Cheung NK, Larson SM (1991) Quantitative imaging of I-124 using positron emission tomography with applications to radioimmunodiagnosis and radioimmunotherapy. Med Phys 18:357–366

Perlmutter JS, Kilbourn MR, Raichle MR, Welch MJ (1987) MPTP-induced up-regulation of in vivo dopaminergic radioligand-receptor binding in humans. Neurology 37:1575–1581

Persson A, Ehrin E, Eriksson L, et al. (1985) Imaging of (C-11)-labelled RO 15-1788 binding to benzodiazepine receptors in the human brain by positron emission tomography. J Psychiatr Res 19:609–615

Phelps ME, Huang SC, Hoffman EJ, Selin C, Sokoloff L, Kuhl DE (1979) Tomographic measurement of local cerebral glucose metabolic rate in humans with (F-18) 2-fluoro-2-deoxy-D-glucose: validation of a method. Ann Neurol 6:371–388

Phelps ME, Barrio JR, Huang SC, Keen RE, Chugani H, Mazziotta JC (1984) Criteria for the tracer kinetic measurement of cerebral protein synthesis in humans with positron CT. Ann Neurol 15:S192–193

Piwnica-Worms D (1995) Making sense out of anti-sense: challenges of imaging gene translation with radiolabeled oligonucleotides. J Nucl Med 35:1064–1066

Plautz G, Nabel EG, Nabel GJ (1991) Selective elimination of recombinant genes in vivo with a suicide retroviral vector. New Biol 3:709–715

Poupeye E, Counsell RE, De Leenheer A, et al. (1989) Synthesis of 11C-labelled thymidine for tumor visualization using positron emission tomography. Int J Radiat Appl Instrum [A] 40:57–61

Raichle ME, Eichling JO, Straatman MG, Welch MJ, Larson DB, Ter-Pogossian MM (1976) Blood brain barrier permeability of ^{11}C labeled alcohols and ^{15}O labeled water. Am J Physiol 230:543–547

Raichle ME, Markham J, Larson K, Grubb RLJ, Welch MJ (1981) Measurement of local cerebral blood flow in man with positron emission tomography. J Cereb Blood Flow Metab 1:S19–20

Rasey JS, Koh W, Grierson JR, et al. (1989) Radiolabeled fluoromisonidazole as an imaging agent for tumor hypoxia. Int J Radiat Oncol Biol Phys 17:985–992

Rasey JS, Nelson NJ, Chin L, et al. (1990) Characteristics of the binding of labeled fluoromisonidazole in cells in vitro. Radiat Res 122:301–308

Ratib O, Phelps ME, Huang SC, Henze E, Selin CE, Schelbert HR (1982) Positron tomography with deoxyglucose for estimating local myocardial glucose metabolism. J Nucl Med 23:577–583

Reeve J, Arlot M, Wooton R, et al. (1988) Skeletal blood flow, iliac histomorphometry, and strontium kinetics in osteoporosis: a relationship between blood flow and corrected apposition rate. J Clin Endocrinol Metab 66:1124–1131

Rege S, Maass A, Chaiken L, et al. (1994) Use of positron emission tomography with fluorodeoxyglucose in patients with extracranial head and neck cancers. Cancer 73:3047–3058

Rege SD, Chaiken L, Hoh CK, et al. (1993) Change induced by radiation therapy in FDG uptake in normal and malignant structures of the head and neck: quantitation with PET. Radiology 189:807–812

Reiman RE, Huvos AG, Benua RS, et al. (1981) Quotient imaging with N-13 glutamate in osteogenic sarcoma. Cancer 48:1976–1981

Reiman RE, Rosen G, Gelbard AS, et al. (1982) Imaging of primary Ewing sarcoma with N-13 L-glutamate. Radiology 142:494–500

Reivich M, Kuhl DE, Wolf A, et al. (1979) The (F-18) fluorodeoxyglucose method for the measurement of local cerebral glucose utilization in man. Circ Res 44:127–137

Rhodes CG, Wise RJS, Gibbs JM, et al. (1983) In vivo disturbance of the oxidative metabolism of glucose in human cerebral gliomas. Ann Neurol 14:614–626

Robinson RG, Mayberg HS, Wong DF, et al. (1987) Relationship of poststroke depression to PET scan asymmetry in cortical serotonin receptors. Soc Neurosci 13:851–853

Rottenberg CA, Ginos JZ, Kearfott KJ, Junck L, Bigner DD (1984) In vivo measurement of regional brain tissue pH using positron emission tomography. Ann Neurol 15:98–102

Rozental JM, Levine RL, Mehta MP, et al. (1991a) Early changes in tumor metabolism after treatment: the effects of stereotactic radiotherapy. Int J Radiat Oncol Biol Phys 20:1053–1060

Rozental JM, Levine RL, Nickles RJ (1991b) Changes in glucose uptake by malignant gliomas: preliminary study of prognostic significance. J Neurooncol 10:75–83

Rozental JM, Cohen JD, Mehta MP, Levine RL, Hanson JM, Nickles RJ (1993) Acute changes in glucose uptake after treatment: the effects of carmustine (BCNU) on human glioblastoma multiforme. J Neurooncol 15:57–66

Rubin SC, Kairemo KJ, Brownell AL, et al. (1993) High-resolution positron emission tomography of human ovarian cancer in nude rats using 124I-labeled monoclonal antibodies. Gynecol Oncol 48:61–67

Sadato N, Yonekura Y, Senda M, et al. (1993) PET and the autoradiographic method with continuous inhalation of oxygen-15-gas: theoretical analysis and comparison with conventional steady-state methods. J Nucl Med 34:1672–1680

Sanchez JF, Urdiales JL, Mates JM, Nunez de Castro I (1992) The induction of ornithine decarboxylase by ornithine takes place at post-transcriptional level in perifused Ehrlich carcinoma cells. Cancer Lett 67:187–92

Sato K, Kameyama M, Ishiwata K, Katakura R, Yoshimoto T (1992) Metabolic changes of glioma following chemotherapy: an experimental study using four PET tracers. J Neurooncol 14:81–89

Sawataishi J, Mineura K, Sasajima T, Kowada M, Sugawara A, Shishido F (1992) Effects of radiotherapy determined by 11C-methyl-L-methionine positron emission tomography in patients with primary cerebral malignant lymphoma. Neuroradiology 34:517–519

Schelbert HR, Schwaiger M (1986) PET studies of the heart. In: Phelps M, Mazziotta J, Schelbert H (eds) Positron emission tomography and autoradiography: principles and applications for the brain and heart. Raven, New York, p 581

Schelstraete K, Simons M, Deman J, et al. (1982) Uptake of 13N-ammonia by human tumors as studies by positron emission tomography. Br J Radiol 55:797–804

Schwaiger M, Muzik O (1991) Assessment of myocardial perfusion by positron emission tomography. Am J Cardiol 67:35–43

Shackney SE, McCormack GW, Cuchural GJ (1978) Growth rate patterns of solid tumors and their relation to responsiveness to therapy. Ann Intern Med 89:107–121

Shah A, Schelbert HR, Schwaiger M, Hansen H, Selin C (1985) Measurement of regional myocardial blood flow with N-13 ammonia and positron-emission tomography in intact dogs. J Am Coll Cardiol 5:92–96

Shibata T, Yamamoto K, Hayashi N, et al. (1988) Dynamic positron emission tomography with 13N-ammonia in liver tumors. Eur J Nucl Med 14:607–611

Shields AF, Kozell LB, Link JM, et al. (1990a) Comparison of PET imaging using (11C)thymidine labeled in the ring-2 and methyl positions (abstr). J Nucl Med 31:794

Shields AF, Lim K, Grierson J, Link J, Krohn KA (1990b) Utilization of labeled thymidine in DNA synthesis: studies for PET. J Nucl Med 31:337–342

Shields AF, Graham MM, Kozawa SM, et al. (1992) Contribution of labeled carbon dioxide to PET imaging of carbon-11-labeled compounds. J Nucl Med 33:581–584

Slosman DO, Pittet N, Donath A, Polla BS (1993) Fluorodeoxyglucose cell incorporation as an index of cell proliferation: evaluation of accuracy in cell culture. Eur J Nucl Med 20:1084–1088

Smith FW, Heys SD, Evans NT, et al. (1992) Pattern of 2-deoxy-2-[18F]-fluoro-D-glucose accumulation in liver tumours: primary, metastatic and after chemotherapy. Nucl Med Commun 13:193–195

Snook DE, Rowlinson BG, Sharma HL, Epenetos AA (1990) Preparation and in vivo study of 124I-labelled monoclonal antibody H17E2 in a human tumour xenograft model. A prelude to positron emission tomography (PET). Br J Cancer Suppl 10:89–91

Sokoloff L, Reivich M, Kennedy C, et al. (1977) The (^{14}C) deoxyglucose method for the measurement of local cerebral glucose utilization: theory, procedure, and normal values in the conscious and anesthetized albino rat. J Neurochem 28:897–916

Strauss LG, Conti PS (1991) The applications of PET in clinical oncology. J Nucl Med 32:623–648

Strauss LG, Clorius JH, Schlag P, et al. (1989) Recurrence of colorectal tumors: PET evaluation. Radiology 170:329–332

Sweeney MJ, Ashmore J, Morris HP (1963) Comparative biochemistry of hepatomas. IV. Isotope studies of glucose and fructose metabolism in liver tumors of different growth rates. Cancer Res 23:995–1002

Syrota A, Dormon D, Berger J, et al. (1983) C-11 ligand binding to adrenergic and muscarinic receptors of the human heart studied in vivo by PET (abstr). J Nucl Med 24:20–21

Syrota A, Castaing M, Rougemont D, et al. (1985) Regional tissue pH and oxygen metabolism in human cerebral infarction studied with positron emission tomography. In: Greitz T, Ingvar DH, Widen L (eds) The metabolism of the human brain studied with positron emissiontomography. Raven, New York, pp 285–303

Taniguchi H, Oguro A, Takeuchi K, et al. (1993) Difference in regional hepatic blood flow in liver segments – noninvasive measurement of regional hepatic arterial and portal blood flow in human by positron emission tomography with H2(15)O. Ann Nucl Med 7:141–145

Tewari DS, Qian Y, Tewari M, et al. (1994) Mechanistic features associated with induction of metalloproteinases in human gingival fibroblasts by interleukin-1. Arch Oral Biol 39:657–664

Thorell JO, Stone ES, von Hilst H, Ingvar M (1993) Synthesis of [1-11C]D-glucosamine and evaluation of its in vivo distribution in rat with PET. Appl Radiat Isot 44:799–805

Tomura N, Kato T, Ogawa T, et al. (1990) The changes of cerebral blood flow and metabolism of normal brain tissue after surgery, radiation, and chemotherapy in brain tumor patients: evaluated by position emission tomography. No To Shinkei 42:1085–1092

Tse KKM, Buchpiguel CA, Mozley PD, Alavi A (1994a) Utility of FDG PET in detecting bony metastases. Comparison with bone scintigraphy. In: ICP, The Institute for Clinical PET. Proceedings of the Sixth Annual International PET Conference, Washington, DC

Tse N, Hoh C, Hawkins R, Phelps M, Glaspy J (1994b) Positron emission tomography diagnosis of pulmonary metastases in osteogenic sarcoma. Am J Clin Oncol 17:22–25

Tsurumi Y, Kameyama M, Ishiwata K, et al. (1990) 18F-fluoro-2'deoxyuridine (FUdR) as a tracer of nucleic acid metabolism in brain tumors. J Neurosurg 72:110–113

Tyler JL, Diksic M, Villemure JG, et al. (1987) Metabolic and hemodynamic evaluation of gliomas using positron emission tomography. J Nucl Med 28:1123–1133

Usdin E, Borchardt RT, Creveling CR (1979) Transmethylation. Elsevier, New York, pp 105–122

Uygur GA, Dogan AS, Hichwa RD, Kirchner PT, Watkins GL, Ponto LL (1994) Evaluation of malignant lymphomas with F-18 FDG and PET. In: ICP, The Institute for Clinical PET. Proceedings of the Sixth Annual International PET Conference. Washington, DC

Vaidyanathan G, Bigner DD, Zalutsky MR (1992) Fluorine-18-labeled monoclonal antibody fragments: a potential approach for combining radioimmunoscintigraphy and positron emission tomography. J Nucl Med 33:1535–1541

Van Brocklin HF, Liu A, Welch MJ, O'Neil JP, Katzenellenbogen JA (1994) The synthesis of 7 alpha-methyl-substituted estrogens labeled with fluorine-18: potential breast tumor imaging agents. Steroids 59:34–45

van Eijkeren M, De Schryver A, Goethals P, et al. (1992) Measurement of short-term 11C-thymidine activity in human head and neck tumours using positron emission tomography (PET). Acta Oncol 31:539–543

van Langevelde A, van der Molen H, Journee de Korver J, Paans AM, Pauwels EK, Vaalburg W (1988) Potential radiopharmaceuticals for the detection of ocular melanoma. III. A study with 14C and 11C labelled tyrosine and dihydroxyphenylalanine. Eur J Nucl Med 14:382–387

Vaupel P (1990) Oxygenation of human tumors. Strahlenther Onkol 166:377–386

Vaupel P, Kallinowski F, Okunieff P (1990) Blood flow, oxygen consumption and tissue oxygenation of human tumors. Adv Exp Med Biol 277:895–905

Velazquez M, Haller J, Amundsen T, Schuster DP (1991a) Regional lung water measurements with PET: accuracy, reproducibility, and linearity. J Nucl Med 32:719–725

Velazquez M, Weibel ER, Kuhn C, Schuster DP (1991b) PET evaluation of pulmonary vascular permeability: a structure-function correlation. J Appl Physiol 70:2206–2216

Verhagen A, Luurtsema G, Pesser JW, et al. (1991) Preclinical evaluation of a positron emitting progestin ([18F]fluoro-16 alpha-methyl-19-norprogesterone) for imaging progesterone receptor positive tumours with positron emission tomography. Cancer Lett 59:125–132

Wagner HN, Burns HD, Dannals RF, et al. (1983) Imaging dopamine receptors in the human brain by positron tomography. Science 221:1264–1267

Wahl RL (1994) Emerging applications of PET in oncology: melanoma, lymphoma and prostate cancers. In: ICP, The Institute for Clinical PET. Proceedings of the Sixth Annual International PET Conference. Washington, DC

Wahl RL, Henry CA, Ethier SP (1992) Serum glucose: effects on tumor and normal tissue accumulation of 2-[F-18]-fluoro-2-deoxy-D-glucose in rodents with mammary carcinoma. Radiology 183:643–647

Wahl RL, Zasadny K, Helvie M, Hutchins GD, Weber B, Cody R (1993) Metabolic monitoring of breast cancer chemohormonotherapy using positron emission tomography: initial evaluation. J Clin Oncol 11:2101–2111

Warburg O (1925) Über den Glukosestoffwechsel der Carcinomzelle. Klin Wochenschr 4:534–536

Watanabe A, Tanaka R, Takeda N, et al. (1989) DNA synthesis, blood flow, and glucose utilization in experimental rat brain tumors. J Neurosurg 70:86–91

Weber MJ, Nakamura KD, Salter DW, et al. (1984) Molecular events leading to enhanced glucose transport in Rous sarcoma virus-transformed cells. Fed Proc Am Soc Exp Biol 1984:2246–2250

Westera G, Reist HW, Buchegger F, et al. (1991) Radioimmuno positron emission tomography with monoclonal antibodies: a new approach to quantifying in vivo tumour concentration and biodistribution for radioimmunotherapy. Nucl Med Commun 12:429–437

Westerberg G, Langstrom B (1994) Labelling of proteins with 11C in high specific radioactivity: [11C]albumin and [11C]transferrin. Appl Radiat Isot 45:773–782

Wieland DM, Hutchins GD, Rosenspire KC, et al. (1989) (C-11) Hydroxyephedrine (HED): a high specific activity alternative to 6-(F-18)fluorometaraminol (FMR) for heart neuronal imaging. J Nucl Med 30:767–772

Wieland DM, Rosenspire KC, Hutchins GD, et al. (1990) Neuronal mapping of the heart with 6-[18F]fluorometaraminol. J Med Chem 33:956–964

Wilson CB, Snook DE, Dhokia B, et al. (1991) Quantitative measurement of monoclonal antibody distribution and blood flow using positron emission tomography and 124iodine in patients with breast cancer. Int J Cancer 47:344–347

Wolf W, Presant CA, Servis KL, et al. (1990) Tumor trapping of 5-fluorouracil: in vivo 19F NMR spectroscopic pharmacokinetics in tumor bearing humans and rabbits. Proc Natl Acad Sci USA 87:492–496

Wong WH, Hicks K (1994) A clinically practical method to acquire parametric images of unidirectional metabolic rates and blood spaces. J Nucl Med 35:1206–1212

Wooten R, Dove C (1986) The single-passage extraction of 18F in rabbit bone. Clin Phys Physiol Meas 7:333–345

Yang D, Emran A, Tansey W, et al. (1990) Radiosynthesis of fluorotamoxifen analogs (abstr). J Nucl Med 31:903

Yang DJ, Li C, Kuang LR, et al. (1994) Imaging, biodistribution and therapy potential of halogenated tamoxifen analogues. Life Sci 55:53–67

Yen CK, Budinger TF, Friedland RP, et al. (1982) Brain tumor evaluation using Rb-82 and positron emission tomography. J Nucl Med 23:532–537

Young AB, Frey KA, Agranoff BW (1986) Receptor assays: in vitro and in vivo. In: Phelps ME, Mazziotta JC, Schelbert HR (eds) Positron emission tomography and autoradiography: principles and applications for the brain and heart. Raven, New York, pp 73–95

Young H, Carnochan P, Zweit J, Babich J, Cherry S, Ott R (1994) Evaluation of copper(II)-pyruvaldehyde bis (N-4-methylthiosemicarbazone) for tissue blood flow measurement using a trapped tracer model. Eur J Nucl Med 21:336–341

16 Modification of Blood Flow

C.W. Song

16.1
Introduction

The development and morphological characteristics of tumor vasculatures and tumor blood flow have been extensively investigated during the past several decades, and considerable new information on tumor blood flow relevant to tumor treatment has been revealed in recent years (ENDRICH and MESSMER 1981; GULLINO 1991; JAIN 1988; JAIN and WARD-HARTLEY 1984; LÜSCHER and VANHOUTTE 1990; REINHOLD 1979; SECCOMBE and SCHAFF 1994; VAUPEL 1994; WARREN 1979). The growth of solid tumors requires an adequate vascular distribution and blood flow (FOLKMAN 1986; GULLINO 1991; WARREN 1979). Ironically, the response of tumors to various treatments, such as radiotherapy and chemotherapy, also requires an adequate blood flow. A limiting factor for complete control of tumors by radiotherapy is believed to be the presence of hypoxic cells in tumors which are about three times more resistant to radiotherapy than normoxic cells. Since the oxygen is supplied to tumor cells through

C. W. SONG, PhD, Department of Therapeutic Radiology, University of Minnesota Medical School, 424 Harvard Street S.E., Box 494 UMHC, Minneapolis, MN 55455, USA

tumor blood flow, the status of tumor blood flow greatly affects the effectiveness of radiotherapy. In chemotherapy, tumor blood flow serves as a vehicle for the delivery of cytotoxic agents to tumor cells. Therefore, adequate vascular supply and blood flow are prerequisites for a good response of tumors to most chemotherapeutic drugs. On the other hand, vascular insufficiency and poor tumor blood flow are desirable in treating tumors with hyperthermia since blood flow is the main heat dissipation route, and thus poor blood flow would facilitate tumor heating (JAIN and WARD-HARTLEY 1984; REINHOLD and ENDRICH 1986; SONG 1984, 1991; VAUPEL 1990). Reduction of tumor blood flow following administration of bioreductive drugs has been reported to enhance the drugs ability to kill hypoxic tumor cells (BROWN and GIACCIA 1994; CHAPLIN and TROTTER 1991; DENEKAMP 1991; DENEKAMP and HILL 1991). In this chapter, modifications of blood flow by various vasoactive compounds for the purpose of improving tumor therapies as well as changes in tumor blood flow by treatments, such as ionizing radiation and hyperthermia, are briefly discussed.

16.2
Tumor Vasculature

Detailed descriptions of the morphological and functional characteristics of the tumor vasculature are beyond the scope of this chapter. However, some features of tumor vasculature directly relevant to the modification of tumor blood flow are briefly described here. The hastily formed tumor blood vessels are composed of single-layered endothelial cells. The endothelial cells in tumor vessels are often separated by gaps, which are occupied by tumor cells. The thin, capillary-like tumor vessels lack smooth muscle layer and innervation, and thus they are unable to autoregulate (MATTSSON et al. 1979; REINHOLD 1979; WARREN 1979). It is not uncommon, however, to find mature blood vessels with a smooth muscle layer and neural junctions inside slowly growing tumors or highly differentiated tumors. Even in the fast growing tumors, blood vessels with normal features can often be found, particularly in the periphery of the tumors (MATTSSON et al. 1979). Most blood vessels having normal features in the tumors are believed to be blood vessels of normal tissues incorporated into the growing tumor mass. Surprisingly, the normal blood vessels which are incorporated into tumors are able to remain intact and retain their contractile properties. Since tumor vessels are unable to autoregulate, as previously mentioned, the changes in tumor blood flow are mainly due to changes in blood flow in adjacent normal tissues and arterial blood pressure. Therefore, the change in tumor blood flow caused by external factors is greatly influenced by the structural relationship between the tumor vascular beds and the vascular beds of the surrounding normal tissues (Fig. 16.1; HIRST 1989). When the vascular beds in tumors and those in normal tissues are in parallel, the change in blood flow in the tumors and that in normal tissues would be opposite for the following reason: an increase in normal tissue blood flow due to vasodilation would shunt away the blood flow from the tumor to the normal tissue, resulting in a decrease in tumor blood flow. Conversely, when the tumor blood flow is in

series with the normal tissue blood flow, the changes in tumor blood flow and that in normal tissue blood flow would be similar because the blood that leaves the normal tissue vascular bed directly flows into the tumors. In many tumors, the "series" and "parallel" types may be mixed or combined. The change in blood flow in such a "combination"-type vascular bed would vary depending on the relative contribution of the series and parallel types.

Another important feature of tumor blood flow, at least in experimental tumors, is that the blood flow spontaneously stops and resumes (CHAPLIN et al. 1989). The duration of such a temporal stoppage of blood flow is as long as several minutes. One of the probable causes for such intermittent blood flow is that either white blood cells or aggregated platelets temporarily plug the narrow tumor vessels. Even a small increase in rigidity of the blood cells for some reason, such as an increase in acidity in tumors, would greatly retard the movement of the cells through microvessels (JAIN 1988; NASH and DORMANDY 1990). Another cause for the intermittent blood flow is the temporal closing of part of the tumor vascular bed due to a decrease in perfusion pressure. As the tumor size increases, the tumor vessels elongate and branch without a concomitant increase in the number of feeding arteries, and thus the intravascular pressure in the elongated vessels progressively declines.

16.3
Mechanisms of Blood Flow Modification in Normal Tissues

The dynamic vascular tone in normal tissues is a consequence of interactions among neuronal influence, local release and action of vasoactive factors, systemically released hormones, and concentrations of ions, notably H^+, K^+ and Ca^{2+}, as well as the pO_2 and pCO_2 in the circulating blood. It is now an established fact that dilation and constriction of the vascular smooth muscle are caused, respectively, mainly by endothelium-derived relaxing factors (EDRF) and endothelium-derived contracting factors (EDCF) synthesized in endothelial cells upon stimulation by vasoactive factors (LÜSCHER and VANHOUTTE 1990; SECCOMBE and SCHAFF 1994). It should be noted, however, that many factors act as both EDRF and EDCF, depending on the vessel type and concentration (TODA 1984; VANHOUTTE et al. 1984; VANHOUTTE 1985; LÜSCHER and VANHOUTTE 1990).

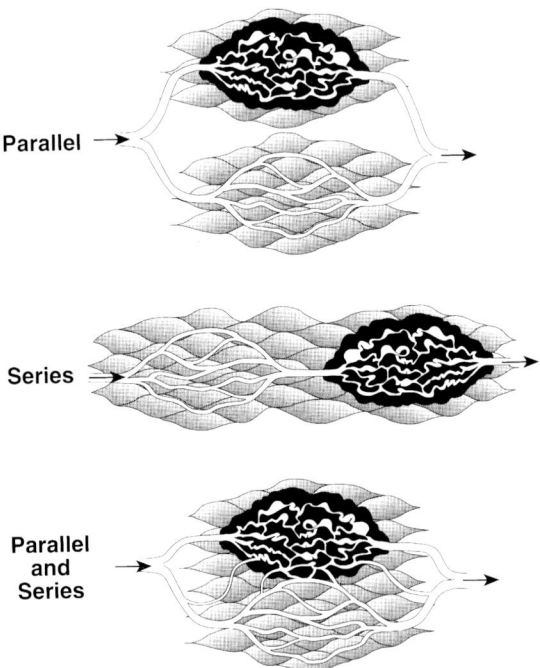

Parallel

Series

Parallel and Series

Fig. 16.1. Geometric relationship between the tumor vascular bed and normal tissue vascular bed. They are either "in parallel", "in series" or "combined." (Modeled after HIRST 1989)

16.3.1
Vasodilation

The evidence that endothelial cells are directly involved in vasodilation was provided by Furchgott and Zawadzki (1980). These investigators observed that the vascular relaxation caused by acetylcholine was due to relaxation of vascular smooth muscle caused by a substance synthesized in endothelial cells. It was then found that the messenger to smooth muscle from endothelial cells, endothelium-derived relaxing factor is none other than nitric oxide (NO) radical (Furchgott et al. 1987; Hutchinson et al. 1987; Ignarro et al. 1987; Palmer et al. 1987; Sakuma et al. 1988; Secrest and Chapnick 1988). Subsequently, the same NO was found to be responsible for the vasodilation caused by a variety of vasoactive compounds (Lüscher and Vanhoutte 1990; Seccombe and Schaff 1994; Fig. 16.2).

Acetylcholine, histamine, vasopressin, adrenaline (epinephrine) and noradrenaline (norepinephrine), bradykinin, adenosine triphosphate (ATP), adenosine diphosphate (ADP), 5-hydroxytryptamine (5-HT, serotonin) and thrombin are all receptor-mediated agonists for the synthesis of NO in endothelial cells (Seccombe and Schaff 1994). Sodium fluoride (NaF), phospholipase C, K^+ ions and the calcium ionophore A23187 are receptor-independent agonists for the synthesis of NO in endothelial cells. Different signal transductions pathways are involved in the synthesis of NO in endothelial cells upon stimulation by the variety of agonists mentioned.

Platelets tend to aggregate for a variety of reasons and retard blood perfusion. However, aggregating platelets release ADP and 5-HT, which activate specific receptors in endothelial cells and trigger the synthesis and release of NO (Cohen et al. 1983; Houston et al. 1985). Prothrombin, which is released from aggregating platelets, reacts with thromboplastin and becomes thrombin. The thrombin then acts on the thrombin-specific receptor on the surface of endothelial cells, evoking release of NO as well as prostacyclin and tissue plasminogen activator. The released NO in turn acts as a potent antithrombogenic factor, not only by inhibiting further aggregation, but by promoting platelet disaggregation.

It should be noted that NO is not the only EDRF. Arachidonic acid is a major component of membrane phospholipids and is metabolized in endothelial cells to several potent vasodilators, such as prostaglandin E_2 (PGE_2), prostacyclin (PGI_2), leukotrienes and thromboxane A_2 (DeMey et al. 1982; Gerritsen and Cheli 1983; Moncada and Vane 1979). All these metabolites of arachidonic acid formed in endothelial cells directly act on smooth muscle cells although some of them may stimulate endothelial cells to release NO. As discussed later, attempts have been made recently to increase tumor blood flow by manipulating the NO content in tumors.

16.3.2
Vasoconstriction

Endothelial cells also synthesize and release EDCF upon stimulation by various vasoactive compounds, many of which are also able to stimulate endothelial cells to release EDRF (Fig. 16.3; Altiere et al. 1986; Katûsic et al. 1988; Lüscher and Vanhoutte 1990; Seccombe and Schaff 1994). The known

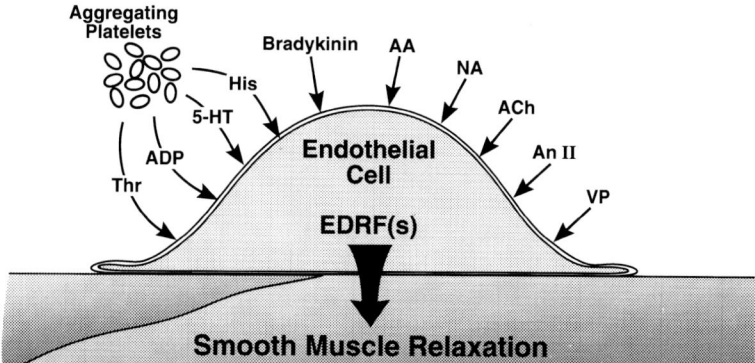

Fig. 16.2. Release of endothelium-derived relaxing factor (*EDRF*) from endothelial cells stimulated by various neurohumoral mediators. *AA*, arachidonic acid; *ACh*, acetylcholine; *ADP*, adenosine diphosphate; *An II*, angiotensin II; *His*, histamine; *5-HT*, 5-hydroxytryptamine (serotonin); *NA*, noradrenaline (norepinephrine); *Thr*, thrombin; *VP*, vasopressin

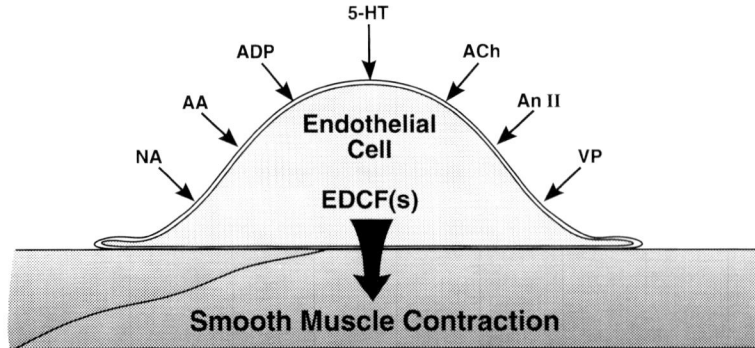

Fig. 16.3. Release of endothelium-derived contraction factor (*EDCF*) from endothelial cells stimulated by various neurohumoral mediators. *AA*, arachidonic acid; *ACh*, acetylcholine;

ADP, adenosine diphospate; *An II*, angiotensin II; *5-HT*, 5-hydroxytryptamine (serotonin); *NA*, noradrenaline (norepinephrine); *VP*, vasopressin

vasoconstrictors which stimulate endothelial cells to release EDCF are acetylcholine, noradrenaline (norepinephrine) and nicotine (ALTIERE et al. 1986; SHIRAHASE et al. 1988; USUI et al. 1987). The vasoconstricting effect of A23187 has also been reported to be endothelium-dependent (KATÛSIC et al. 1988) The nature of EDCF has not yet been completely revealed, but it appears to be composed of several major components, such as thromboxane A_2, angiotensin II, superoxide ions and endothelins.

The inhibitors of cyclooxygenase, an enzyme involved in the metabolism of arachidonic acid, block the vasocontraction caused by arachidonic acid, indicating that the metabolites of arachidonic acid are important components of the EDCF. A major metabolite of arachidonic acid is thromboxane A_2, which has potent vasoconstriction activity, suggesting that thromboxane A_2 is an EDCF (ALTIERE et al. 1986). However, it should be noted that thromboxane A_2 is synthesized by aggregating platelets in addition to endothelial cells.

Angiotensin II is another potent vasoconstrictor which is formed from angiotensin I (SAYE et al. 1984). Angiotensin II causes vasoconstriction by directly acting on smooth muscle cells and has been reported to potentiate adrenergic neurotransmission (KIFOR and DZAU 1987). Superoxide ions (O_2^-) are formed during activation of cyclooxygenase, and it has been known that O_2^- is a potent vasoconstrictor (KATÛSIC and VANHOUTTE 1989; KONTOS 1985). Therefore, it has been suggested that O_2^- might be an EDCF. In support of this notion, superoxide dismutase prevents endothelium-dependent contractions caused by A23187. Endothelin is a peptide released from endothelial cells and possesses vasoconstrictive activity (GILLESPIE et al. 1986; HICKEY et al. 1985). The levels of endothelin in circulating

blood are extremely low, indicating that the role of this peptide as a vasoconstrictor may be minimal under normal conditions. Endothelin-1 has been known to potentiate other vasoconstrictor hormones such as norepinephrine and serotonin.

16.4
Chemical Modification of Tumor Blood Flow

Numerous compounds have been investigated for their effect on tumor blood flow. However, only those that have some clinical relevance are discussed herein.

16.4.1
Reversible or Short-Term Changes

The effects of agents reported to cause reversible or short-term changes in tumor blood flow by affecting blood vessel walls are briefly described. It should be emphasized that many of the reversible changes in tumor blood flow by vasoactive factors may not result from changes in tumor vessels, but result indirectly from the changes in normal tissue blood flow.

16.4.1.1
Nitric Oxide

As discussed in the previous section, NO is believed to be the main EDRF. Several groups of investigators have recently studied the effect of NO on tumor blood flow. WOOD et al. (1993, 1994) reported that systemic administration of the NO donor SIN-1 to mice bearing the SCCVII/Ha tumor improved the energy level in tumors. An injection of SIN-1 also increased the response of tumors to X rays, indicating that SIN-1 increased tumor blood flow, thereby increasing tumor oxygenation. In view of the fact that only 10% of vessels in the SCCVII tumor possess smooth muscle (CHAPLIN and TROTTER 1991), how NO could increase the blood flow in the tumor is unclear. SONG et al. (1994, unpublished observation) investigated the effect of NO on the blood flow in R3230 Ac tumors of Fischer rats. An i.v. injection of the NO donor DEA/NO decreased tumor blood flow in the rat tumors by 30%. An i.v. injection of DEA/NO to C3H mice bearing FSaII tumors decreased the tumor pO_2 for about 30 min. It appeared that NO caused a vasodilation in normal tissues and shunted blood away from the tumors to normal tissues. Interestingly, inhibition of NO synthesis also decreases tumor blood flow probably by causing constriction in blood vessels including the arteries that feed the tumors. It has been known that synthesis of NO from L-arginine can be effectively inhibited by an analogue of L-arginine, L-NAME (N-nitro-L-arginine methyl ester). SONG et al. (1995) observed that an i.v. injection of L-NAME also decreased the pO_2 in FSaII tumors, suggesting that inhibition of NO synthesis decreased tumor blood flow. In this context, ANDRADE et al. (1992) reported that the blood flow in a murine adenocarcinoma and melanoma grown in sponge implants in mice significantly decreased by local injections of L-NAME.

16.4.1.2
Angiotensin II

Angiotensin, a known vasoconstrictor, was found to increase the filling of the tumor vasculature with contrast medium in angiographic examination of dog and human tumors (ELKIN and MENG 1966; EKELUND et al. 1972; EKELUND 1973). It appeared that constriction of normal tissue vasculature by angiotensin resulted in an increase in blood flow in tumor vasculature lacking constrictive properties. However, the increase in tumor blood flow by angiotensin II appeared to be short-lived since the improvement of angiographic imaging could be obtained only when the contrast medium was injected within a short time after angiotensin injection (EKELUND and GÖTHLIN 1977). EKELUND and LUNDERQUIST (1974) reported that the administration of angiotensin increased the diagnostic value of angiography in 70% of 80 human tumors in the kidneys, liver, pancreas, bone and soft tissues. SUZUKI et al. (1981) also reported that angiotensin II increased the ratio of tumor blood flow to normal tissue blood flow in human hepatic tumors. It was probable that the improvement in angiographic contrast between tumors and normal tissues in the studies mentioned above has resulted from smaller constriction in the arteries feeding the tumors relative to those in normal tissue, and not from an increase in tumor blood flow. JIRTLE et al. (1978) infused angiotensin II into rats bearing MW-9B mammary tumors and observed that blood flow in both tumors and the surrounding mammary gland tissue decreased. The decrease in blood flow, however, was greater in tumors than in normal tissues. HIRST et al. (1991) studied the effect of angiotensin II on the blood flow in NT carcinoma of CBA mouse. Systemic injection of angiotensin II increased the blood flow in tumors grown intradermally for the first several minutes followed by a return to control level, while it decreased the blood flow in tumors grown intraabdominally. A direct administration of angiotensin II to the arteries feeding the tumors caused vasoconstriction. It is probable that the tumor vessels that were constricted by angiotensin II in the study by HIRST et al. (1991) were of host origin having constrictive properties.

16.4.1.3
Vasopressin

As alluded to previously, vasopressin is an endogenous endothelium-dependent vasoactive peptide which dilates large cerebral and coronary blood vessels while it constricts renal and peripheral blood vessels (VANHOUTTE 1985). CARLSSON and ERIKSSON (1970) reported that vasopressin constricted the blood vessels in the kidney while it exerted no effect on the blood vessels in kidney tumors. GÖTHLIN (1976) reported that vasopressin caused vasoconstriction followed by a compensatory increase in blood flow in the normal kidney while it induced temporal vasodilation in hypervascularized kidney tumors.

16.4.1.4
Norepinephrine

The results of a number of investigations using a variety of experimental tumors indicated that norepinephrine decreases tumor blood flow (CATER et al. 1962; GULLINO and GRANTHAM 1961; RANKIN and PHERNETTON 1976). ACKERMAN and HECHMER (1977) reported that in a Walker tumor implanted in the rat liver, norepinephrine appeared to open up capillary-like vessels in the central parts of the tumor. MATTSSON et al. (1978) reported that norepinephrine reduced the blood flow in hepatoma as well as in normal tissues of rats and concluded that the change in blood flow by norepinephrine was due either to a direct effect on tumor blood vessels or to an indirect effect on normal vessels in the transplantation area. JIRTLE et al. (1978) reported that the blood flow in a mammary carcinoma of rats was significantly larger than that in surrounding mammary gland tissues and that continuous infusion of norepinephrine into host rats preferentially reduced the tumor blood flow. As a consequence of such differential effects, the blood flow in the tumors and normal tissues became almost equal after norepinephrine injection.

16.4.1.5
5-Hydroxytryptamine

5-hydroxytryptamine (5-HT, serotonin) is a potent endogenous vasoactive agent which stimulates endothelial cells to release NO, and thus causes relaxation of smooth muscle (Fig. 2). Conversely, 5-HT causes vasoconstriction when it acts directly on the vascular smooth muscle (LÜSCHER and VANHOUTTE 1990). 5-HT has been reported to reduce blood flow in a variety of experimental tumors (JIRTLE 1988; KNAPP et al. 1985). A local decrease in blood pressure due to a dilation in blood vessels in normal tissues surrounding the tumors appeared to be the major cause for the decline in tumor blood flow. HIRST et al. (1991) observed that an i.p. injection of 5-HT to tumor-bearing mice caused an abrupt decrease in tumor blood flow, more so in the tumors transplanted in muscle than those in the abdominal fat pad. CHAPLIN (1986) reported that reduction of blood flow by 5-HT enhanced the response of tumors to the bioreductive drug RSU-1069.

16.4.1.6
Hydralazine

Hydralazine has attracted considerable attention in recent years because of its potential usefulness in reducing tumor blood flow and thus increasing the response of tumors to bioreductive drugs and hyperthermia. A marked decrease in blood flow was observed in a variety of experimental rodent and dog tumors by hydralazine (CHAPLIN and TROTTER 1991; HORSMAN et al. 1989; KALMUS et al. 1990; VOORHEES and BABBS 1982). Such a decrease in tumor blood flow by hydralazine was apparently due to dilation of normal tissue blood vessels accompanied by an increase in normal tissue blood flow and a decline in local and systemic blood pressure. ROWELL and CLARK (1990) observed that hydralazine caused an increase rather than a decrease in blood flow in 20 human lung tumors and concluded that hydralazine may have caused dilation in blood vessels feeding the tumors. An important fact is that hydralazine does not dilate blood vessels in all normal tissues. HASEGAWA and SONG (1991) clearly demonstrated that the magnitude of the increase in blood flow caused by hydralazine varied markedly in different normal tissues. In fact, hydralazine decreased the blood flow in the liver, while it caused varying degrees of increases in the blood flow in other normal tissues. Consequently, the blood flow ratio in tumors to that in normal tissues markedly decreased for the tumors grown in the muscle, while it did not change for the tumors grown in other sites. The effect of hydralazine on tumor blood flow may be dose dependent. KALMUS et al. (1990) observed that whereas doses of hydralazine of $1.0\,\mu g/kg$ or more reduced the microcirculation in FSall tumors of mice, $0.25\,\mu g/kg$ hydralazine caused a transitional increase in blood flow in the tumors.

16.4.1.7
Nicotinamide

Intermittent tumor blood flow has been demonstrated to produce microregional and acute hypoxia in tumors and render the tumors resistant to radiotherapy. Nicotinamide was first reported to be a radiosensitizer of tumors by CALCUTT et al. in 1970, but the mechanism of such nicotinamide-induced radiosensitization was unclear. It is now believed that nicotinamide prevents the intermittent blood flow in tumors (HONESS and BLEEHEN 1993; HORSMAN et al. 1990; LEE and SONG 1992). KELLEHER and VAUPEL (1994) reported that a subtle increase in tumor oxygenation, probably due to prevention of intermittent blood flow, accounts for the radiosensitization of in vivo tumors by nicotinamide. In addition to acute hypoxia caused by the limitation of blood flow, chronically hypoxic cells exist in regions beyond the limitation of oxygen diffusion in tumors. The chronically hypoxic cells may be reoxygenated by increasing the oxygen transport capacity of circulating blood. The feasibility of increasing the radioresponse of tumors with the use of nicotinamide to reduce acute hypoxia in combination with other agents such as carbogen (CHAPLIN et al. 1991), perfluoro compounds (KJELLEN et al. 1991; ONO et al. 1993) or pentoxifylline (LEE et al. 1992) to lessen chronic hypoxia has been investigated.

16.4.2
Irreversible or Long-Term Changes

Tumor vasculature may serve as a potent target for tumor treatment because prolonged ischemia as a result of vascular damage would lead to massive tumor cell death (DENEKAMP 1991). Flavone acetic acid and some cytokines, such as interleukin-1 and TNF, belong to a group of agents that are able to cause longterm stoppage of tumor blood flow or nearly irreversible vascular damage in tumors.

16.4.2.1
Flavone Acetic Acid

Flavone acetic acid (FAA), a synthetic flavonoid, is only slightly cytotoxic in vitro. However, when injected into tumor-bearing mice, it causes a rapid and marked reduction in tumor blood flow and induces ischemic cell death in the tumors within several hours (CAPOLONGO et al. 1987; CORBETT et al. 1986). SMITH et al. (1991) reported that in addition to the vascular shutdown, the number of platelets in circulation rapidly dropped and the blood clotting time became shorter after FAA injection into tumor-bearing mice. The reduction of blood flow and the decline in platelet number were correlated. It was suggested that some tumor factor(s) in the circulation potentiate the effect of FAA to deplete the platelet pool and cause blood coagulation. There are some indications that the antitumor effect of FAA is related to immunogenicity of tumors (BIBBY et al. 1991). HILL et al. (1991) concluded that while it is clear that antitumor activity of FAA involves the vasculature in solid tumors, it is also clear that unknown tumor-host factors modify the action of FAA. Unfortunately, for reasons which are still unclear, the anti-

tumor activity or antivasculature activity of FAA is not evident in human tumors (KERR et al. 1989).

16.4.2.2
Cytokines

It has been known that the antitumor activity of interleukin-1α (IL-1α) and tumor necrosis factor-α (TNF-α) is much more marked in vivo than in vitro. Such a disparity between the effects of these cytokines in vitro and in vivo has been attributed to their antivascular activity and also to stimulation of immune reactions in vivo (BEVILACQUA et al. 1986; NAWROTH and STERN 1986). IL-1α and TNF-α are different cytokines, but they share a number of similar biological properties. BELARDELLI et al. (1989) concluded that alterations of tumor blood vessels may be the primary events leading to early changes in tumor metabolism and subsequent tumor cell death in solid tumors treated with IL-1β and TNF. BRAUNSCHWEIGER and his colleagues (1988) reported that IL-1α caused vascular damage in mouse tumors, accompanied by a decrease in tumor blood flow, vasodilation, vascular congestion, hemorrhagic necrosis and a decline in energy status. Other biological effects of these cytokines are an increase in the endothelial-cell-surface expression of leukocyte adhesion molecules and an increase in procoagulants, plasminogen activator inhibitor and prostaglandins, which are related to vascular damage. KALLINOWSKI et al. (1989) observed that both single and repeated application of TNF-α at $<10\,\mu g/kg$ increased the blood flow in rat tumors, whereas at $>100\,\mu g/kg$, it decreased tumor blood flow. When combined, IL-1α and TNF-α exert their antitumor activity additively. By decreasing tumor blood flow, IL-1α and TNF-α appear to induce hypoxia and lower the energy supply in tumors (KLUGE et al. 1992) and thus enhance the effect of bioreductive drugs (BRAUNSCHWEIGER et al. 1991) and hyperthermia (SONG et al. 1993).

16.4.3
Modifiers of Blood Viscosity

Blood flow rate in tissues is a function not only of vessel diameter but also of blood viscosity, which is in turn dependent on the number of red and white cells, deformability of red and white cells, tendency of RBCs to aggregate, activity and aggregation of platelets and plasma viscosity (NASH and DORMANDY 1990; SEVICK and JAIN 1988). The blood flow rate itself alters blood viscosity; as the blood flow rate declines, the aggregation of red blood cells and thus the viscosity of blood increases. The tumor vessels are very permeable, probably due to the immature nature of these vessels characterized by large gaps between endothelial cells, and lack a smooth muscle layer. Blood may become increasingly viscous as a result of extravasation of plasma or plasma water as the blood passes through the long, immature tumor vessels. In addition, the biochemical composition of blood may change, including a decrease in plasma pO_2 and an increase in lactic acid content. The combination of these changes in the blood may increase the rigidity of red blood cells. It is probable that such local rheological changes are among the causes for the intermittent blood flow in tumors, as discussed previously. Various drugs that are able to decrease the viscosity of blood have been used in treating patients with circulatory disorders. Calcium antagonists or Ca^{2+} entry blockers and pentoxifylline are known to decrease red blood cell rigidity and have been tested for their usefulness in increasing tumor blood flow. Conversely, an increase in blood viscosity and a decrease in tumor blood flow has some therapeutic potential in enhancing the effects of hyperthermia or bioreductive drugs. The most well known means of increasing plasma viscosity is hyperglycemia.

16.4.3.1
Calcium Antagonists

Calcium channel blockers or calcium antagonists have characteristics of inhibiting vasoconstriction and thus increasing blood flow in tumors. The increase in blood flow in some tumors by calcium antagonists, however, may be attributed to dilation of normal blood vessels incoporated into the tumors (VAUPEL and MENKE 1989) as well as to increases in red cell fluidity (DECREE et al. 1979; SCOTT et al. 1980). JIRTLE (1988) demonstrated that flunarizine, Ca^{2+} entry blocker, could increase the blood flow in the SMT-2A tumor of rats. Interestingly, the increase in blood flow in SMT-2A tumors by flunarizine occurred in both the outer viable and the inner necrotic regions, which was different from the preferential increase in blood flow in the peripheral area by other vasodilating compounds. Further studies indicated that flunarizine reduced intermittent blood flow in

SMT-2A tumors, resulting in a reduction of acutely hypoxic tumor cells. HILL and STIRLING (1987) have shown that flunarizine increased blood flow in the KHT murine sarcoma, and VAUPEL and MENKE (1987) demonstrated that flunarizine improved the oxygenation status in the DS carcinosarcoma of rats by increasing blood flow. WOOD and HIRST (1989) investigated the effects of four calcium antagonists, flunarizine, verapamil, nisoldipine and diltiazem, on the blood flow and the radioresponse of SCCVII/St mouse tumors. Variable increases in blood perfusion and radiosensitization in tumors were obtained by these compounds. At high doses, however, verapamil, nisoldipine and diltiazem caused marked reductions in blood flow and an increase in radioresistance. VAUPEL and MENKE (1989) reported that the changes in blood flow in DS carcinosarcoma of rats effected by flunarizine, verapamil and diltiazem were very dependent on the changes in effect of the drugs on the mean arterial blood pressure and on the proportions of normal vessels in the tumors. A noticeable increase in tumor blood flow was effected by flunarizine, which appeared to reduce the flow resistance in the host arterioles incorporated into the tumors without reducing the mean arterial blood pressure.

16.4.3.2
Pentoxifylline

Pentoxifylline (PTX) is a derivative of methylxanthine and is being used clinically for patients with intermittent claudication. PTX has been known to increase the fluidity of both red and white blood cells, resulting in a decrease in blood viscosity (ARMSTRONG et al. 1990); AVIADO and PORTER 1984; LOWE 1990). EHRLY (1978), LEE et al. (1992), and SONG et al. (1992, 1994) reported that PTX increased blood flow and pO_2 in R3230 Ac tumors of rats and SCK and FSall tumors of mice. Such an increase in blood flow and pO_2 by PTX enhanced the radioresponse of tumors. HONESS et al. (1993, 1995) reported that PTX increased both blood flow and HpO_2 in RIF-1 tumors of mice and also increased tumor radiosensitivity, and PRICE et al. (1995) reported similar findings. The time course of the increase in tumor oxygenation corresponded well with that of the increase in blood flow. HIRST et al. (1992) also studied the effect of PTX on blood flow in NT mouse tumors and observed that PTX significantly increased blood flow in tumors grown in skin, muscle or gut. The effect of PTX in increasing tumor blood flow was greater in large tumors than in small tumors.

16.4.3.3
Hyperglycemia

Artificial induction of a hyperglycemic state by administration of high doses of glucose causes marked physiological changes, including decreased blood flow and increased acidity in tumors of various species (VAUPEL et al. 1989; WARD-HARTLEY and JAIN 1987; SEVICK and JAIN 1988). The possibility that the aforementioned physiological changes in tumors might increase the response of tumors to hyperthermia attracted a great deal of attention by investigators. CALDERWOOD and DICKSON (1980) reported that an i.p. injection of glucose reduced blood flow in three different rat tumors by more than 90% for a few hours. Indications are that the magnitude of the decline in tumor blood flow depends on the site of tumor growth. Ross et al. (1989) observed that the decrease in blood flow by hyperglycemia in the C6 glioma implanted intracerebrally was far less than that in the subcutaneous C6 glioma, suggesting that the blood flow in brain tumors may be resistant to hyperglycemic stress.

A number of mechanisms responsible for the decline in tumor blood flow by hyperglycemia has been proposed. In hyperglycemic animals, the cardiac output significantly decreased, which may be related to an increase in blood viscosity probably caused by an increase in red cell rigidity by glucose (WARD-HARTLEY and JAIN 1987). VAUPEL et al. (1989) observed that an i.p. injection of glucose induced a shift of intravascular water to the abdominal cavity, resulting in a marked increase in blood viscosity and a decrease in tumor blood flow. Subsequently, KALMUS et al. (1989) reported that an i.v. injection of glucose reduced blood flow in FSall tumors and normal skin of mice although the i.v. injection was less effective than the i.p. injection. These investigators concluded that the decline in blood flow by i.v. injected glucose was again due to a decrease in cardiac output and an increase in viscous resistance to flow, which did not seem to be glucose specific because similar effects occurred by i.v. galactose. The acidic intratumor environment caused by the reduction of blood flow might further increase the rigidity of red blood cells and further decrease tumor blood flow. Small decreases in pH by hyperglycemia were observed in human tumors (THISTLETHWAITE et al. 1987). Whether such a decline in pH in human tu-

mors by hyperglycemia was due to a decline in blood flow is not clear.

16.5
Physical Modification of Tumor Blood Flow

16.5.1
Radiation

There are numerous reports on the effects of radiation on blood flow in a variety of experimental tumors as well as in human tumors (e.g., CLEMENT et al. 1976; DEWHIRST et al. 1990; HILMAS and GILLETTE 1974; KALLMAN et al. 1972; SONG et al. 1972, 1974; TING et al. 1991; TOZER et al. 1991). The vascular changes in irradiated tumors are largely dependent on tumor type, tumor size, site of tumor growth and radiation dose and fractionation schedule. The conclusions on the radiation effects on tumor vasculature also tend to vary depending on what methods are used for the analysis of results. For example, tumors may appear to be well vascularized when studied using histologic methods after they begin to shrink because the parenchymal cells are dead and removed by that time. However, not all blood vessels that are visible in histologic preparations might function, and thus the blood flow might not be as large as the histologic preparations indicate. In general, irradiation of tumors with high doses of radiation in a single exposure causes vascular damage and decreases tumor blood flow. On the other hand, irradiation with intermediate doses, i.e., 5–10 Gy in a single exposure, appears to increase blood flow in some but not all tumors, while a single exposure to 1–2 Gy radiation causes no significant change in tumor vasculatures. The ideal treatment schedule in radiotherapy is supposed to preserve the functional integrity of the tumor vasculature, thereby preventing the surviving tumor cells from becoming radioresistant hypoxic cells.

16.5.2
Hyperthermia

With increasing interest in the hyperthermic treatment of tumors, the effect of heat on tumor blood flow has been intensely investigated, as reviewed by JAIN and WARD-HARTLEY (1984), REINHOLD and ENDRICH (1986), SONG (1984, 1991), VAUPEL et al. (1988) and VAUPEL (1990). The efficacy of hyperthermic treatment of tumors is greatly influenced by blood flow for primarily two reasons. First, the temperature distribution in tumors during heating depends mostly on the heat dissipation by blood flow. Second, the thermosensitivity of tumor cells is greatly influenced by environmental factors, such as pH and nutritional condition, which are dependent on blood flow. In rodent tumors grown subcutaneously or intramuscularly, the vasculature is more vulnerable to heat than the vasculatures in surrounding normal tissues. It has been observed that heating at 41°–42°C for 30–60 min induces small increases in tumor blood flow, usually less than two-fold. After heating at these temperatures, tumor blood flow returns to its original level. Upon heating to temperatures of 43°–44°C or higher, the blood flow in most rodent tumors begins to decrease soon after the initiation of heating and eventually stops completely. The mechanisms for the heat-induced decrease in tumor blood flow may vary depending on the heat dose. The blood viscosity might increase upon heating at relatively low temperatures due to a variety of reasons, such as extravasation of plasma, and the blood vessels might be plugged by hardened and aggregated white and red blood cells and also by aggregated platelets. When heated at high temperatures, e.g., 43°–44°C, hemorrhagic vascular damage prevails throughout the tumors, indicating that direct vascular damage occurs. Contrary to tumor blood flow, the blood flow in the skin and skeletal muscle of rodents increases as much as five- to tenfold upon heating at 42°–44°C. The temperature at which the blood flow starts to decline in the skin and muscle is 1°–2°C higher than that in the tumors (Fig. 16.4). Song et al. (1989a) observed that human skin blood flow increases as much as ten- to 15-fold at 43°–44°C. The heat dissipation through blood flow would be much more efficient in normal tissues than in tumors because of the significantly greater increase in blood flow in normal tissues relative to that in tumors. In situations in which tumor blood flow is markedly greater than that in surrounding normal tissues before heating, it may still remain larger than the normal tissue blood flow despite the large increase in normal tissue blood flow. The blood flow in internal organs has been reported to increase only two- to fourfold. However, the blood flow in internal organs is much larger than that in skin and muscle and usually greater than the blood flow in tumors grown in the organs. Therefore, the normal tissue blood flow in internal organs may remain larger than the tumor blood flow during heating. In general, the blood vessels in human tumors are more mature

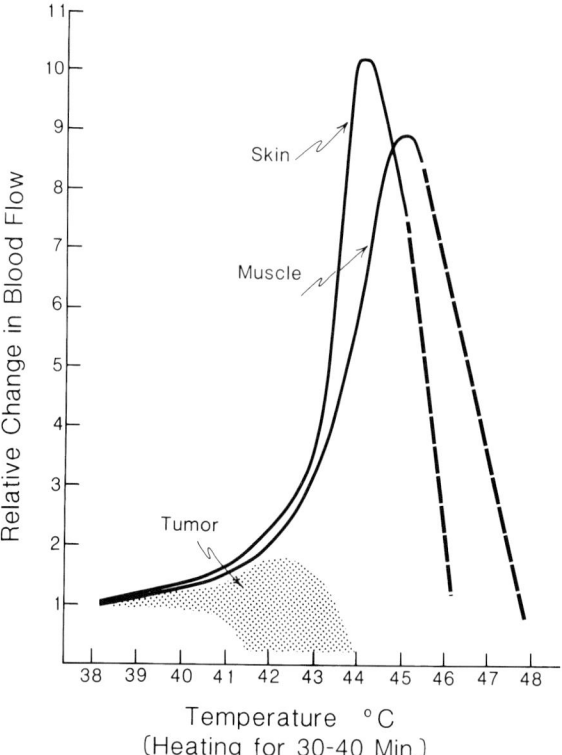

Fig. 16.4. Relative changes in blood flow in the skin or muscle and in tumors of rodents upon heating at various temperatures for 30–40 min. (From Song 1984)

than the blood vessels in rapidly growing rodent tumors, and thus the vasculature in human tumors may be heat resistant compared to that in rodent tumors. Nevertheless, human tumors treated with hyperthermia demonstrated significantly greater vascular damage than surrounding normal tissues, indicating, as in rodents, that the tumor vasculature is more vulnerable to heat than the normal tissue vasculatures (Lyng et al. 1991). Such differential responses of blood vessels in tumors and normal tissues to heat are believed to be the reason, at least in part, for the preferential damage to tumors by hyperthermia. A potentially important fact is that the vasculature in tumors and normal tissues is able to adapt to heat stress, so that they become heat resistant or thermotolerant after heating. Song et al. (1989b) observed that reheating at temperatures as high as 43°–44°C following preheating at relatively low temperatures, e.g., 42°C, resulted in a marked increase in tumor blood flow. The increase in tumor blood flow is undesirable for raising the tumor temperature to a cytotoxic level. On the other hand, a heat-induced increase in tumor blood flow in tumors with vascular thermotolerance would improve tumor oxygenation and thus increase the response of tumors to radiotherapy. The increase in tumor blood flow caused by heating at relatively low temperatures of 40°–42°C without preheating may also increase tumor pO$_2$ and thus enhance the radioresponse of tumor cells.

16.6
Summary

The vascular tone in normal tissues is controlled mainly by endothelium-derived relaxing factors (EDRF) and endothelium-derived contracting factors (EDCF). The concentrations of various ions, such as H$^+$, K$^+$ and Ca^{2+} in the circulating blood and also the tissue pO$_2$ and pCO$_2$ influence blood flow by changing the flexibility and aggregation of blood cells and by affecting synthesis of EDRF and EDCF. The EDRF and EDCF which are formed in endothelial cells upon stimulation by various vasoactive factors freely diffuse to smooth muscle cells and cause relaxation or constriction of the smooth muscle. Some vasoactive factors directly act on the smooth muscle. Since the newly formed tumor vessels are devoid of smooth muscle cells, the tumor blood vessels do not respond to the EDRF or EDCF or other vasoactive humoral factors. Therefore, the changes in tumor blood flow by vasoactive compounds are believed to be mostly due to the changes in blood flow in surrounding normal tissues. When the blood flow in normal tissues increases and blood pressure drops due to the action of vasodilators, the blood is shunted away from the tumors to normal tissues. Vasoconstrictors reduce the blood flow in normal tissues and raise the blood pressure, which leads to an increase in tumor blood flow. However, such an increase in tumor blood flow by some vasoconstrictors lasts for only a short period since subsequent constriction of arteries feeding the tumors causes a decrease in tumor blood flow. Consequently, in general, both vasodilators such as hydralazine and vasoconstrictors such as angiotensin II can decrease tumor blood flow. The intermittent tumor blood flow can be decreased by nicotinamide by as yet unknown mechanisms. Reduction of blood viscosity by increasing the deformability of red and white blood cells by pentoxifylline or calcium entry blockers such as flunarizine has been reported to increase tumor blood flow. In contrast, an increase in blood viscosity by hyperglycemia decreases tumor blood flow. FAA and TNF cause a prolonged decline in tumor blood flow mainly by damaging the endothelium. An exposure to moderate doses of ionizing radiation may increase tumor blood flow, while exposure to high doses of radiation decreases tumor blood flow. Tumor blood flow increases and thus tumor pO$_2$ increases by heating at moderate temperatures, i.e., <42°C, while tumor blood flow decreases at higher temperatures.

Acknowledgements. The author would like to thank Ms. Peggy Evans for her editorial assistance. This work was supported in part by NCI grant numbers CA13353 and CA44114.

References

Ackerman NB, Hechmer PA (1977) Effects of pharmacological agents on the microcirculation of tumor implanted in the liver. Bibl Anat 15:301–303

Altiere RJ, Kiritsy-Roy JA, Catravas JD (1986) Acetylcholine-induced contractions in isolated rabbit pulmonary arteries: role of thromboxane A2. J Pharmacol Exp Ther 236:535–541

Andrade SP, Hart IR, Piper PJ (1992) Inhibitors of nitric oxide synthase selectively reduce flow in tumour-associated neovasculature. Br J Pharmacol 107:1092–1095

Armstrong M, Needham D, Hatchell DL, Nunn RS (1990) Effect of pentoxifylline on the flow of polymorphonuclear leukocytes through a model capillary. Angiology 41:253–262

Aviado DM, Porter JM (1984) Pentoxifylline: a new drug for the treatment of intermittent claudication. Pharmacotherapy 4:297–307

Belardelli F, Proietti E, Ciolli V, et al. (1989) Interleukin-1β induces tumor necrosis and early morphologic and metabolic changes in transplantable mouse tumors. Similarities with the anti-tumor effects of tumor necrosis factor α and B. Int J Cancer 44:116–123

Bevilacqua MP, Pober JS, Majeau GR, Fiers W, Cotran R, Gimbrone MA (1986) Recombinant tumour necrosis factor induces procoagulant activity in cultured human vascular endothelium. Characterization and comparison with the actions of interleukin 1. Proc Natl Acad Sci USA 83:4533–4537

Bibby MC, Double JA, Phillips RM, Quinn PM (1991) Flavone acetic acid: is vascular shutdown the crucial mechanism of action? Int J Radiat Biol 60:395–399

Braunschweiger PG, Johnson CS, Kumar N, Ord V, Furmanski P (1988) Antitumor effects of recombinant human interleukin-1α on RIF-1 and PancO$_2$ solid tumors. Cancer Res 48:6011–6016

Braunschweiger PG, Jones SA, Johnson CS, Furmanski P (1991) Interleukin-1α-induced tumour pathophysiologies can be exploited with bioreductive alkylating agents. Int J Radiat Biol 60:369–372

Brown JM, Giaccia AJ (1994) Tumor hypoxia: the picture has changed in the 1990s. Int J Radiat Biol 65:95–102

Calcutt G, Ting SM, Preece AW (1970) Tissue NAD levels and the response to irradiation or cytotoxic drugs. Br J Cancer 24:380–388

Calderwood SK, Dickson JA (1980) Effect of hyperglycemia on blood flow, pH and response to hyperthermia (42°C) of the Yoshida sarcoma in the rat. Cancer Res 40:4728–4733

Capolongo LS, Balconi G, Ubezio P, Glavassi R, Taraboletti G, Regonesi A, Yoder C, D-Incalci M (1987) Antiproliferative properties of flavone acetic acid (NSC347512) (LM957): a new anticancer agent. Eur J Cancer Clin Oncol 23:1529–1535

Carlsson B, Eriksson U (1970) Renal angiography under the influence of vasopressin and bradykinin. Am J Roentgenol 109:161–166

Cater DB, Grigson CMB, Watkinsson DA (1962) Changes of oxygen tension in tumours induced by vasoconstrictor and vasodilator drugs. Acta Radiol 58:401–436

Chaplin DJ (1986) Potentiation of RSU-1069 tumour cytotoxicity by 5-hydroxytryptamine. Br J Cancer 54:727–731

Chaplin DJ, Trotter MJ (1991) Chemical modifiers of tumor blood flow. In: Vaupel P, Jain RK (eds) Tumor blood supply and metabolic microenvironment. Fischer, Stuttgart, pp 65–85 (Funktionsanalyse biologischer Systeme 20)

Chaplin DJ, Acker B, Olive PL (1989) Potentiation of the tumor cytotoxicity of melphalan by vasodilatory drugs. Int J Radiat Oncol Biol Phys 16:1131–1135

Chaplin DJ, Horsman MR, Aoki DS (1991) Nicotinamide, Fluosol-DA and carbogen: strategy to reoxygenate acutely and chronically hypoxic cells in vivo. Br J Cancer 63:109–113

Clement JJ, Song CW, Levitt SH (1976) Changes in functional vascularity and cell number following X-irradiation of murine carcinoma. Int J Radiat Oncol Biol Phys 1:671–678

Cohen RA, Shepherd JT, Vanhoutte PM (1983) Inhibitory role of the endothelium in the response of isolated coronary arteries to platelets. Science 221:273–274

Corbett TH, Bissery MC, Wozniak A, et al. (1986) Activity of flavone acetic acid (NSC-347512) against solid tumours of mice. Invest New Drugs 4:207–220

DeCree J, DeCock W, Guekens H, DeClerk F, Beerens M, Verhaegen H (1979) The rheological effects of cinnarizine and flunarizine in normal and pathological conditions. Angiology 30:505–515

DeMey JG, Claeys M, Vanhoutte PM (1982) Endothelium-dependent inhibitory effects of acetylcholine, adenosine diphosphate, thrombin and arachidonic acid in the canine femoral artery. J Pharmacol Exp Ther 222:166–173

Denekamp J (1991) The current status of targeting tumour vasculature as a means of cancer therapy: an overview. Int J Radiat Biol 60:401–408

Denekamp J, Hill S (1991) Angiogenic attack as a therapeutic strategy for cancer. Radiother Oncol Suppl 20:103–112

Dewhirst MW, Oliver R, Tso CY, Gustafson C, Secomb T, Gross JF (1990) Heterogeneity in tumor microvascular response to radiation. Int J Radiat Oncol Biol Phys 18:559–568

Ehrly A (1978) The effect of pentoxifylline on the human blood. Curr Med Res Opin 5:608–613

Ekelund L (1973) Pharmako-Angiographie der Niere. Radiologe 13:279–282

Ekelund L, Göthlin J (1977) Effect of angiotensin on normal renal circulation determined by angiography and a dye dilution technique. Acta Radiol Diagn 18:39–48

Ekelund L, Lunderquist A (1974) Pharmacoangiography with angiotension. Radiology 110:533–540

Ekelund L, Göthlin J, Lunderquist A (1972) Diagnostic improvement with angiotensin in renal angiography. Radiology 105:33–37

Elkin M, Meng C-H (1966) The effect of angiotensin on renal vascularity in dogs. Am J Roentgenol 98:927–934

Endrich B, Messmer K (1981) Microcirculation in transplanted tumors. Drug Res 31:2007–2011

Folkman J (1986) How is blood vessel growth regulated in normal and neoplastic tissue? Cancer Res 46:467–473

Furchgott RF, Zawadzki JV (1980) The obligatory role of endothelial cells in the relaxation of arterial smooth muscle by acetylcholine. Nature 299:373–376

Furchgott RF, Khan MT, Jothianandan D (1987) Comparison of endothelium-dependent relaxation and nitric oxide-induced relaxation in rabbit aorta (abstr). Fed Proc 46:385

Gerritsen ME, Cheli CD (1983) Arachidonic acid metabolism and prostaglandin endoperoxide metabolism in isolated rabbit coronary microvessel and isolated and cultivated coronary microvessel endothelial cells. J Clin Invest 72:1658–1671

Gillespie MN, Owasoyo JO, McMurtry IF, O'Brien RF (1986) Sustained coronary vasoconstriction provoked by a peptidergic substance released from endothelial cells in culture. J Pharmacol Exp Ther 236:339–343

Göthlin J (1976) Effect of vasopressin on human renal circulation investigated by angiography and a dye dilution technique. Acta Radiol Diagn 17:763–772

Gullino PM (1991) Tumor angiogenesis 1990: comments on the state of the art. In: Vaupel P, Jain RK (eds) Tumor blood supply and metabolic microenvironment. Fischer, Stuttgart, pp 11–26 (Funktionsanalyse biologischer Systeme 20)

Gullino PM, Grantham FH (1961) Studies on the exchange of fluids between host and tumor. II. The blood flow of hepatomas and other tumors in rats and mice. J Natl Cancer Inst 27:1465–1491

Hasegawa T, Song CW (1991) Effect of hydralazine on the blood flow in tumors and normal tissues in rats. Int J Radiat Oncol Biol Phys 20:1001–1007

Hickey KA, Rubanyi GM, Paul RJ, Highsmith RF (1985) Characterization of a coronary vasoconstrictor produced by cultured endothelial cells. Am J Physiol 248:550–556

Hill RP, Stirling D (1987) Oxygen and tumor response. In: Fielden EM, Fowler IF, Hendry JH, Scott D (eds) Proceedings of the 8th International Congress on Radiation Research, vol 2. Taylor and Francis, London, pp 725–730

Hill SA, Williams KB, Denekamp J (1991) Studies with a panel of tumours having a variable sensitivity to FAA, to investigate its mechanisms of action. Int J Radiat Biol 60:379–384

Hilmas DE, Gillette EL (1974) Microvasculature of C3H/Bi mouse mammary tumors after X-irradiation. Radiat Res 61:128–143

Hirst DG (1989) Tumor blood flow modification therapeutic benefit: is this approach ready for clinical application? Review. In: Michael B, Hance M (eds) Gray Laboratory 1989 Annual Report. Cancer Research Campaign, London, pp 14–17

Hirst DG, Hirst VK, Shaffi KM, Prise VE, Joiner B (1991) The influence of vasoactive agents on the perfusion of tumours growing in three sites in the mouse. Int J Radiat Biol 60:211–218

Hirst DG, Tozer GM, Sensky PL, et al. (1992) Tumor vascular physiology. In: Denekamp J (ed) Gray Laboratory 1992 Annual Report. Cancer Research Campaign, London, pp 40–46

Honess DJ, Bleehen NM (1993) Effects of the radiosensitising agent nicotinamide on relative tissue perfusion and kidney function in C3H mice. Radiother Oncol 27:140–148

Honess DJ, Dennis IF, Bleehen NM (1993) Pentoxifylline: its pharmcacokinetics and ability to improve tumour perfusion and radiosensitivity in mice. Radiother Oncol 28:208–218

Honess DJ, Andrews MS, Ward R, Bleehen NM (1995) Pentoxifylline increases RIF-1 tumour pO_2 in a manner compatible with its availability to increase relative tumour perfusion. Acta Oncol 34:385–389

Horsman MR, Christensen KL, Overgaard J (1989) Hydralazine-induced enhancement of hyperthermic damage in C3H mammary carcinoma in vivo. Int J Hyperthermia 5:123–136

Horsman MR, Wood PJ, Chaplin DJ, Brown JM, Overgaard J (1990) The potentiation of radiation damage by nicotinamide in the SCCVII tumour in vivo. Radiother Oncol 18:49–57

Houston DS, Shepherd JT, Vanhoutte PM (1985) Adenine nucleotides, serotonin and endothelium-dependent relaxation to platelets. Am J Physiol 248:H389–H395

Hutchinson PJA, Palmer RMJ, Moncada S (1987) Comparative pharmacology of EDRF and nitric oxide on vascular strips. Eur J Pharmacol 141:445–451

Ignarro LJ, Buga JM, Wood KS, Byrns RE, Chaudhuri G (1987) Endothelium-derived relaxing factor produced and released from artery and vein is nitric oxide. Proc Natl Acad Sci USA 84:9265–9269

Jain RK (1988) Determination of tumor blood flow: a review. Cancer Res 48:2641–2658

Jain RK, Ward-Hartley K (1984) Tumor blood flow: characterization, modifications and role in hyperthermia. IEEE Trans Sonics Ultrasonics 31:504–526

Jirtle RL (1988) Chemical modification of tumour blood flow. Int J Hyperthermia 4:355–371

Jirtle RL, Clifton KH, Rankin JHG (1978) Effects of several vasoactive drugs on the vascular resistance of MTW-9B tumors in W/Fu rats. Cancer Res 38:2385–2390

Kallinowski F, Schaefer C, Tyler G, Vaupel P (1989) In vivo targets of recombinant human tumour necrosis factor-α: blood flow, oxygen consumption and growth of isotransplanted rat tumours. Br J Cancer 60:555–560

Kallman RF, DeNardo GL, Stasch MJ (1972) Blood flow in irradiated mouse sarcoma as determined by the clearance of xenon-133. Cancer Res 32:483–490

Kalmus J, Okunieff P, Vaupel P (1989) Effect of intraperitoneal versus intravenous glucose administration on laser Doppler flow in murine FSall tumor and normal skin. Cancer Res 49:6313-6317

Kalmus J, Okunieff P, Vaupel P (1990) Dose-dependent effects of hydralazine on microcirculatory function and hyperthermic response of murine FSall tumors. Cancer Res 50:15–19

Katûsic ZS, Vanhoutte P (1989) Superoxide anion is an endothelium-derived contracting factor. Am J Physiol 257:H33–H37

Katûsic ZS, Shepherd JT, Vanhoutte PM (1988) Endothelium-dependent contractions to calcium ionophore A23187, arachidonic acid and acetylcholine in canine basilar arteries. Stroke 19:476–479

Kelleher DK, Vaupel PW (1994) Possible mechanisms involved in tumor radiosensitization following nicotinamide administration. Radiother Oncol 32:47–53

Kerr DJ, Maughan T, Newlands E, Rustin G, Bleehen NM, Lewis C, Kaye SB (1989) Phase II trials of flavone acetic acid in advanced malignant melanoma and colorectal cancer. Br J Cancer 60:104–106

Kifor I, Dzau VJ (1987) Endothelial renin-angiotensin pathway: evidence for intracellular synthesis and secretion of angiotensin. Circ Res 60:422–428

Kjellen E, Joiner MC, Collier JM, Johns H, Rojas A (1991) A therapeutic benefit from combining normobaric carbogen or oxygen with nicotinamide in fractionated X-ray treatments. Radiother Oncol 22:81–91

Kluge M, Elger B, Engel T, Schaefer C, Segga J, Vaupel P (1992) Acute effects of tumor necrosis factor α or lymphotoxin on global blood flow, laser Doppler flux, and bioenergetic status of subcutaneous rodent tumors. Cancer Res 52:2161–2173

Knapp WH, Debatin J, Layer K, Helus F, Altmann A, Sinn HJ, Ostertag H (1985) Selective drug induced reduction of blood flow in tumor transplants. Int J Radiat Oncol Biol Phys 11:1357–1366

Kontos HA (1985) Oxygen radicals in cerebral vascular injury. Circ Res 57:508–516

Kukovetz WR, Holzmann S, Wurm A, Pöch G (1979) Evidence of cyclin GMP-mediated relaxing effects of nitro-compounds in coronary smooth muscle. Naunyn Schmiedebergs Arch Pharmacol 310:129–138

Lee I, Song CW (1992) The oxygenation of murine tumor isografts and human tumor xenografts by nicotinamide. Radiat Res 130:65–71

Lee I, Kim JH, Levitt SH, Song CW (1992) Increase in tumor response by pentoxifylline alone or in combination with nicotinamide. Int J Radiat Oncol Biol Phys 22:425–429

Lowe GDO (1990) Drugs that modify red blood cell characteristics. In: Fleming JS (ed) Drugs and the delivery of oxygen to tissue. CRC Press, Boca Raton, pp 253–264

Lüscher TF, Vanhoutte PM (1990) The endothelium: modulator of cardiovascular function. CRC Press, Boca Raton

Lyng H, Monge OR, Bohler PJ, Rofstad EK (1991) The relevance of tumour and surrounding normal tissue vascular density in clinical hyperthermia of locally advanced breast carcinoma. Int J Radiat Biol 60:189–193

Mattsson J, Appelgren L, Karlsson L, Peterson H-I (1978) Influence of vasoactive drugs and ischemia on intra-tumour blood flow distribution. Eur J Cancer 14:761–764

Mattsson J, Appelgren L, Hamberger B, Peterson H-I (1979) Tumor vessel innervation and influence of vasoactive drugs on tumor blood flow. In: Peterson H-I (ed) Tumor blood circulation. CRC Press, Boca Raton, pp 129–135

Moncada S, Vane JR (1979) Pharmacology and endogenous roles of prostaglandin endoperoxide, thromboxane A_2 and prostacyclin. Pharmacol Rev 30:293–331

Nash GB, Dormandy JA (1990) The involvement of red cell aggregation and blood cell rigidity in impaired microcirculatory efficiency and oxygen delivery. In: Fleming JS (ed) Drugs and the delivery of oxygen to tissues. CRC Press, Boca Raton, pp 227–252

Nawroth P, Stern D (1986) Modulation of endothelial cell hemostatic properties by tumor necrosis factor. J Exp Med 163:740–733

Ono K, Masunaga S, Akuta K, Akaboshi M, Abe M (1993) Radiosensitization of SCCVII tumours and normal tissues by nicotinamide and carbogen: analysis by micronucleus assay. Radiother Oncol 28:162–167

Palmer RMJ, Ferrige AG, Moncada S (1987) Nitric oxide release accounts for the biological activity of endothelium-derived relaxing factor. Nature 327:524–526

Palmer RMJ, Ashton DS, Moncada S (1988) Vascular endothelial cells synthesize nitric oxide from L-arginine. Nature 333:664–666

Price MJ, Li LT, Tward JD, McBride WH, Lavey RS (1995) Effect of nicotinamide and pentoxifylline on normal tissue and FSa tumor oxygenation. Acta Oncol 34:391–395

Rankin JHG, Phernetton T (1976) Effects of prostaglandin E_2 on blood flow to the VX_2 carcinoma (abstract). Fed Proc 35:297

Rapoport RM, Murad F (1983) Antagonist-induced endothelium-dependent relaxation in rat thoracic aorta may be mediated through cGMP. Circ Res 52:352–357

Reinhold HS (1979) In vivo observations of tumor blood flow. In: Peterson HI (ed) Tumor blood circulation. CRC Press, Boca Raton, pp 115–128

Reinhold HS, Endrich B (1986) Tumor microcirculation as target for hyperthermia. Int J Hyperthermia 2:111–137

Ross BD, Mitchell SL, Merkle H, Garwood M (1989) In vivo ^{31}P and ^{1}H NMR studies of rat brain tumor pH and blood flow during acute hyperglycemia: differential effects between subcutaneous and intracerebral locations. Magn Res Med 12:219–234

Rowell NP, Clark K (1990) The effect of oral hydralazine on blood pressure, cardiac output and peripheral reistance with respect to dose, age and acetylation status. Radiother Oncol 18:293–298

Sakuma I, Stuehr D, Gross SS, Nathaan C, Levi R (1988) Identification of arginine as a precursor of endothelium-derived relaxing factor. Proc Natl Acad Sci USA 85:8664–8667

Saye JA, Singer HA, Peach MJ (1984) Role of endothelium in conversion of angiotensin I to angiotensin II in rabbit aorta. Hypertension 6:216–221

Scott CK, Persico FJ, Carpenter K, Chasin M (1980) The effects of flunarizine, a new calcium antagonist, on human red blood cells in vitro. Angiology 31:320–330

Seccombe JF, Schaff HV (1994) Vasoactive factors produced by the endothelium. Landes, Austin

Secrest RJ, Chapnick BM (1988) Endothelial-dependent relaxation induced by leukotrienes C_4, D_4 and E_4 in isolated canine arteries. Circ Res 62:983–991

Sevick EM, Jain RK (1988) Blood flow, and venous blood pH of tissue-isolated Walker 256 carcinoma during hyperglycemia. Cancer Res 48:1201–1207

Shirahase H, Usui H, Kurahashi K, Fujiwara M, Fukui K (1988) Endothelium-dependent contraction induced by nicotine in isolated canine basilar artery – possible involvement of thromboxane A_2 (TXA_2) like substance. Life Sci 42:437–445

Smith KA, Thurston G, Murray JC (1991) Systemic effects of FAA are enhanced by implanted tumours. Int J Radiat Biol 60:389–393

Song CW (1984) Effect of local hyperthermia on blood flow and microenvironment: a review. Cancer Res 44:4721s–4730s

Song CW (1991) Role of blood flow in hyperthermia. In: Urano M, Douple E (eds) Hyperthermia and oncology, vol 3. VSP, Utrecht, pp 275–315

Song CW, Payne JT, Levitt SH (1972) Vascularity and blood flow in X-irradiated Walker carcinoma 256 of rats. Radiology 104:693–697

Song CW, Sung JH, Clement JJ, Levitt SH (1974) Vascular changes in neuroblastoma of mice following X-irradiation. Cancer Res 34:2344–2350

Song CW, Chelstrom LM, Levitt SH, Haumschild DJ (1989a) Effects of temperature on blood circulation measured with the laser Doppler method. Int J Radiat Oncol Biol Phys 17:1041–1047

Song CW, Lin J-C, Chelstrom LM, Levitt SH (1989b) The kinetics of vascular thermotolerance in SCK tumors of A/J mice. Int J Radiat Oncol Biol Phys 17:799–802

Song CW, Hasegawa T, Kwon H, Lyons JC, Levitt SH (1992) Increase in tumor oxygenation and radiosensitivity by pentoxifylline. Radiat Res 130:205–210

Song CW, Lin J-C, Lyons JC (1993) Antitumor effect of interleukin 1-α in combination with hyperthermia. Cancer Res 53:324–328

Song CW, Makepeace CM, Griffin RJ, Hasegawa T, Osborn JL, Choi I-B, Nah BS (1994) Increase in tumor blood flow by pentoxifylline. Int J Radiat Oncol Biol Phys 29:433–437

Song CW, Makepeace CM, Griffin RJ, Keefer LJ (1995) Modification of tumor oxygenation by nitric oxide and inhibitors of nitric oxide synthase. In: Vaupel P, Kelleher DK, Gunderoth M (eds) Tumor oxygenation. Akademie der Wissenschaften und der Literatur, Mainz, pp 119–124

Suzuki M, Hori K, Abe I, Sito S, Sato H (1981) A new approach to cancer chemotherapy: selective enhancement of tumor blood flow with angiotensin II. J Natl Cancer Inst 67:663–669

Thistlethwaite AJ, Alexander GA, Moylan DJ III, Leeper DB (1987) Modification of human tumor pH by elevation of blood glucose. Int J Radiat Oncol Biol Phys 13:603–610

Ting LL, Belfi CA, Tefft M, Ngo FWH (1991) KHT sarcoma blood perfusion change after single-dose x-ray irradiation. Int J Radiat Biol 60:335–339

Toda N (1984) Endothelium-dependent relaxation induced by angiotension II and histamine in isolated arteries of dog. Br J Pharmacol 81:301–307

Tozer GW, Myers R, Cunningham VJ (1991) Radiation-induced modification of blood flow distribution in a rat fibrosarcoma. Int J Radiat Biol 60:327–334

Usui H, Kurahasi K, Shirahase H, Fukuki K, Fujiwara M (1987) Endothelium-dependent vasocontraction in response to noradrenaline in the canine cerebral artery. Jpn J Pharmacol 44:226–231

Vanhoutte PM (1985) Automatic nerve, endothelium, aggregating platelets and contraction of coronary arterial smooth muscle. In: Bevan JA, Godfraind T, Maxwell RA, Stoclet JC, Worcel M (eds) Vascular neuroeffector mechanisms. Elsevier, New York, pp 205–210

Vanhoutte PM, Katûsic ZS, Shepherd JT (1984) Vasopressin induces endothelium-dependent relaxations of cerebral and coronary, but not of systemic arteries. J Hypertens Suppl 3:421–422

Vaupel PW (1990) Pathophysiological mechanisms of hyperthermia in cancer therapy. In: Gautherie M (ed) Biological basis of oncologic thermotherapy. Springer, Berlin Heidelberg New York, pp 73–134

Vaupel PW (1994) Blood flow, oxygenation, tissue pH distribution, and bioenergetic status of tumors. Schering Research Foundation, Berlin

Vaupel P, Menke H (1987) Blood flow, vascular resistance and oxygen availability in malignant tumors upon intravenous flunarizine. Adv Exp Med Biol 215:393–398

Vaupel P, Menke H (1989) Effect of various calcium antagonists on blood flow and red blood cell flux in malignant tumors. Prog Appl Microcirc 14:88–103

Vaupel PW, Kallinowski F, Kluge M (1988) Pathophysiology of tumors in hyperthermia. Recent Results Cancer Res 107:65–75

Vaupel P, Okunieff P, Kluge M (1989) Response of tumour red blood cell flux to hyperthermia and/or hyperglycemia. Int J Hyperthermia 5:199–210

Voorhees WD, Babbs CF (1982) Hydralazine-enhanced selective heating of transmissible venereal tumor implants in dogs. Eur J Cancer Clin Oncol 19:1027–1033

Ward-Hartley KA, Jain RK (1987) Effect of glucose and galactose on microcirculatory flow in normal and neoplastic tissues in rabbits. Cancer Res 47:371–377

Warren BA (1979) The vascular morphology of tumors. In: Peterson HI (ed) Tumor blood circulation. CRC Press, Boca Raton, pp 1–47

Wood PJ, Hirst DG (1989) Modification of tumor response to calcium antagonists in the SCCVII/st tumor implanted at two different sites. Int J Radiat Biol 56:355–367

Wood PJ, Stratford IJ, Adams GE, Szabo C, Thiemermann C, Vane JR (1993) Modification of energy metabolism and radiation response of a murine tumor by changes in nitric oxide availability. Biochem Biophys Res Commun 192: 505–510

Wood PJ, Sansom JM, Stratford IJ, Adams GE, Szabo C, Thiemermann C, Vane JR (1994) Modification of metabolism of transplantable and spontaneous murine tumors by the nitric oxide synthase inhibitor nitro-L-arginine. Int J Radiat Oncol Biol Phys 29:433–437

17 Modification of Oxygen Supply

D.W. Siemann[1], M.R. Horsman[2], and D.J. Chaplin[3]

CONTENTS

17.1
Introduction

As a tumor grows, the vascular development often cannot keep pace with the rapid and uncontrolled proliferation of the malignant cell population. Consequently, solid tumor masses typically exhibit abnormal blood vessel networks which, unlike vessels in normal tissues, fail to provide adequate and homogeneous nutritional support to the tumor cells (Warren 1979; Vaupel et al. 1989). Such spatial and temporal heterogeneity of the microcirculation within a tumor can result in an expansion of intercapillary space and a decrease in vessel density. These factors may consequently lead to increased oxygen (O_2) diffusion distances. Solid tumors therefore can be comprised of cell subpopulations existing under a variety of microenvironmental con-

ditions. Indeed, areas of low oxygen tensions (pO_2) have been well documented in both rodent and human tumors (Kolstad 1968; Gatenby et al. 1988; Vaupel et al. 1989, 1991; Kallinowski et al. 1990; Höckel et al. 1993; Horsman et al. 1995). Frequency distributions of measured pO_2 values in malignancies have shown, in general, pO_2 values that are far lower than those in the surrounding normal tissues (Vaupel et al. 1989).

The hypoxia associated with tumors was originally considered to be strictly chronic in nature and based on the diffusion-limited model of hypoxia as described by Thomlinson and Gray (1955). In this model, large intercapillary distances resulting from rapid tumor cell proliferation lead to hypoxic cells existing at the rim of the O_2 diffusion distance. More recently it was suggested (Brown 1979; Sutherland and Franko 1980) that another type of hypoxia, namely acute or perfusion-limited hypoxia, might also exist in tumors. Why this intermittent blood flow occurs in tumors has not been established, although suggestions for the causative factors include vessel plugging by blood cells or circulating tumor cells (Jain 1988), collapse of vessels in regions of high tumor interstitial pressure (Sevick and Jain 1989), or spontaneous vasomotor activity in normal tissue vessels incorporated into the tumor which subsequently affects flow in downstream tumor microvessels (Intaglietta et al. 1977; Reinhold et al. 1977). Using a dual fluorescent stain technique coupled with histological evaluations, this type of hypoxia has been demonstrated in experimental tumors and shown to result from transient stoppages of tumor blood flow (Chaplin et al. 1986, 1987; Trotter et al. 1989). In view of these findings it is currently commonly believed that both types of hypoxia may exist in solid tumors (Coleman 1988).

Heterogeneity in oxygenation within tumors has been implicated as a major contributing factor for failure to cure neoplastic disease by radiation or chemotherapy. In the early 1900s it was recognized that interference with the blood supply could reduce

[1] D.W. Siemann, PhD, Department of Radiation Oncology, Shands Cancer Center, University of Florida, P.O.Box 100385, Gainesville, FL 32610, USA
[2] M.R. Horsman, PhD, Danish Cancer Society, Department of Experimental Clinical Oncology, Nørrebrogade 44, DK-8000 Aarhus C, Denmark
[3] D.J. Chaplin, PhD, Tumour Microcirculation Group, CRC Gray Laboratory Cancer Research Trust Mount Vernon Hospital, Northwood, Middlesex HA6 2JR, UK

radiation-induced skin reactions. Subsequently it was demonstrated that hypoxic cells were approximately two to three times more resistant to radiation than were normoxic cells. However, it was not until the 1955 paper of THOMLINSON and GRAY that the full magnitude of the potential impact of hypoxia on clinical radiation therapy was realized. These authors utilized tumor histology measurements and O_2 diffusion distance calculations to predict that viable tumor cells might exist in regions of radiobiological hypoxia. This hypothesis was confirmed experimentally, and it is now well established (1) that hypoxic cells exist in most rodent tumor models and (2) that these cells lead to refractory radiation responses (MOULDER and ROCKWELL 1984). Evidence for the presence of hypoxic cells in human tumors and their ability to affect the outcome of radiation therapy in patients is less readily available. However, (a) the clinical gains observed with hyperbaric O_2, (b) the recognition from both retrospective and prospective trials that anemia represents a poor prognostic factor often associated with local radiation failures, (c) the success with hypoxic cell sensitizers reported in the DAHANCA studies, and (d) the apparent significance of the intratumor pO_2 distributions as a determinant in patients receiving standard radiotherapy all support the notion that hypoxic cells are probably one of the major reasons for failure to control certain tumor types with conventional radiotherapy (DISCHE 1991; GATENBY et al. 1988; OVERGAARD 1989; HÖCKEL et al. 1993; BUSH 1986).

If inadequate tumor oxygenation represents a major reason for the failure to cure some human cancers by radiation therapy, then one approach to overcome this problem is the development of methods aimed at increasing the quantity of O_2 delivered to the tumor. A number of strategies to improve tumor oxygenation by altering O_2 delivery have been considered. These include: (1) the administration of artificial blood substitutes, (2) right-shifting the oxyhemoglobin (HbO_2) dissociation curve, (3) the use of agents that increase tumor blood flow or inhibit cellular respiration, and (4) high O_2 content gas breathing either alone or coupled with the agent nicotinamide.

17.2
Increase in Oxygen Transport Capacity of the Blood

Historically, one aspect of improving tumor oxygenation that has received considerable attention is the practice of correcting the detrimental effects of anemia in cancer patients undergoing radiotherapy. This approach was undertaken because retrospective clinical studies had indicated that anemic patients had a poorer prognosis than patients with normal Hb levels (BUSH 1986; OVERGAARD 1989). Transfusion of red blood cells before irradiation subsequently was shown to lead to reductions in (1) the fraction of hypoxic tumor cells in preclinical investigations (HILL et al. 1971; HIRST and WOOD 1987a; SIEMANN et al. 1989) and (2) the number of local recurrences in a few prospective clinical trials (BUSH 1986). However, given the potential problems associated with blood transfusions, other manipulations of the O_2 transport capacity, particularly artificial blood substitutes, have been proposed for use in cancer patients.

The artificial blood substitutes that have received the greatest interest in cancer research as O_2 carriers are the perfluorochemical (PFC) emulsions. PFCs are small in size (<0.2 μm) and have a high O_2 carrying capacity when combined with high O_2 tensions. PFC emulsions have been tested in both the laboratory and the clinic and have demonstrated effectiveness in a number of clinical settings. For example, PFCs may be used as O_2 transport agents to minimize ischemic damage after myocardial infarction (NUNN et al. 1983). These properties of high O_2 carrying capacity along with small size also have made the PFCs potential candidates to overcome hypoxia-related treatment resistance in solid tumors. Consequently the PFC emulsions have been studied extensively for possible cancer treatment application (TEICHER and ROSE 1984; SONG et al. 1985; ROCKWELL et al. 1986). Although there exists evidence that this approach can produce some improvements in the radiation response of certain experimental rodent tumors, results from clinical investigations of the leading PFC, Fluosol-DA, as an adjuvant to radiation therapy have not been overly encouraging (ROSE et al. 1986; SASAI et al. 1989).

An alternative approach to blood transfusion is the use of erythropoietin (rhEPO) administration to correct anemia. Recent preclinical investigations (KELLEHER et al. 1995) have shown that repeated administration of rhEPO to tumor-bearing anemic rats could lead to an increase in the median tumor pO_2. Yet despite this increase the degree of tumor oxygenation was still observed to be significantly lower than that measured in nonanemic animals. These findings nevertheless suggest that this agent may have the potential to improve the response of tumors to radiotherapy. Clearly more work with this treatment strategy is warranted.

17.3
Modification of Hemoglobin Affinity

Aside from the quantity of Hb available for O_2 transport, the amount of O_2 released at the tissues is strongly dependent on the shape or position of the HbO_2 dissociation curve (Hb affinity for O_2). Although normally only ~25% of the O_2 is released from the Hb between the arterial and venous ends of a capillary, a shift to the left (increased Hb affinity) adversely affects tissue oxygenation, whereas a shift to the right (reduced Hb affinity) will enhance O_2 release to the tissue. The position of the HbO_2 dissociation curve is normally modified by several allosteric factors, including pH, serum phosphates, 2,3-diphosphoglycerate (2,3 DPG), CO_2, and carbon monoxide (CO).

Factors that alter the O_2 binding capacity of Hb also can significantly influence the extent of tumor hypoxia. For example, CO levels similar to those found in smokers increase the Hb affinity for O_2 and shift the HbO_2 dissociation curve to the left (SIEMANN et al. 1978). At constant Hb levels this results in either a reduction in the quantity of O_2 being delivered to the tissues or a drop in the venous pO_2 in order for the Hb to release a normal amount of O_2 to the tissues. In tumor-bearing mice the presence of such HbCO levels resulted in both reduced tumor oxygenation, as assessed by computerized pO_2 histography (HORSMAN et al. 1994a), and larger fractions of radiobiologically hypoxic tumor cells (SIEMANN et al. 1978; GRAU et al. 1992). Consequently, increased Hb affinity for O_2 led to refractory responses to both single and multiple dose radiation treatments (SIEMANN et al. 1978; GRAU et al. 1992). These preclinical investigations support the recent clinical findings that smoking and/or increased HbCO levels are poor prognostic indicators for head and neck cancer patients treated for cure with radiation therapy (OVERGAARD and GRAU 1995).

Alternatively, a shift to the right of the HbO_2 dissociation curve will enhance the O_2 release from Hb such that, for the same arterial-venous O_2 difference, the venous O_2 tensions will be maintained at higher values. Such reductions in the Hb affinity for O_2 have been shown in high altitude adaptation studies to improve O_2 delivery to tissues (LENFANT et al. 1971). In addition, successful right-shifting of the HbO_2 dissociation curve has been possible in situ through the

infusion of 2,3-diphosphoglycerate (2,3 DPG) enriched red blood cells or the direct administration of glycolytic precursors (SIEMANN et al. 1989; VALERI 1982). Clinically, right-shifting of the HbO_2 dissociation curve has been shown to have therapeutic implications for ischemia and hypoxemia (VALERI 1982). Thus a shift to the right of the HbO_2 dissociation curve in a tumor-bearing host might be expected to improve tumor oxygenation and, hence, reduce the number of hypoxic cells in a tumor. Indeed, a reduction in the Hb affinity for O_2, mediated through an increase in the red blood cell 2,3 DPG levels, has been shown to sensitize hypoxic cells to radiation in a number of preclinical tumor models (SIEMANN et al. 1979; SIEMANN and MACLER 1986; HIRST and WOOD 1987b).

However, since changes in Hb affinity brought on by the manipulation of allosteric factors (such as 2,3 DPG) normally modifying the position of the HbO_2 dissociation curve may be subject to autoregulation, other agents that shift the HbO_2 dissociation curve to the right have also received attention. For example,

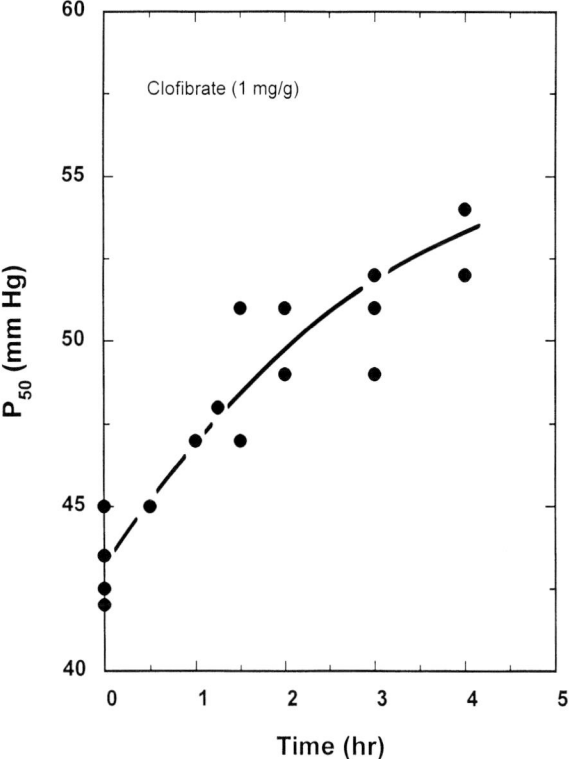

Fig. 17.1. P_{50} values (pO_2 values at which the Hb is 50% saturated with O_2) determined in C3H/HeJ mice at various times after injecting 1 mg/g clofibrate

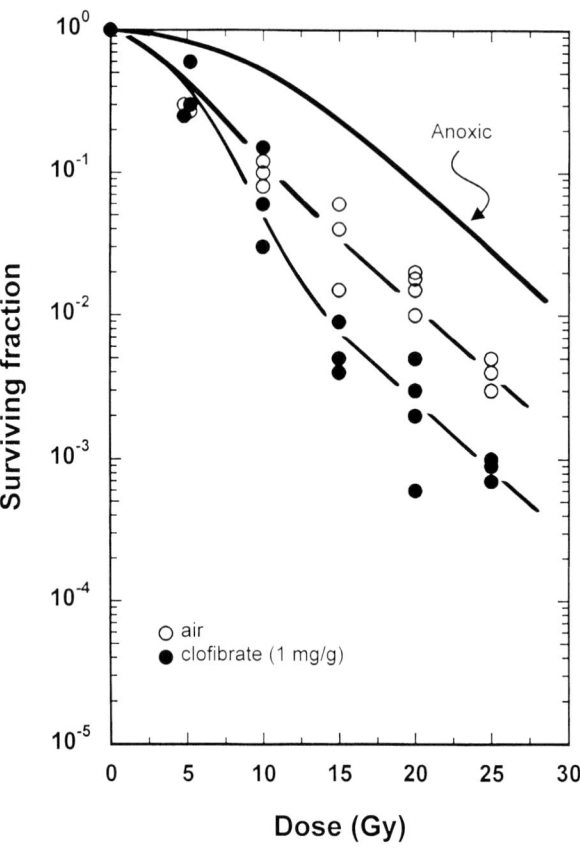

Fig. 17.2. Clonogenic cell survival in KHT sarcomas irradiated in mice under air-breathing conditions with various doses of radiation either along (○) or 4 hr after the administration of a 1 mg/g dose of clofibrate (●). The curve labeled anoxic was obtained by measuring the survival of tumor cells irradiated in dead animals and is shown for comparison

17.4
Inspiratory Hyperoxia

17.4.1
Hyperbaric Hyperoxia

One potential solution to overcoming tumor hypoxia is to increase the amount of O_2 carried in the blood by the breathing of gas mixtures with high O_2 content. Thus it was suggested in 1955 by CHURCHILL-DAVIDSON et al. that the breathing of O_2 at hyperbaric pressure (HPO), which increases the arterial O_2 tensions, might increase the O_2 diffusion distance sufficiently to oxygenate some of the hypoxic tumor cells. Under experimental conditions there is indeed good evidence that HPO is an efficient tool for improving the O_2 supply to the tumor. This has been demonstrated by direct measurements of tissue pO_2 values and of HbO_2 saturations of single red blood cells within tumor microvessels using cryospectrophotometry (MÜLLER-KLIESER et al. 1983). HPO has also proved successful in reducing the fraction of radiobiologically hypoxic tumor cells in a number of animal studies in which the effectiveness of tumor irradiation in animals breathing air or HPO was compared (MILNE et al. 1973; SUIT et al. 1981). Although a number of clinical trials with hyperbaric O_2 (HPO) have been performed, only a few have given improved results (DISCHE 1991). One explanation for the failure of this method may be the vasoconstrictive action of O_2, which can counteract the increase in blood pO_2 by reducing blood flow (MILNE et al. 1973; MÜLLER-KLIESER et al. 1983).

17.4.2
Normobaric Hyperoxia

Since vasoconstriction is less severe with high O_2 content gases at ambient pressure, both atmospheric O_2 and carbogen (5% CO_2: 95% O_2) have been considered as alternatives to HPO. Preclinical animal model studies have indicated that such gas mixtures may be effective at oxygenating tumors (HORSMAN et al. 1994a, 1995; FENTON and SIEMANN 1995) and sensitizing them to radiation (MILNE et al. 1973; KJELLEN et al. 1991; ROJAS 1991; CHAPLIN et al. 1993; SIEMANN et al. 1977, 1994; HORSMAN et al. 1994a, 1995; FENTON and SIEMANN 1995). However, the tumor sensitization efficacy of such gases can be strongly dependent on the preirradiation gas breathing time (SIEMANN et al. 1977, 1994; CHAPLIN et al.

the application of antilipidemics such as clofibrate, which bind the Hb molecule and alter O_2 release by reducing the Hb affinity (Fig. 17.1), has been shown to lead to radiosensitization of hypoxic cells in several animal tumor models (Fig. 17.2; see also HIRST et al. 1987). Although the antilipidemics are not the ideal candidates for improving the oxygenation of solid tumors in patients, the preclinical oncology investigations with these compounds not only indicate the validity of this approach but also support the notion that therapeutic interventions based on manipulating the Hb affinity for O_2 warrant further investigation.

1993). Yet, despite considerable experience with atmospheric O_2 and carbogen in experimental rodent tumors, few clinical studies have been performed using these gases (RUBIN et al. 1979).

17.5
Low-Dose Hyperthermia

Another possible approach to improve tumor oxygenation is the use of low-dose hyperthermia. The O_2 status of any tissue is very dependent on the blood supply to that tissue, and any increase in blood flow should improve oxygenation. Indeed, it has been well documented that heating tumors can modify blood flow (STEWART and BEGG 1983; SONG 1984; VAUPEL and KALLINOWSKI 1987). However, this effect is strongly dependent on both the time and temperature of heating. In general, at mild hyperthermia levels ($\leq 41°C$) and/or short exposure times ($\leq 30\,min$), there is a transient increase in tumor blood flow, while at higher temperatures and/or longer treatment times flow drops below control levels due to vascular occlusion that results from heat-induced vascular damage.

As was discussed earlier, the amount of O_2 released at the tissue is dependent on the HbO_2 dissociation curve. There exists some evidence that heat may modify the position of this curve (VAUPEL and KALLINOWSKI 1987). Again, this effect appears to be both exposure-time and temperature dependent, such that while a treatment at 40°C for 30 min will right-shift the dissociation curve, longer time intervals or higher temperatures actually left-shift the curve.

Although low-dose hyperthermia can improve the oxygenation of certain animal tumors (BICHER et al. 1980; VAUPEL and KALLINOWSKI 1987), some human tumor xenografts (ROSZINSKI et al. 1991), and even human tumors in patients (BICHER et al. 1980; BICHER and MITAGVARIA 1984), the exact relevance of this finding is not clear. Numerous experimental studies have examined the effect of combining heat and radiation to treat tumors. However, in general, the results show little or no enhancement of radiation damage when low-dose hyperthermia, such as might improve oxygenation, is used. Rather, improvements are seen only with higher heat doses that typically decrease blood flow and left-shift the HbO_2 dissociation curve (OVERGAARD 1978; FIELD and BLEEHEN 1979; HORSMAN and OVERGAARD 1989).

17.6
Vasoactive Agents

Blood flow is an important determinant of tissue oxygenation. Hence one method to overcoming the hypoxic cell problem in tumors is through the use of vasoactive agents directed at improving oxygenation by increasing blood flow. Foremost amongst the vasoactive compounds that can increase tumor blood flow are the calcium antagonists, particularly verapamil, flunarizine, and cinnarizine (JIRTLE 1988; KAELIN et al. 1982; VAUPEL and MENKE 1989). Calcium antagonists are known to inhibit the respiration rate of cancer cells under in vitro conditions (BIAGLOW et al. 1986; VAUPEL and MÜLLER-KLIESER 1986). Most critically, in several animal models these agents have been shown to be capable of effecting the microcirculation sufficiently to increase tumor blood flow and enhance the hypoxic cell radiation response. However, this topic is the subject of another chapter in this book.

Another interesting agent that has received considerable attention is nicotinamide (NIC). This compound has been shown to improve tumor blood flow and to enhance tumor radiation sensitivity. Perhaps most interestingly, it has been shown that the ability of NIC to increase the radiation response of tumors is at least in part a consequence of a reduction in the proportion of perfusion-limited or acutely hypoxic cells (HORSMAN et al. 1990b; CHAPLIN et al. 1990).

17.7
Targeting Acute and Chronic Hypoxia in Tumors

The manipulations described above are all directed at increasing the quantity of O_2 carried in the blood. Under these circumstances, tissue oxygenation is improved because the higher blood O_2 tensions (pO_2) result in an increase in the O_2 diffusion distance. As such it would appear that manipulations of this sort exert their influence primarily on the chronic or O_2-diffusion-limited hypoxic cells. Furthermore, since such manipulations have been demonstrated to improve tumor response to therapy, these results also imply that diffusion-limited hypoxia can critically impact therapy outcome. However, the vascular architecture and its response varies among tumor types, and it is likely that the *type of hypoxia* present may vary significantly between tumors. Indeed, it is currently commonly believed that two types of hypoxia may exist in tumors: (1) chronic or diffusion-limited hypoxia in which hypoxic cells exist at the rim of the oxygen diffusion distance and (2) acute or perfusion-limited hypoxia, perhaps arising as a consequence of transient ischemia. Although our understanding and detailed characterizations of the hypoxic cell populations in solid tumors to date have been based primarily on rodent tumor models and selected human tumor xenografts, it is clear from the available broad base of preclinical models that the *type of hypoxia* present in a tumor may significantly influence the choice of treatment strategy (COLEMAN 1988; CHAPLIN et al. 1993; HORSMAN et al. 1994b; SIEMANN et al. 1994).

17.8
Carbogen Breathing plus Nicotinamide

If different types of hypoxic cells exist in human tumors, attempts to eliminate hypoxia in tumors must involve treatments targeted against both chronic and acute hypoxia. Strategies such as PFC emulsions, respiratory hyperoxia, and changes in Hb affinity would prove ineffective at eliminating all of the radiobiologically hypoxic cells. One of the few agents that can reduce the temporal microregional

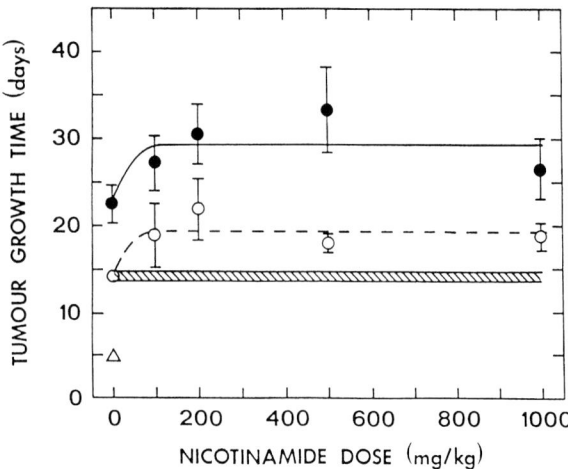

Fig. 17.4. The effect of nicotinamide dose and carbogen breathing on the response of a C3H mammary carcinoma to radiation (15 Gy). Various doses of nicotinamide (NIC) were injected 20 min prior to irradiation. Carbogen breathing was for 10 min prior to and during the radiation treatment. Data shown are control (△), radiation (▨), NIC + radiation (○), and NIC + carbogen + radiation (●). (Results modified from HORSMAN et al. 1994b)

fluctuations in blood flow (acute hypoxia) in tumors is NIC (HORSMAN et al. 1990b; CHAPLIN et al. 1990). An approach to improve the effectiveness of the radiation therapy therefore would be to combine an agent that attacks chronic hypoxia with one that can reduce intermittent changes in tumor blood flow.

Based on this hypothesis, several studies have now shown that NIC with carbogen breathing can lead to both improvements in the tumor oxygenation status (HORSMAN et al. 1994b; Fig. 17.3) and significant enhancements of the tumor response (HORSMAN et al. 1990a, 1994b; CHAPLIN et al. 1991, 1993; KJELLEN et al. 1991; ROJAS 1991; SIEMANN et al. 1994; Figs. 17.4, 17.5). Most critically, the combination of NIC plus carbogen breathing has shown improvements in the tumor radiation response in clinically relevant multifraction treatments (KJELLEN et al. 1991; ROJAS 1991) and with clinically achievable NIC doses (HORSMAN et al. 1994b). Another potential bonus of this combination of agents is that the use of NIC appears to reduce the influence that the preirradiation breathing time of high O_2 content gases can exert on the degree of tumor radiosensitization (CHAPLIN et al. 1993; SIEMANN et al. 1994).

A putative mechanism for the improved tumor responses observed with this combination of agents is that breathing carbogen increases the distance O_2 diffuses from the blood vessels into the tumor, thus

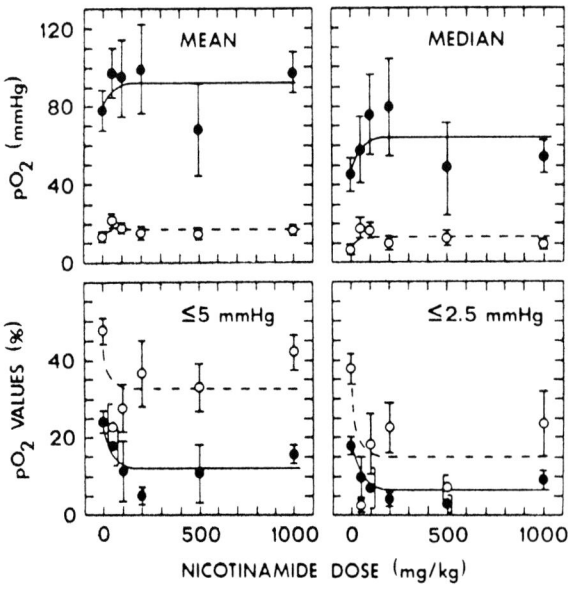

Fig. 17.3. The effect of nicotinamide dose and carbogen breathing on the oxygenation status of a C3H mammary carcinoma. Data shown are the mean and median pO_2 values as well as the percentage ≤2.5 or 5.0 mmHg determined under air (○) or carbogen (●) breathing conditions. (Data taken from HORSMAN et al. 1994b)

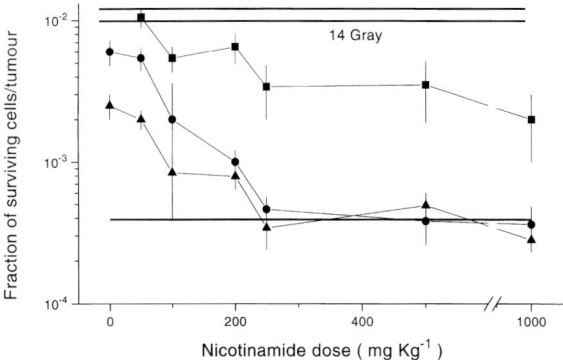

Fig. 17.5. The effect of treating SCCVII tumors with radiation (=), NIC 60 min prior to irradiation (■), NIC + carbogen 60 min prior to irradiation (●), or NIC + carbogen 15 min prior to irradiation (▲). *Lower solid line* indicates survival expected for fully aerobic cells. (Data taken from CHAPLIN et al. 1993)

reducing the fraction of diffusion-limited hypoxic cells. In contrast, NIC may reduce the temporal fluctuations in regional tumor blood flow and thus affect the proportion of perfusion-limited hypoxic cells. However, it should be recognized that neither carbogen breathing nor NIC dosing result in effects strictly confined to one hypoxic cell subpopulation. For example, carbogen therapy not only increases the tissue pO_2 (Fig. 17.3) but also can increase tumor blood flow (KRUUV et al. 1967). Similarly, while NIC probably acts predominantly by reducing intermittent changes in blood flow, this agent has also been shown to sensitize diffusion-limited hypoxic cells (CHAPLIN et al. 1990; HORSMAN et al. 1990b). Nevertheless, the observations that NIC combined with carbogen breathing can dramatically increase the tumor response to radiation has resulted in considerable interest in the evaluation of this combination in a clinical setting where hypoxia may be one of the factors limiting radiotherapy. Indeed, the combination of NIC and carbogen breathing is currently undergoing preliminary clinical testing in Europe.

17.9 Summary

Evidence exists to strongly support the contention that certain human malignancies fail curative radiation therapy because a proportion of the neoplastic cell population is receiving an inadequate supply of O_2 during the radiation treatment. Approaches to improving O_2 delivery include manipulating normally occurring physiological parameters in a man-

ner favorable to improving tissue oxygenation, using chemical agents to supplement normal O_2 transport, or specifically affecting parameters controlling O_2 delivery. Our current experience related to the hypoxia problem in cancer therapy further suggests that O_2-deficient cell populations may arise in tumors from both O_2 diffusion limitations and intermittent fluctuations in blood flow. While increasing the O_2-carrying capacity of the blood would serve to reduce diffusion-limited hypoxic cells, such therapeutic manipulations would have little impact on eliminating hypoxia resulting from transient variations in microregional blood flow. However, strategies which improve blood flow could conceivably impact the latter hypoxic cell type. Obviously, if both perfusion- and diffusion-limited hypoxic cells exist in tumors, one approach might be to combine different therapies such that each is aimed at eliminating a particular hypoxic cell subpopulation. This hypothesis is currently under active investigation in both preclinical laboratory and phase I/II clinical settings.

Acknowledgements. This work was supported by grants from the American NIH (numbers CA 36858 and CA 55300), the Danish Cancer Society, and the British Cancer Research Campaign.

References

Biaglow JE, Varnes ME, Jacobson B, Suit HD (1986) Effect of calcium channel blocking drugs on tumor cell oxygen utilization. Adv Exp Med Biol 200:583–589

Bicher HI, Mitagvaria NP (1984) Changes in tumor oxygenation during microwave hyperthermia: clinical relevance. In: Overgaard J (ed) Hyperthermic oncology, vol 1. Taylor and Francis, London, pp 169–172

Bicher HI, Hetzel FW, Sandhu TS, Frinak S, Vaupel P, O'Hara MD, O'Brien T (1980) Effects of hyperthermia on normal and tumour microenvironment. Radiology 137:523–530

Brown JM (1979) Evidence for acutely hypoxic cells in mouse tumours, and a possible mechanism of reoxygenation. Br J Radiol 52:650–656

Bush RS (1986) The significance of anemia in clinical radiation therapy. Int J Radiat Oncol Biol Phys 12:2047–2050

Chaplin DJ, Durand RE, Olive PL (1986) Acute hypoxia in tumors: implications for modifiers of radiation effects. Int J Radiat Oncol Biol Phys 12:1279–1282

Chaplin DJ, Olive PL, Durand RE (1987) Intermittent blood flow in a murine tumor. Radiobiological effects. Cancer Res 47:597–601

Chaplin DJ, Horsman MR, Trotter MJ (1990) Effect of nicotinamide on the microregional heterogeneity of oxygen delivery within a murine tumor. J Natl Cancer Inst 82:672–676

Chaplin DJ, Horsman MR, Aoki DS (1991) Nicotinamide, fluosol DA and carbogen: a strategy to reoxygenate acutely and chronically hypoxic cells in vivo. Br J Cancer 63:109–113

Chaplin DJ, Horsman MR, Siemann DW (1993) Further evaluation of nicotinamide and carbogen as a strategy to reoxygenate hypoxic cells in vivo: importance of nicotinamide dose and pre-irradiation breathing time. Br J Cancer 68:269–273

Churchill-Davidson I, Sanger C, Thomlinson RH (1955) High pressure oxygen and radiotherapy. Lancet 1:1091–1095

Coleman CN (1988) Hypoxia in tumors: a paradigm for the approach to biochemical and physiologic heterogeneity. J Natl Cancer Inst 80:310–317

Dische S (1991) A review of hypoxic cell radiosensitization. Int J Radiat Oncol Biol Phys 20:147–152

Fenton BM, Siemann DW (1995) Are direct measures of tumor oxygenation reflective of changes in tumor radiosensitivity following oxygen manipulation? Acta Oncol 34:307–311

Field SB, Bleehen NM (1979) Hyperthermia in the treatment of cancer. Cancer Treat Rev 6:63–94

Gatenby RA, Kessler HB, Rosenblum JS, Coia LR, Moldofsky PJ, Hartz WH, Broder GJ (1988) Oxygen distribution in squamous cell carcinoma metastases and its relatonship to outcome of radiation therapy. Int J Radiat Oncol Biol Phys 14:831–838

Grau C, Horsman MR, Overgaard J (1992) Influence of carboxyhemoglobin level on tumor growth, blood flow and radiation response in an experimental model. Int J Radiat Oncol Biol Phys 22:421–424

Hill RP, Bush RS, Yeung P (1971) The effect of anemia on the fraction of hypoxic cells in an experimental tumor. Br J Radiol 44:299

Hirst DG, Wood PJ (1987a) The adaptive response of mouse tumours to anaemia and retransfusion. Int J Radiat Biol 51:597–609

Hirst DG, Wood PJ (1987b) The influence of hemoglobin affinity for oxygen on tumor radiosensitivity. Br J Cancer 55:487–491

Hirst DG, Wood PJ, Schwartz HC (1987) The modification of hemoglobin affinity for oxygen and tumor radiosensitization by antilipidemic drugs. Radiat Res 112:164–172

Höckel M, Knoop C, Schlenger K, et al. (1993) Intratumoral pO$_2$ predicts survival in advanced cancer of the uterine cervix. Radiother Oncol 26:45–50

Horsman MR, Overgaard J (1989) Thermal radiosensitization in animal tumors: the potential for therapeutic gain. In: Urano M, Double E (eds) Hyperthermia and oncology, vol 2. VSP, Utrecht, pp 113–145

Horsman MR, Chaplin DJ, Overgaard J (1990a) Combination of nicotinamide and hyperthermia to eliminate radioresistant chronically and acutely hypoxic tumor cells. Cancer Res 50:7430–7436

Horsman MR, Wood PJ, Chaplin DJ, Brown JM, Overgaard J (1990b) The potentiation of radiation damage by nicotinamide in the SCCVII tumour in vivo. Radiother Oncol 18:49–57

Horsman MR, Khalil AA, Siemann DW, et al. (1994a) Relationship between radiobiological hypoxia in tumors and electrode measurement of tumor oxygenation. Int J Radiat Oncol Biol Phys 29:439–442

Horsman MR, Nordsmark M, Khalil AA, Hill SA, Chaplin DJ, Siemann DW, Overgaard J (1994b) Reducing acute and chronic hypoxia in tumours by combining nicotinamide with carbogen breathing. Acta Oncol 33:371–376

Horsman MR, Khalil AA, Nordsmark M, et al. (1995) The use of oxygen electrodes to predict radiobiological hypoxia in

tumours. In: Vaupel P, Kelleher DK, Günderoth M (eds) Tumor oxygenation. Fischer, Stuttgart, pp 49–57

Intaglietta M, Myers RR, Gross JF, Reinhold HS (1977) Dynamics of microvascular flow in implanted mouse mammary tumors. Bibl Anat 15:237–246

Jain RK (1988) Determinants of tumor blood flow: a review. Cancer Res 48:2641–2658

Jirtle RL (1988) Chemical modification of tumour blood flow. Int J Hyperthermia 4:355–371

Kaelin WGJ, Shrivastav S, Shand DG, Jirtle RL (1982) Effect of verapamil on malignant tissue blood flow in SMT-2A tumor-bearing rats. Cancer Res 42:3944–3949

Kallinowski F, Zander R, Hoeckel M, Vaupel P (1990) Tumor tissue oxygenation as evaluated by computerized-pO$_2$-histography. Int J Radiat Oncol Biol Phys 19:953–961

Kelleher DK, Matthiensen U, Thews O, Vaupel P (1995) Tumor oxygenation in anemic rats: effects of erythropoietin treatment vs blood transfusion. Acta Oncol, in press

Kjellen E, Joiner MC, Collier JM, Johns H, Rojas AF (1991) A therapeutic benefit from combining normobaric carbogen or oxygen with nicotinamide in fractionated X-ray treatments. Radiother Oncol 22:81–91

Kolstad P (1968) Intercapillary distance, oxygen tension and local recurrence in cervix cancer. Scand J Clin Lab Invest 106:145–157

Kruuv JA, Inch WR, McCredie JA (1967) 1. Effects of breathing gases containing carbon dioxide at atmospheric pressure. 2. Effects of vasodilator drugs. Cancer 20:51–65

Lenfant C, Torrance JD, Reynafarje C (1971) Shift in the O$_2$-Hb dissociation curve at altitude: mechanism and effect. J Appl Physiol 30:625–631

Milne N, Hill RP, Bush RS (1973) Factors affecting hypoxic KHT tumor cells in mice breathing O$_2$, O$_2$ + CO$_2$, or hyperbaric oxygen with or without anaesthesia. Radiology 106:663–671

Moulder JE, Rockwell S (1984) Hypoxic fraction of solid tumors: experimental techniques, methods of analysis, and a survey of existing data. Int J Radiat Oncol Biol Phys 10:695–712

Müller-Klieser W, Vaupel P, Manz R (1983) Tumour oxygenation under normobaric and hyperbaric conditions. Br J Radiol 50:559–564

Nunn GR, Dance G, Peters J, Cohn LH (1983) Effect of fluorocarbon exchange transfusion on myocardial infarction size in dogs. Am J Cardiol 52:203–205

Overgaard J (1978) The effect of local hyperthermia alone and in combination with radiation on solid tumors. In: Streffer C (ed) Cancer therapy by hyperthermia and radiation. Urban and Schwartzen berg, Baltimore, pp 49–62

Overgaard J (1989) Sensitization of hypoxic tumour cells – clinical experience. Int J Radiat Biol 56:801–811

Overgaard J, Grau C (1995) Significance of hemoglobin concentration for treatment outcome. In: Molls E, Vaupel P (eds) Medical radiology: blood flow, oxygenation and microenvironment of human tumors – implications for clinical radio-oncology. Springer, Berlin Heidelberg New York

Reinhold HS, Blackiewicz B, Block A (1977) Oxygenation and reoxygenation in "sandwich" tumours. Bibl Anat 15:270–272

Rockwell S, Mate TP, Irvin CG, Nierenburg M (1986) Reactions of tumors and normal tissues in mice to irradiation in the presence and absence of a perfluorochemical emulsion. Int J Radiat Oncol Biol Phys 12:1315–1318

Rojas A (1991) Radiosensitization with normobaric oxygen and carbogen. Radiother Oncol 20:65–70

Rose C, Lustig R, Mclntosch N, Teicher B (1986) A clinical trial of Fluosol-DA 20% in advanced squamous cell carcinoma of the head and neck. Int J Radiat Oncol Biol Phys 12:1325–1327

Roszinski S, Wiedemann G, Jiang SZ, Baretton G, Wagner T, Weiss C (1991) Effects of hyperthermia and/or hyperglycemia on pH and pO_2 in well oxygenated xenografted human sarcoma. Int J Radiat Oncol Biol Phys 20:1273–1280

Rubin P, Hanley J, Keys HM, Marcial V, Brady L (1979) Carbogen breathing during radiation therapy. The radiation therapy oncology group study. Int J Radiat Oncol Biol Phys 5:1963–1970

Sasai K, Ono K, Nishidai T (1989) Variation in tumor response to fluosol-DA (20%). Int J Radiat Oncol Biol Phys 15:1149–1152

Sevick EM, Jain RK (1989) Geometric resistance to blood flow in solid tumors perfused ex vivo – effects of tumor size and perfusion pressure. Cancer Res 49:3506–3512

Siemann DW, Macler LM (1986) Tumor radiosensitization through reductions in hemoglobin affinity. Int J Radiat Oncol Biol Phys 12:1295–1297

Siemann DW, Hill RP, Bush RS (1977) The importance of the pre-irradiation breathing times of oxygen and carbogen (5% CO_2:95% O_2) on the in vivo radiation response of a murine sarcoma. Int J Radiat Oncol Biol Phys 2:903–911

Siemann DW, Hill RP, Bush RS (1978) Smoking: the influence of carboxyhemoglobin (HbCO) on tumor oxygenation and response to radiation. Int J Radiat Oncol Biol Phys 4:657–662

Siemann DW, Hill RP, Bush RS, Chhabra P (1979) The in vivo radiation response of an experimental tumor: the effect of exposing tumor-bearing mice to a reduced oxygen environment prior to but not during radiation. Int J Radiat Oncol Biol Phys 5:61–68

Siemann DW, Alliet KL, Macler LM (1989) Manipulations in the oxygen transport capacity of blood as a means of sensitizing tumors to radiation therapy. Int J Radiat Oncol Biol Phys 15:1169–1172

Siemann DW, Horsman MR, Chaplin DJ (1994) The radiation response of KHT sarcomas following nicotinamide treatment and carbogen breathing. Radiother Oncol 31:117–122

Song CW (1984) Effect of local hyperthermia on blood flow and microenvironment: a review. Cancer Res Suppl 44:4721–4730

Song CW, Zhang WL, Pence DM, Lee I, Levitt SH (1985) Increased radiosensitivity of tumours by perfluorochemicals and carbogen. Int J Radiat Oncol Biol Phys 11:1833–1836

Stewart F, Begg A (1983) Blood flow changes in transplanted mouse tumours and skin after mild hyperthermia. Br J Radiol 56:477–482

Sutherland RM, Franko AJ (1980) On the nature of the radiobiologically hypoxic fraction in tumors. Int J Radiat Oncol Biol Phys 5:117–120

Suit HD, Maimonis P, Michaelis HB, Sedlacek RS (1981) Comparison of hyperbaric oxygen and misonidazole in fractionated irradiation of murine tumors. Radiat Res 87:360–367

Teicher BA, Rose CM (1984) Oxygen-carrying perfluorochemical emulsion as an adjuvant to radiation therapy in mice. Cancer Res 44:4285–4288

Thomlinson RH, Gray LH (1955) The histological structure of some human lung cancers and the possible implications for radiotherapy. Br J Cancer 9:539–549

Trotter MJ, Chaplin DJ, Durand RE, Olive PL (1989) The use of fluorescent probes to identify regions of transient perfusion in murine tumours. Int J Radiat Oncol Biol Phys 16:931–935

Valeri CR (1982) Use of rejuvenation solutionsin blood preservation. Crit Rev Clin Lab Sci 17:299–374

Vaupel P, Kallinowski F (1987) Physiological effects of hyperthermia. In: Streffer C (ed) Recent results in cancer research, vol 4. Springer, Berlin Heidelberg New York, pp 71–109

Vaupel P, Menke H (1989) Effect of various calcium antagonists on blood flow and red cell flux in malignant tumors. Prog Appl Microcirc 14:88–103

Vaupel P, Müller-Klieser W (1986) Verapamil inhibits the respiration rate of cancer cells. Adv Exp Med Biol 200:645–648

Vaupel P, Kallinowski F, Okunieff P (1989) Blood flow, oxygen and nutrient supply, and metabolic microenvironment of human tumors: a review. Cancer Res 49:6449–6465

Vaupel P, Schlenger K, Knoop C, Hoeckel M (1991) Oxygenation of human tumors: evaluation of tissue oxygen distribution in beast cancers by computerized O_2 tension measurements. Cancer Res 51:3316–3322

Warren BA (1979) The vascular morphology of tumors. In: Petersen HI (ed) Tumor blood circulation. CRC Press, Boca Raton, pp 1–47

18 The Potential Benefit of Hypoxic Cytotoxins in Radio-Oncology

J.M. BROWN[1]

CONTENTS

18.1
Why Use Hypoxic Cytotoxins?

18.1.1
Introduction

Hypoxic cytotoxins – also known as bioreductive drugs – are compounds that are reductively metabolized under low oxygen conditions to produce toxic products. The reason that they can be of great benefit in cancer therapy is because their reductive metabolism to toxic products can be far greater in tumors – because of tumor hypoxia – than it is in dose-limiting normal tissues. Furthermore, the hypoxic cells are often resistant to nonsurgical treatments of cancer, primarily radiation and anticancer drugs, either because of low oxygen levels per se, or because of inadequate drug diffusion or because of reduced cellular proliferation (Fig. 18.1). Thus, the potential of hypoxic cytotoxins is that not only can they be selectively activated in tumors, but that the generation of the toxic products will be in, or adjacent to,

the cells in those tumors resistant to conventional anticancer therapies.

Hypoxic cytotoxins drugs constitute an entirely new and largely untested class of anticancer agents principally because the target cells within the solid tumor are different from those of conventional drugs and radiation. Typically, hypoxic cytotoxins have their greatest cytotoxicity to the cells at maximum distance from tumor blood vessels, thereby complementing the pattern of cytotoxicity for both radiation and anticancer drugs, which is maximum for the cells immediately adjacent to the blood vessels (Fig. 18.1). Thus, bioreductive drugs have the potential of overcoming a major cause of resistance of solid tumors to conventional therapies, namely, that resulting from the inadequate oxygenation and drug delivery to tumor cells distant to blood vessels.

18.1.2
How Do These Drugs Preferentially Kill Hypoxic Cells?

Figure 18.2 shows the general mechanism of action of bioreductive drugs. The bioreductive drug D is typically enzymatically reduced to a free radical by the addition of a single electron. These enzymes, of which cytochrome p450, cytochrome b5, NADPH-cytochrome C reductase, and NADH-cytochrome b5 reductase are the most important, transfer a single electron to the bioreductive drug, the source of which comes from a reduced pyridine nucleotide cofactor. Because of the high affinity of molecular oxygen for free radicals (in Fig. 18.2), they are highly reactive with oxygen, which will "back oxidize" the radical to the parent drug. In the absence of oxygen, however, this radical anion can exhibit toxicity itself or be further metabolized to a toxic species. All bioreductive drugs showing greater toxicity to hypoxic cells demonstrate this enzymatic reduction by a 1-electron step as the first reduction process. Also shown in Fig. 18.2 is reduction by 2-electron transfer (by DT diaphorase), a reaction which is not oxygen-

[1] Department of Radiation Oncology, Division of Radiation Biology, Stanford University School of Medicine, Stanford, CA 94305-5468, USA

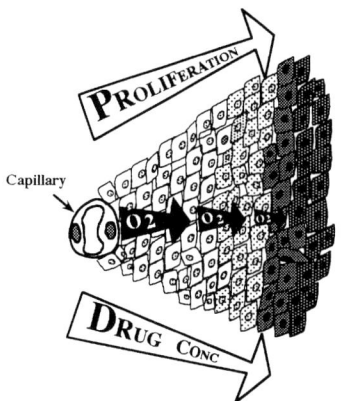
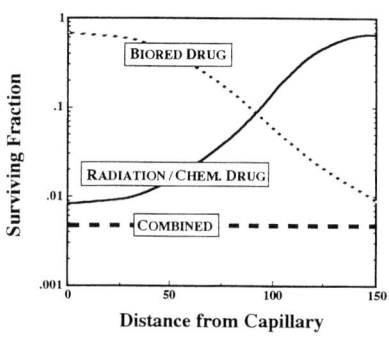

Fig. 18.1. The effect of increasing distance from a tumor capillary on the concentration of oxygen, cellular proliferation, and the concentration of an anticancer drug. These changing concentrations and proliferation cause decreasing cell killing as a function of distance from the capillary for both radiation and the majority of chemotherapeutic drugs. Hypoxic cytotoxic drugs, however, since they are activated only under hypoxic conditions, show a complementary profile with maxi-

mum cytotoxicity for the hypoxic cells distant from capillaries. The combination of radiation + bioreductive drug or of a chemotherapeutic agent + bioreductive drug would be expected to give a more uniform level of cell killing as a function of distance from the capillary, as shown. This would markedly increase the tumor cell kill by radiation or the anticancer drug without increasing toxicity to normal tissues

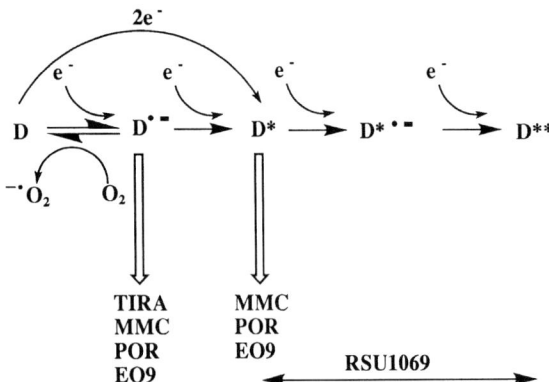

Fig. 18.2. General reduction scheme for a bioreductive drug *D* enzymatically reduced by an oxygen-sensitive 1-electron reduction to the radical anion and then by a further series of 1-electron reductions to other species. Also shown is reduction by 2-electron transfer (by DT diaphorase), which is not oxygen-sensitive. The reduction metabolites thought to be toxic for the individual bioreductive drugs are shown. *TIRA*, tirapazamine; *MMC*, mitomycin C; *POR*, porfiromycin; *EO9*, specific quinone antibiotic. (Redrawn from RAUTH et al. 1993)

sensitive. Nonetheless, this reaction can lead to a more toxic product than the original parent compound. Of course, employing this strategy would only be useful for a tumor with very high levels of DT diaphorase with a drug that was a good substrate for DT diaphorase and that was metabolized to a more toxic product than the parent drug.

18.1.3
Will Hypoxic Cytotoxins Be Any Better than Hypoxic Radiosensitizers or Techniques to Improve Tumor Oxygenation?

Combining a hypoxic cytotoxin with fractionated irradiation is fundamentally different from strategies of either oxygenating the tumor cells or sensitizing them with a hypoxic radiosensitizer, since these approaches can only overcome the problem of hypoxia in tumors and cannot create more cell killing than if the tumor were fully oxygenated. However, when a hypoxic cytotoxin is added to fractionated radiation therapy, a considerably *greater* killing of the tumor cell population can be produced than if the tumor were fully oxygenated (BROWN and KOONG 1991). Under these conditions, hypoxia becomes an advantage that can be exploited. Figure 18.3 shows a theoretical analysis of combining a hypoxic cytotoxin with every fraction of a typical radiotherapy regimen of 30 × 2 Gy with three conditions of tumor oxygenation: 0%, 30%, and 50% hypoxic cells in the tumor. It is apparent that if the tumor contains any hypoxic cells and if the hypoxic cytotoxin can kill 50% of these cells each time it is given, then the resistance of the tumor caused by hypoxia can be overcome. However, if more than 50% of the hypoxic tumor cells are killed each time the bioreductive drug is given, then greater cell killing is achieved than if the tumor were fully oxygenated. Under these conditions the addi-

Fig. 18.3. Calculations of the influence of adding a hypoxic cytotoxin with each dose of a fractionated course of 30 × 2 Gy to a tumor with different proportions of hypoxic cells. Surviving fractions after 30 × 2 Gy with the hypoxic cytotoxin are calculated for different fractions of the hypoxic cells killed with each drug dose

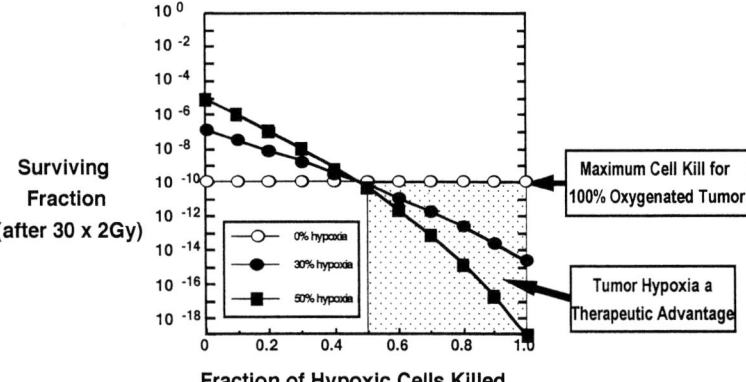

tion of the hypoxic cytotoxin to fractionated irradiation must be superior to any technique of improving tumor oxygenation or the use of even a "perfect" hypoxic radiosensitizer since the best that these can do is to sensitize the tumor to the level of full oxygenation.

Two important points should be noted in this analysis. First, it assumes that reoxygenation/rehypoxiation (BROWN and LEMMON 1992) occurs prior to each dose of radiation or bioreductive drug, i.e., the surviving cells in the tumor population return to their original proportions of aerobic and hypoxic cells. There is ample evidence both from experimental and clinical studies that reoxygenation occurs during a course of radiation therapy, and if this is at least partially related to fluctuating blood flow (BROWN 1979), then we would expect rehypoxiation also to occur, as suggested by experimental evidence for one tumor (KIM and BROWN 1994). The possibility that no reoxygenation/rehypoxiation could occur during a 6-week course of radiation therapy is highly unlikely, since hypoxic cells cannot remain viable for such long periods of time within a tumor.

Second, the above analysis assumes that the hypoxic cytotoxin is given with each of the radiation doses. Calculations show that the benefit of adding the hypoxic cytotoxin increases with the number of times it is given with fractionated radiation (BROWN and KOONG 1991). Although it is generally recognized that it would be of maximum benefit to give a hypoxic radiosensitizer with as many of the radiation doses as possible, this has so far not been the practice with cytotoxic agents. For example, in the only clinical trial attempting to test a bioreductive drug with fractionated radiation, the patients received only one dose of mitomycin C at the beginning of radiotherapy and one after the end of the 6-week course (HAFFTY et al. 1993; WEISSBERG et al.

1989). The theoretical analysis indicates that giving the drug only two times would have a negligible effect on the overall level of tumor cell kill, irrespective of what proportion of the hypoxic cells were killed by the drug (BROWN and KOONG 1991). The only way for this not to be the case would be for no rehypoxiation (regeneration of hypoxic cells) to occur during the 6-week course of radiation therapy, and as discussed above, this seems improbable.

18.1.4
Experimental Studies of the Combination of a Hypoxic Cytotoxin with Fractionated Irradiation

The development in the past decade of two agents that have high selective toxicity for hypoxic cells, both in vitro and in vivo, namely, RSU 1069 (or its prodrug, RB 6145), a dual-function agent containing both nitroimidazole and aziridine moieties, and tirapazamine (SR 4233), a benzotriazine di-N-oxide, has allowed these calculations in the theoretical analysis discussed above to be tested experimentally. Figure 18.4 shows the results of adding a dose of tirapazamine to each dose of a fractionated radiation regimen to the SCCVII tumor in vivo and then determining their response using clonogenic assay of the cells removed following different numbers of the treatments. The response of the combined treatment is much greater than the response to radiation alone, to the bioreductive drug alone, or to the predicted "additive" response. However, the calculation of additivity assumes a homogeneous tumor population. If, on the other hand, it is assumed that the tumor contains both aerobic and hypoxic cells and that both reoxygenation and rehypoxiation occur following each treatment, then results obtained are very similar to those predicted (BROWN and LEMMON

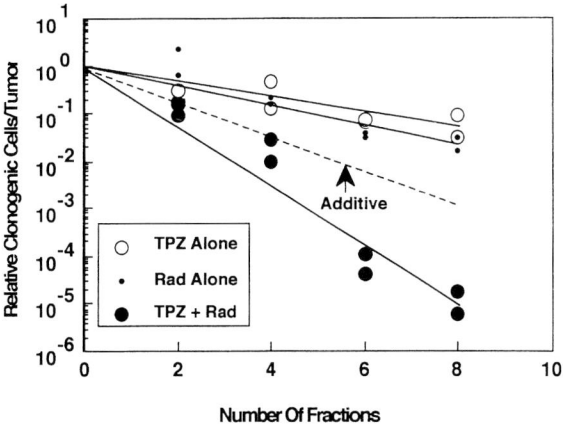

Fig. 18.4. Effect of adding tirapazamine (*TPZ*; 0.11 mmol/kg) 30 min before each 2.5 Gy fraction (*Rad*) given twice daily to the SCCVII tumor. (Redrawn from BROWN and LEMMON 1990)

1990). Both for this tumor and for other murine tumors, we have also shown that the combination of tirapazamine and fractionated irradiation produced more cell killing than was produced by a large dose of the hypoxic radiosensitizer etanidazole preceding each radiation dose for a number of different tumors (BROWN and LEMMON 1990). It was also superior to oxygenating the tumor prior to each radiation dose with carbogen combined with nicotinamide (DORIE et al. 1994). We have recently completed a series of experiments with human tumors transplanted into SCID mice, giving tirapazamine prior to each of eight doses of 2.5 Gy. A summary of the dose modifying factors (DMFs) for the combination compared to radiation alone is shown in Table 18.1 for these tumors as well as for mouse tumors treated similarly in the same series of experiments. It is apparent that except for one of the human tumors (HT 1080), there is substantial enhancement of the tumor cell kill compared to radiation only.

Thus, the results obtained experimentally are consistent with the theoretical modeling that adding a hypoxic cytotoxin to fractionated irradiation can produce more tumor cell killing than if the tumor were fully oxygenated. Caution has to be exercised, however, in extrapolating these results directly to the clinical situation. The modeling, for example, assumes that reoxygenation and rehypoxiation occur throughout the whole course of treatment and that all cells in the tumor population are potentially able to undergo these processes. Such aspects of tumor physiology are in need of investigation.

18.2
Hypoxic Cytotoxins in or near Clinical Testing

18.2.1
Quinones

The three bioreductive drugs of current clinical interest in this class are mitomycin C, porfiromycin, and E09. All are structurally similar (Fig. 18.5) and require reductive metabolism for activity. They are reductively metabolized to bifunctional alkylating agents, and probably produce their major activity through the formation of DNA interstrand crosslinks. Despite these similarities, the drugs have interesting differences, probably as a result of their differing substrate specificities for the various activating reductive enzymes in the cell and the cytotoxicities of some of their metabolites. Although, in general, the differential toxicity to hypoxic cells is not as large for this class as for RSU1069 and tirapazamine, they can be highly active as cytotoxic drugs, both in preclinical models and, in the case of mitomycin C, in the clinic.

Table 18.1. Logs of cell kill produced by 8 × 2.5 Gy in 4 days with or without tirapazamine (0.12 mmol/kg) given 30 min prior to each radiation dose

Treatment	Tumor						
	SCCVII	KHT	RIF-1	EMT-6	A549	HT1080	HT29
Radiation only	2.8	2.4	2.8	2.5	1.5	3.1	2.0
Radiation + tirapazamine	4.8	3.7	4.7	3.5	2.0	3.3	3.0
DMF	1.7	1.5	1.7	1.4	1.3	1.1	1.5

DMF, dose modifying factors.

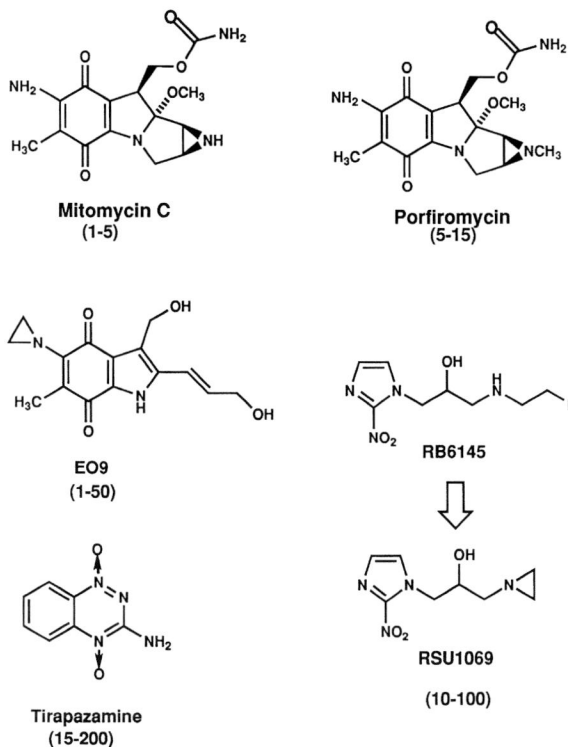

Fig. 18.5. Structures of the bioreductive cytotoxic agents of current clinical interest. The *numbers in parentheses* associated with each drug show the approximate range of the ratios of drug concentrations to produce equal cell kill for hypoxic and aerobic cells ("hypoxic cytotoxicity ratios") for a variety of different tumor cell lines

18.2.1.1
Mitomycin C

Mitomycin C (MMC) was introduced into the clinic in 1958 and has demonstrated its clinical efficacy towards a number of different tumors in combination with other chemotherapeutic drugs and with radiation. Reviews of its mechanism of action, pharmacology, preclinical, and clinical activity have been published (CROOKE and BRADNER 1976; RAUTH et al. 1993; ROCKWELL et al. 1993). We will discuss here only certain aspects of the preferential cytotoxicity of this compound towards hypoxic cells.

Selective cytotoxicity towards hypoxic cells has been demonstrated for the mouse tumor cell line EMT6 (FRACASSO and SARTORELLI 1986; LUDWIG et al. 1984; ROCKWELL et al. 1982), mouse KHT cells, hamster CHO and V79 cells, human HeLa cells (RAUTH et al. 1983), and human fibroblasts (MARSHALL et al. 1989). However, no differential cytotoxicity was demonstrable for the human WiDR colon cancer cell line or for ten fresh human tumor

biopsies tested using the human tumor clonogenic assay (LUDWIG et al. 1984). Even in those cases in which a clear differential hypoxic cytotoxicity was seen, the difference in drug concentrations to achieve equal cytotoxicity under aerobic and hypoxic conditions is not large, varying from one (no differential) to a maximum of five to six.

One factor that may contribute both to the low differential toxicity between hypoxic and aerobic cells and to the variability between different laboratories is the extreme sensitivity of the toxicity of MMC to inhibition by small levels of oxygen. This was demonstrated by MARSHALL and RAUTH (1986), who showed that the maximum change in selective toxicity of Chinese hamster cells in vitro under hypoxic conditions occurred at solution of oxygen concentrations less than 0.02% ($\sim$0.15 mmHg). This level is already at maximum radiation resistance, and it is unusual for investigators to obtain levels lower than this in producing hypoxic cells without taking extreme measures. Thus, although MMC clearly exhibits preferential toxicity under hypoxic conditions in some (though not all) cell lines, extremely low levels of oxygen are required to produce the maximum differential. Indeed, at oxygen levels at which the cells already exhibit maximum, or close to maximum, resistance to radiation, there could be very little differential toxicity of MMC for aerobic and hypoxic cells. A consequence of this is that if radiation and MMC are combined, there will be a population of cells at intermediate oxygen concentrations that is sensitive neither to radiation nor to MMC (MARSHALL and RAUTH 1986).

The question of whether MMC shows selective toxicity to hypoxic cells in vivo has been addressed using transplanted tumors in mice by combining the drug with a large single dose of radiation. This radiation dose will selectively kill the aerobic cells of the tumor, leaving the hypoxic cells as the principal survivors. Thus, if MMC were selectively toxic to hypoxic cells in vivo, it should produce greater cell kill of an irradiated tumor than a control tumor (which will consist largely of aerobic cells). In general, this has either not been found, or the difference between what would be expected from the cytotoxicity to the whole population of tumor cells and that which was observed for the irradiated tumors was very small (ADAMS et al. 1992; RAUTH et al. 1983; ROCKWELL and KENNEDY 1979; ROCKWELL et al. 1989).

In summary, though MMC is justifiably considered to be the prototype bioreductive drug, the experimental data show that it is far from ideal in this role. Its selective toxicity towards hypoxic cells is extremely modest, and requires extremely low levels of oxygen to obtain maximum cytotoxicity. This will minimize the number of cells in the tumor for which it will express preferential cytotoxicity and, indeed, it will not kill all of the cells resistant to ionizing radiation. In agreement with this prediction based on in vitro data, the in vivo experiments with mouse tumors have shown little or no *preferential* cytotoxicity for hypoxic cells. Nonetheless, the drug is toxic to hypoxic cells in vivo (though no more than for aerobic cells), and this might help to explain the positive result of the clinical trials obtained with a combination of radiation and mitomycin C (HAFFTY et al. 1993; WEISSBERG et al. 1989).

18.2.1.2
Porfiromycin

Porfiromycin (POR) is the *n*-methyl aziridine analogue of MMC (Fig. 18.5) and appears to be metabolized in a similar way as is MMC. However, it has a greater differential to hypoxic cells than does MMC as a result of its lowered aerobic cytotoxicity (KEYES et al. 1985). When combined with irradiation in single doses, it has also been shown to produce supra-additive effects, indicating a preferential cytotoxicity to hypoxic cells in EMT6 tumors. These studies have encouraged the Yale group to replace mitomycin C with POR in their studies of the combination of radiation with a bioreductive alkylating agent. A clinical trial with this combination is presently in progress. This trial will be a somewhat better test of the principle that selective killing of hypoxic cells with minimal killing of aerobic cells can potentiate the effects of radiotherapy. However, the limitation of the need to achieve very low levels of oxygenation for maximum hypoxic cell killing is the same for POR as it is for MMC (MARSHALL and RAUTH 1986), so this compound is still far from ideal. Also, the fact that it still produces cytotoxicity to aerobic cells at the doses delivered will make it difficult to interpret the results of any trial as related strictly to the killing of hypoxic cells.

18.2.1.3
E09

The quinone antibiotic E09 (Fig. 18.5) has emerged as a promising new bioreductive drug and has recently entered phase I clinical trials in Europe. Although E09 has the indoquinone backbone of MMC,

its activity against tumor models in vitro and in vivo showed the drugs to have different patterns of antitumor activity (HENDRIKS et al. 1993). Under some circumstances E09 appears to act as a hypoxia selective drug and in others to be an example of a bioreductive drug fulfilling the "enzyme-directed" model (WORKMAN and WALTON 1990). E09 is a much more efficient substrate for DT diaphorase than are either MMC or POR (WALTON et al. 1991), and this appears in large measure to be responsible for its activity. Thus, in cells possessing moderate to high levels of DT diaphorase, the oxygen-insensitive 2-electron reduction pathway is dominant, and for these cells, E09 would not be expected to have any preferential cytotoxicity under hypoxic conditions. The fact that the compound is a highly efficient substrate for DT diaphorase would suggest that its aerobic cytotoxicity should be highly dependent upon cellular DT diaphorase levels, which can vary by several orders of magnitude in different cell types. Experimental data have confirmed these expectations (WALTON et al. 1992a; PLUMB et al. 1994; ROBERTSON et al. 1994). Since under hypoxic conditions quinone metabolism can follow both the 1- and 2-electron pathways, it would be expected that for cells having high levels of DT diaphorase, the 2-electron pathway would continue to dominate under hypoxic conditions and, therefore, there would be little or no difference in the aerobic and hypoxic cytotoxicities for cells with high DT diaphorase levels. This has been shown to be the case (PLUMB et al. 1994; ROBERTSON et al. 1994).

In summary, E09 would appear to be a promising drug for use against tumors that have very high levels of DT diaphorase. For these tumors, killing of aerobic and hypoxic cells would be expected to be similar (ignoring drug distribution effects). For the tumors with low DT diaphorase, the strategy for use of E09 would be different, because under these conditions, it would be preferentially toxic to the hypoxic cells. Thus, it would be preferable to combine it with an agent selectively toxic to aerobic cells, such as radiation or chemotherapeutic drugs with activity limited by diffusion or cellular proliferation.

18.2.2
Nitroaromatics

A number of nitroheterocyclic compounds have been investigated as bioreductive hypoxic cytotoxic agents. The roots of this work began in attempts to overcome the resistance of hypoxic cells to killing by ionizing radiation (ADAMS and COOKE 1969). Two nitroimidazole compounds emerged from these studies, metronidazole and misonidazole, both of which showed activity as radiosensitizers of hypoxic cells, not only in vitro, but also to hypoxic cells in transplanted tumors in mice. It was then demonstrated that these compounds showed preferential toxicity towards hypoxic cells in the absence of radiation (HALL and ROIZIN-TOWLE 1975; MOHINDRA and RAUTH 1976). The principal observation that preferential cytotoxicity to hypoxic cells was also produced in vivo by metronidazole and misonidazole came from studies showing that greater cell killing could be observed if the drug was given *following* a large dose of radiation than with radiation alone (BROWN 1977; DENEKAMP and MCNALLY 1978).

Although the preferential toxicity of the nitroimidazoles metronidazole, misonidazole and the less neurotoxic etanidazole towards hypoxic cells can be demonstrated in vivo, the fact that the effect is considerably less than that of radiosensitization and requires a dose very close to the lethal dose to the animal has not encouraged the use of these compounds as hypoxic cytotoxins.

18.2.2.1
RSU 1069 (RB 6145, PD 144872)

In an effort to increase the potency of misonidazole, ADAMS et al. (1984) synthesized a compound with an aziridine moiety in the N1 side chain of a 2-nitroimiazole. The compound produced, RSU 1069 (1[2-nitro-1-imidazolyl]-3-aziridinyl-2-propynol; Fig. 18.5), was shown to have greater efficiency as a hypoxic cell radiosensitizer and as a hypoxia-selective cytotoxin than misonidazole, both in vitro and in vivo (ADAMS et al. 1984; CHAPLIN et al. 1986; HILL et al. 1986; WHITMORE and GULYAS 1986; WONG et al. 1991). Studies in vivo indicated that the potentiation of the radiation effect was the result primarily of its killing rather than radiosensitization of the hypoxic tumor cells (HILL et al. 1986).

Studies with repair deficient cell lines suggest that under aerobic conditions, RSU 1069 behaves primarily as a monofunctional alkylating agent (resulting from the aziridine ring), while under hypoxic conditions, reduction of the nitro group converts it to a bifunctional alkylating agent (WHITMORE and GULYAS 1986). The mechanism of cell killing of RSU 1069 under hypoxic conditions appears to be the production of DNA interstrand crosslinks (OLIVE

1995; O'NEILL et al. 1987; WHITMORE and GULYAS 1986).

In an attempt of overcome the problem of severe emesis in patients observed with RSU 1069 (HORWICH et al. 1986), a large number of analogues were investigated, both in vitro and in vivo (AHMED et al. 1986). The best of these compounds was RB 6145, a prodrug of RSU 1069 (Fig. 18.5). Under physiological conditions, the bromoethylamino group of RB 6145 cyclizes to the aziridine group of RSU 1069. More recently, the optical isomers of the racemic mixture of RB 6145 have been synthesized for direct comparison of their biological efficacy and toxicity (SEBOLT-LEOPOLD et al. 1993). It appears that the R isomer of RB 6145, PD 144872, is considerably less emetic than the L isomer and RB 6145, but maintains the same radiation potentiation in vivo. Unfortunately an unexpected occular toxicity of the drug has halted its advancement into clinical trials (SEBOLT-LEOPOLD, personal communication).

18.2.3
N-Oxides

18.2.3.1
Tirapazamine

Tirapazamine (SR 4233, WIN 59075, 3-amino-1,2,4-benzotriazine 1,4-dioxide) is the first representative of a third class of oxygen-inhibtitable bioreductive compounds to be developed for clinical use. The compounds has electron-affinity slightly higher than that of misonidazole ($E1/2 = -332\,mV$ cf $-389\,mV$ for MISO) and has comparable activity as a hypoxic cell radiosensitizer. However, it is cytotoxic to hypoxic cells at concentrations much lower than those needed to produce radiosensitization and shows a hypoxic cytotoxicity ratio of between 20 and 300 for the vast majority of human and rodent cell lines (BROWN 1993; ZEMAN et al. 1986).

The profile of cytotoxicity for tirapazamine as a function of oxygen concentration is different from that of both the quinone antibiotic and nitroimidazole bioreductive drugs. Essentially, tirapazamine maintains its "hypoxic cytotoxicity" at oxygen concentrations approximately ten times higher than do RSU 1069 and other nitroimidazole bioreductive drugs (KOCH 1993). This results in the calculated additive killing produced by X-rays and tirapazamine being more uniform over the whole range of oxygen concentrations than for X-rays combined with other bioreductive drugs (KOCH 1993). However, these results have been obtained with single cells in vitro, and it is not yet clear that they will apply to cells in solid tumors. If the results do apply to tumors, then it would suggest that the profile exhibited by tirapazamine and, presumably, other benzotriazine N-oxides would be advantageous in vivo, since tumors have oxygen concentrations spanning the range from extreme hypoxia to fully aerobic (Höckel et al. 1991; Vaupel et al. 1991).

Several groups have investigated the mechanism for the selective hypoxic cytotoxicity of tirapazamine, and there is general agreement on the broad outlines of the mechanism. Figure 18.6 shows a diagram of the proposed mechanism. It shows that the DNA-damaging species is an oxidizing radical which is "back-oxidized" to the parent drug in the presence of oxygen. In the absence of oxygen, this oxidizing radical is shown as abstracting hydrogen from cellular target, leaving an oxidized target molecule along with the stable 2-electron reduction product, SR 4317.

The principal evidence for this mechanism is first that SR 4317, which is only produced in tissue culture under hypoxic conditions, is cytotoxic neither to aerobic nor to hypoxic cells, despite the fact that it is taken up by these cells (Baker et al. 1988). Second, the fact that free radicals are involved in the cytotoxicity is supported by the finding that DMSO, a potent radical scavenger, substantially reduces hypoxic cytotoxicity (Brown 1991). Third, definitive evidence of the free radical has been obtained using electron spin resonance (Lloyd et al. 1991). These same investigators also identified the superoxide radical under aerobic conditions. The evidence that DNA is the likely target for cell killing by tirapazamine under hypoxic conditions comes from findings that cell

lines deficient in DNA double-strand break repair are hypersensitive to tirapazamine cytotoxicity under hypoxic conditions (Biedermann et al. 1991), and also that for equal cell killing, the same final number of chromosome breaks have been found both for tirapazamine and for ionizing radiation (Wang et al. 1993).

Tirapazamine can be reductively metabolized by a number of enzymes, of which cytochrome P450 and NADPH cytochrome P450 reductase are considered to be the most important (Walton et al. 1992b; Wang et al. 1993). In mouse liver microsomes where high levels of both cytochrome P450 and NADPH cytochrome P450 reductase occur, these enzymes account for approximately 70% and 30%, respectively, of metabolism of tirapazamine (Walton et al. 1992). Cahill and White (1990) showed that purified rat liver NADPH cytochrome P450 reductase could alone convert tirapazamine to its 2-electron reduction product, and Lloyd et al. (1991) argued that this enzyme is likely to be the most important for reduction of tirapazamine by rat liver microsomes. However, hypoxic metabolism by tumor cells is of greatest relevance, and Wang et al. (1993) showed that both P450 and NADPH cytochrome P450 reductase were involved in the reduction of tirapazamine by tumor cell homogenates. Since cytochrome P450 is dependent on the presence of NADPH cytochrome P450 reductase for its metabolic action, it would thus appear that cytochrome P450 reductase can have both a direct and indirect role in the metabolism of tirapazamine.

The fact that tirapazamine preferentially kills hypoxic cells in vitro is no guarantee that it also preferentially kills hypoxic cells in solid tumors, where problems such as inadequate diffusion, too

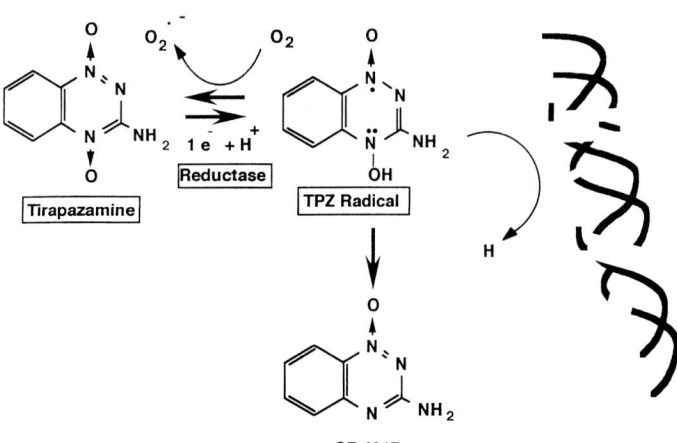

Fig. 18.6. A representation of the mechanism of action of tirapazamine (*TPZ*). (Redrawn from Brown 1993)

rapid metabolism, and a prohibitive systemic toxicity may occur. However, there is considerable evidence that selective killing of hypoxic cells does in fact occur in transplanted tumors in mice. ZEMAN et al. (1988) found a roughly 20-fold increase in cell kill over that expected from the toxicity of radiation alone and that of tirapazamine alone when the two were combined. Preferential killing of hypoxic cells is also consistent with the date obtained for combined fractionated irradiation with tirapazamine for a variety of transplanted tumors (BROWN and LEMMON 1990). It has also been shown that following treatment of mice bearing SCCVII tumors with tirapazamine, the hypoxic fraction (determined by the paired survival curve method) falls to approximately 8% of pretreatment levels by 1h (KIM and BROWN 1994). These experiments strongly suggest that tirapazamine selectively kills hypoxic cells in tumors, although some contribution by radiosensitization cannot be eliminated (ZEMAN and BROWN 1989).

A number of phase I, II and III clinical studies of tirapazamine have now been completed. The drug has been given either as a single dose every 3 weeks (with chemotherapy) or multiple times per week (with radiotherapy). Data obtained to date indicate that maximum tolerated doses of the drug produce total exposure profiles (area under the curve) that would produce significant enhancement of the response of tumors to radiation and chemotherapy.

In summary, tirapazamine is the first drug to enter clinical trials purely as a bioreductive agent with toxicity only to hypoxic cells. All of the indications to date are that the drug has the promise of reversing some of the resistance of solid tumors, both to radiotherapy and to some anticancer drugs.

18.2.3.2
Other N-Oxides

The aliphatic bis-N-oxide AQ4N was developed by PATTERSON (1993) as a prodrug of AQ4. Reductive activation of the weakly DNA-affinic AQ4N produces the tight DNA binder AQ4. AQ4N has about tenfold in vitro selectivity towards hypoxic cells, although, in order to achieve substantial cell killing, it is necessary to add rat liver microsomes to the cell cultures. Despite this, AQ4N does appear to be active against T50/80 mammary carcinomas (McKEOWN et al. 1995), although it is not clear whether bioreductive activation is involved.

18.3
Summary

Agents specifically cytotoxic towards hypoxic cells constitute a new and, as yet, clinically untested way not only of overcoming the therapeutic disadvantage of the presence of hypoxic cells in solid tumors, but in addition, of exploiting these cells for additional benefit. The biological rationale for the use of these agents for combination with fractionated irradiation is compelling – more so than that for the use of hypoxic cell radiosensitizers or for various ways of oxygenating tumors. For fractioned radiotherapy this is primarily because, whereas reoxygenation between doses reduces the radiosensitization both from hypoxic sensitizers and from oxygen, it *increases* the potentiation by a hypoxic cytotoxin given with each radiation dose. The rationale for combining hypoxic cytotoxins with standard chemotherapeutic agents is somewhat different. For these agents it is the distance of the tumor cells from the blood vessels and their slow proliferation rates, rather than hypoxia per se, that are responsible for resistance. A hypoxic cytotoxin would be expected to overcome tumor resistance to these chemotherapeutic drugs by preferentially killing the cells furthest from the blood vessels, thereby producing complementary cell killing with the standard anticancer drug. The preferentially studies performed to date, particularly with tirapazamine, have so far confirmed these expectations. Early clinical investigations with this drug are promising. What is also promising is that there is every expectation that better hypoxic cytotoxins will be developed.

Acknowledgement. Supported in part by grants CA 15201 and CA 25990 from the US National Cancer Institute, DHHS.

References

Adams GE, Cooke MS (1969) Electron-affinic sensitization. I. A structural basis for chemical radiosensitizers in bacteria. Int J Radiat Biol 15:457–471

Adams GE, Ahmed I, Sheldon PW, Stratford IJ (1984) RSU 1069, a 2-nitroimidazole containing an alkylating group: high efficiency as a radio- and chemosensitizer in vitro and in vivo. Int J Radiat Oncol Biol Phys 10:1653–1656

Adams GE, Stratford IJ, Edwards HS, Bremner JC, Cole S (1992) Bioreductive drugs as post-irradiation sensitizers: comparison of dual function agents with SR 4233 and the mitomycin C analogue EO9. Int J Radiat Oncol Biol Phys 22:717–720

Ahmed I, Jenkins TC, Walling JM, Stratford IJ, Sheldon PW, Adams GE, Fielden EM (1986) Analogues of RSU-1069:

radiosensitization and toxicity in vitro and in vivo. Int J Radiat Oncol Biol Phys 12:1079–1081

Baker MA, Zeman EM, Hirst VK, Brown JM (1988) Metabolism of SR 4233 by Chinese hamster ovary cells: basis of selective hypoxic cytotoxicity. Cancer Res 48:5947–5952

Biedermann KA, Wang J, Graham RP, Brown JM (1991) SR 4233 cytotoxicity and metabolism in DNA repair-competent and repair-deficient cell cultures. Br J Cancer 63:358–362

Brown JM (1977) Cytotoxic effects of the hypoxic cell radiosensitizer Ro 7-0582 to tumor cells in vivo. Radiat Res 72:469–486

Brown JM (1979) Evidence for acutely hypoxic cells in mouse tumours, and a possible mechanism of reoxygenation. Br J Radiol 52:650–656

Brown JM (1991) Redox activation of benzotriazine N-oxides: mechanisms and potential as anticancer drugs. Plenum, Fermo, Italy

Brown JM (1993) SR 4233 (tirapazamine): a new anticancer drug exploiting hypoxia in solid tumours. Br J Cancer 67:1163–1170

Brown JM, Koong A (1991) Therapeutic advantage of hypoxic cells in tumors: a theoretical study. J Natl Cancer Inst 83:178–185

Brown JM, Lemmon MJ (1990) Potentiation by the hypoxic cytotoxin SR 4233 of cell killing produced by fractionated irradiation of mouse tumors. Cancer Res 50:7745–7749

Brown JM, Lemmon MJ (1992) Fractionation increases the antitumor effect of adding a hypoxic cytotoxin to irradiation. In: Dewey WC, Edington M, Fry RJM, Hall EJ, Whitmore GF (eds) Radiation research: a twentieth century perspective. Academic, San Diego, pp 807–812

Cahill A, White INH (1990) Reductive metabolism of 3-amino-1,2,4-benzotriazine-1,4-dioxide (SR 4233) and the induction of unscheduled DNA synthesis in rat and human derived cell lines Carcinogenesis 11:1407–1411

Chaplin DJ, Durand RE, Stratford IJ, Jenkins TC (1986) The radiosensitizing and toxic effects of RSU-1069 on hypoxic cells in a murine tumor. Int J Radiat Oncol Biol Phys 12:1091–1095

Crooke ST, Bradner WT (1976) Mitomycin C: a review. Cancer Treat Rev 3:121–139

Denekamp J, McNally NJ (1978) The magnitude of hypoxic cell cytotoxicity of misonidazole in human tumours [letter]. Br J Radiol 51:747–748

Dorie MJ, Menke D, Brown JM (1994) Comparison of the enhancement of tumor responses to fractionated irradiation by SR 4233 (tirapazamine) and by nicotinamide with carbogen. Int J Radiat Oncol Biol Phys 28:145–150

Fracasso PM, Sartorelli AC (1986) Cytotoxicity and DNA lesions produced by mitomycin C and porfiromycin in hypoxic and aerobic EMT6 and Chinese hamster ovary cells. Cancer Res 46:3939–3944

Haffty BG, Son YH, Sasaki CT, et al. (1993) Mitomycin C as an adjunct to postoperative radiation therapy in squamous cell carcinoma of the head and neck: results from two randomized clinical trials [see comments]. Int J Radiat Oncol Biol Phys 27:241–250

Hall EJ, Roizin-Towle L (1975) Hypoxic sensitizers: radiobiological studies at the cellular level. Radiology 117:453–457

Hendriks HR, Pizao PE, Berger DP, et al. (1993) EO9: a novel bioreductive alkylating indoloquinone with preferential solid tumour activity and lack of bone marrow toxicity in preclinical models. Eur J Cancer 897–906

Hill RP, Gulyas S, Whitmore GF (1986) Studies of the in vivo and in vitro cytotoxicity of the drug RSU-1069. Br J Cancer 53:743–751

Höckel M, Schlenger K, Knoop C, Vaupel P (1991) Oxygenation of carcinomas of the uterine cervix: evaluation by computerized O_2 tension measurements. Cancer Res 51:6098–6102

Horwich A, Holliday SB, Deacon JM, Peckham MJ (1986) A toxicity and pharmacokinetic study in man of the hypoxic-cell radiosensitiser RSU-1069. Br J Radiol 59:1238–1240

Keyes SR, Rockwell S, Sartorelli AC (1985) Porfiromycin as a bioreductive alkylating agent with selective toxicity to hypoxic EMT6 tumor cells in vivo and in vitro. Cancer Res 45:3642–3645

Kim IH, Brown JM (1994) Reoxygenation and rehypoxiation in the SCCVII mouse tumor. Int J Radiat Oncol Biol Phys 29:493–497

Koch CJ (1993) Unusual oxygen concentration dependence of toxicity of SR-4233, a hypoxic cell toxin. Cancer Res 53:3992–3997

Lloyd RV, Duling DR, Rumyantseva GV, Mason RP, Bridson PK (1991) Microsomal reduction of 3-amino-1,2,4-benzotriazine 1,4-dioxide to a free radical. Mol Pharmacol 40:440–445

Ludwig CU, Peng YM, Beaudry JN, Salmon SE (1984) Cytotoxicity of mitomycin C on clonogenic human carcinoma cells is not enhanced by hypoxia. Cancer Chemother Pharmacol 12:146–150

Marshall RS, Rauth AM (1986) Modification of the cytotoxic activity of mitomycin C by oxygen and ascorbic acid in Chinese hamster ovary cells and a repair-deficient mutant. Cancer Res 46:2709–2713

Marshall RS, Paterson MC, Rauth AM (1989) Deficient activation by a human cell strain leads to mitomycin resistance under aerobic but not hypoxic conditions. Br J Cancer 59:341–346

McKeown SR, Hejmadi MV, McAleer JJA, Patterson LH (1995) Alkylaminoanthraquinone N-oxides: bioreductive potential and positive interaction with radiation in vivo. Br J Cancer, in press

Mohindra JK, Rauth AM (1976) Increased cell killing by metronidazole and nitrofurazone of hypoxic compared to aerobic mammalian cells. Cancer Res 36:930–936

Olive PL (1995) Detection of hypoxia by measurement of DNA damage in individual cells from spheroids and murine tumours exposed to bioreductive drugs. I. Tirapazamine. Br J Cancer 71:529–536

O'Neill P, McNeil SS, Jenkins TC (1987) Induction of DNA crosslinks in vitro upon reduction of the nitroimidazole-aziridines RSU-1069 and RSU-1131. Biochem Pharmacol 36:1787–1792

Patterson LH (1993) Rationale for the use of aliphatic-N-oxides of cytotoxic anthraquinones as prodrug DNA binding agents: a new class of bioreductive agent. Cancer Metastasis Rev 12:119–134

Plumb JA, Gerritsen M, Milroy R, Thomson P, Workman P (1994) Relative importance of DT-diaphorase and hypoxia in the bioactivation of EO9 by human lung tumor cell lines. Int J Radiat Oncol Biol Phys 29:295–299

Rauth AM, Mohindra JK, Tannock IF (1983) Activity of mitomycin C for aerobic and hypoxic cells in vitro and in vivo. Cancer Res 43:4154–4158

Rauth AM, Marshall RS, Kuehl BL (1993) Cellular approaches to bioreductive drug mechanisms. Cancer Metastasis Rev 12:153–164

Robertson N, Haigh A, Adams GE, Stratford IJ (1994) Factors affecting sensitivity to EO9 in rodent and human tumour cells in vitro: DT-diaphorase activity and hypoxia. Eur J Cancer 30:1013–1019

Rockwell S, Kennedy KA (1979) Combination therapy with radiation and mitomycin C: preliminary results with EMT6 tumor cells in vitro and in vivo. Int J Radiat Oncol Biol Phys 5:1673–1676

Rockwell S, Kennedy KA, Sartorelli AC (1982) Mitomycin-C as a prototype bioreductive alkylating agent: in vitro studies of metabolism and cytotoxicity. Int J Radiat Oncol Biol Phys 8:753–755

Rockwell S, Keyes SR, Sartorelli AC (1989) Modulation of the antineoplastic efficacy of mitomycin C by dicoumarol in vivo. Cancer Chemother Pharmacol 24:349–353

Rockwell S, Sartorelli AC, Tomasz M, Kennedy KA (1993) Cellular pharmacology of quinone bioreductive alkylating agents. Cancer Metastasis Rev 12:165–176

Sebolt-Leopold JS, Elliott WL, Showalter HD, Leopold WR (1993) Rationale for selection of PD 144872, the R isomer of RB 6145, for clinical development as a radiosensitizer. Proc Am Assoc Cancer Res 34:362

Siemann DW (1994) In vitro cytotoxicity and chemo-sensitizing activity of the dual function nitroimidazole RB 6145. Int J Radiat Oncol Biol Phys 29:301–306

Siim BG, van Zijl PL, Brown JM (1995) Tirapazamine-induced DNA damage measured using the comet assay correlates with cytotoxicity towards hypoxic tumour cells in vitro. Br J Cancer, to be published

Vaupel P, Schlenger K, Knoop C, Höckel M (1991) Oxygen-ation of human tumors: evaluation of tissue oxygen distri-bution in breast cancers by computerized O_2 tension measurements. Cancer Res 51:3316–3322

Walton MI, Smith PJ, Workman P (1991) The role of NAD(P)H: quinone reductase (EC 1.6.99.2, DT-diaphorase) in the reductive bioactivation of the novel indoloquinone antitumor agent EO9. Cancer Commun 3:199–206

Walton MI, Bibby MC, Double JA, Plumb JA, Workman P (1992a) DT-diaphorase activity correlates with sensitivity to the indoloquinone EO9 in mouse and human colon car-cinomas. Eur J Cancer 1597–1600

Walton MI, Wolf CR, Workman P (1992b) The role of cyto-chrome P450 and cytochrome P450 reductase in the reduc-tive bioactivation of the novel benzotriazine di-N-oxide hypoxic cytotoxin 3-amino-1,2,4-benzotriazine-1,4-dioxide (SR 4233, WIN 59075) by mouse liver. Biochem Pharmacol 44:251–259

Wang J, Biedermann KA, Wolf CR, Brown JM (1993) Metabo-lism of the bioreductive cytotoxin SR 4233 by tumour cells: enzymatic studies. Br J Cancer 67:321–325

Weissberg JB, Son YH, Papac RJ, et al. (1989) Randomized clinical trail of mitomycin C as an adjunct to radiotherapy in head and neck cancer. Int J Radiat Oncol Biol Phys 17:3–9

Whitmore GF, Gulyas S (1986) Studies on the toxicity of RSU-1069. Int J Radiat Oncol Biol Phys 12:1219–1222

Wong KH, Koch CJ, Wallen CA, Wheeler KT (1991) Pharma-cokinetics and cytotoxicity of RSU-1069 in subcutaneous 9L tumours under oxic and hypoxic conditions. BR J Can-cer 63:484–488

Workman P, Walton MI (1990) Enzyme-directed bioreductive drug development. In: Adams GE, Breccia A, Fielden EM, Wardman P. Selective activation of drugs by redox pro-cesses. Plenum, New York, pp 173–191

Zeman EM, Brown JM (1989) Pre- and post-irradiation radiosensitization by SR 4233. Int J Radiat Oncol Biol Phys 16:967–971

Zeman EM, Brown JM, Lemmon MJ, Hirst VK, Lee WW (1986) SR-4233: a new bioreductive agent with high selective tox-icity for hypoxic mammalian cells. Int J Radiat Oncol Biol Phys 12:1239–1242

Zeman EM, Hirst VK, Lemmon MJ, Brown JM (1988) Enhancement of radiation-induced tumor cell killing by the hypoxic cell toxin SR 4233. Radiother Oncol 12:209–218

Subject Index

List of Contributors

J. MARTIN BROWN, Prof. PhD
Department of Radiation Oncology
Division of Radiation Biology
Stanford University School of Medicine
Stanford, CA 94305-5468
USA

DAVID J. CHAPLIN, PhD
Tumour Microcirculation Group
Gray Laboratory Cancer Research Trust
Mount Vernon Hospital
Northwood, Middlesex HA6 2JR
UK

BERNHARD ENDRICH, PD Dr. med.
Krankenhaus St. Elisabeth
Chirurgische Abteilung
Ziegelstrasse 38
D-89407 Dillingen/Donau
Germany

E. FAIT, Mag. rer. nat.
Anatomisches Institut
Johannes Gutenberg-Universität Mainz
D-55099 Mainz
Germany

HORST J. FELDMANN, PD Dr. med.
Klinik und Poliklinik für
Strahlentherapie und Radiologische Onkologie
Klinikum rechts der Isar
der Technischen Universität München
Ismaninger Strasse 22
D-81675 Munich
Germany

CAI GRAU, MD
Department of Experimental Clinical Oncology
Århus University Hospital
44 Nørrebrogade
DK-8000 Århus C
Denmark

MICHAEL HÖCKEL, Prof. Dr. med., Dr. rer. nat.
Department of Obstetrics and Gynecology
University of Mainz Medical School
Langenbeckstrasse 1
D-55101 Mainz
Germany

MICHAEL R. HORSMAN, PhD
Danish Cancer Society
Department of Experimental Clinical Oncology
44 Nørrebrogade
DK-8000 Århus C
Denmark

RAINER JUND, Dr. med.
Klinik und Poliklinik für
Strahlentherapie und Radiologische Onkologie
Klinikum rechts der Isar
der Technischen Universität München
Ismaninger Strasse 22
D-81675 Munich
Germany

M.A. KONERDING, Prof. Dr. med.
Anatomisches Institut
Johannes Gutenberg-Universität Mainz
D-55099 Mainz
Germany

CHRISTIAN LAUBENBACHER, Dr. med.
Klinik für Nuklearmedizin
Klinikum rechts der Isar
der Technischen Universität München
Ismaningerstrasse 22
D-81675 Munich
Germany

ROSS J. MAXWELL, PhD
Århus University Hospitals
NMR Research Centre
Skejby Sygehus
DK-8200 Århus N
Denmark

MICHAEL MOLLS, Prof. Dr. med.
Klinik und Poliklinik für
Strahlentherapie und Radiologische Onkologie
Klinikum rechts der Isar der Technischen Universität
München
Ismaninger Strasse 22
D-81675 Munich
Germany

W. MUELLER-KLIESER, Prof. Dr. rer. nat.
Institute of Physiology and Pathophysiology
University of Mainz
Duesbergweg 6
D-55099 Mainz
Germany

JENS OVERGAARD, Prof. MD
Department of Experimental Clinical Oncology
Århus University Hospital
44 Nørebrogade
DK-8000 Århus C
Denmark

MATTHEW PARLIAMENT, MD
Department of Radiation Oncology
Cross Cancer Institute
11560 University Avenue
Edmonton, Alberta
Canada T6G 1Z2

MARKUS SCHWAIGER, Prof. Dr. med.
Klinik für Nuklearmedizin
Klinikum rechts der Isar
der Technischen Universität München
Ismaningerstrasse 22
D-81675 Munich
Germany

DIETMAR W. SIEMANN, Prof. PhD
Department of Radiation Oncology
Shands Cancer Center
University of Florida
P.O. Box 100385
Gainesville, FL 32610
USA

CHANG W. SONG, Prof. PhD
Department of Therapeutic Radiology
University of Minnesota Medical School
424 Harvard Street S.E, Box 494 UMHC
Minneapolis, MN 55455
USA

PETER STADLER, Dr. med.
Klinik und Poliklinik für
Strahlentherapie und Radiologische Onkologie
Klinikum rechts der Isar der
Technischen Universität München
Ismaninger Strasse 22
D-81675 München
Germany

F. STEINBERG, Dr. med.
Institut für Medizinische Strahlenbiologie
Universitätsklinikum Essen
Hufelandstrasse 55
D-45122 Essen
Germany

C. STREFFER, Prof. Dr. rer. nat., Dr. h.c.
Institut für Medizinische Strahlenbiologie
Universitätsklinikum Essen
Hufelandstrasse 55
D-45122 Essen
Germany

MARION STUBBS, PhD
Cancer Research Campaign Biomedical Magnetic Resonance
Research Group
Division of Biochemistry
St. George`s Hospital Medical School
Cranmer Terrace
London SW17 ORE
UK

M.J. TROTTER, MD
Department of Pathology
Vancouver General Hospital
855 West 12th Avenue
Vancouver
British Columbia
Canada V5Z 1M9

R. URTASUN, Prof. MD
Department of Radiation Oncology
Cross Cancer Institute
11560 University Avenue
Edmonton, Alberta
Canada T6G 1Z2

C. VAN ACKERN, Dr. med.
Anatomisches Institut
Johannes Gutenberg-Universität Mainz
D-55099 Mainz
Germany

PETER VAUPEL, Prof. Dr. med.
Institute of Physiology and Pathophysiology
University of Mainz
Duesbergweg 6
D-55099 Mainz
Germany

MEDICAL RADIOLOGY
Diagnostic Imaging and Radiation Oncology

Titles in the series already published

MEDICAL RADIOLOGY
Diagnostic Imaging and Radiation Oncology

Titles in the series already published

Print and Binding: Kösel GmbH & Co., Kempten